T0329805

Extraction Techniques for Environmental Analysis

Extraction Techniques for Environmental Analysis

John R. Dean

Northumbria University
Newcastle, UK

Registered Offices
John Wiley & Sons, Inc., 111 River Street, Hoboken, NJ 07030, USA
John Wiley & Sons Ltd, The Atrium, Southern Gate, Chichester, West Sussex, PO19 8SQ, UK

Editorial Office
The Atrium, Southern Gate, Chichester, West Sussex, PO19 8SQ, UK

For details of our global editorial offices, customer services, and more information about Wiley products visit us at www.wiley.com.

Wiley also publishes its books in a variety of electronic formats and by print-on-demand. Some content that appears in standard print versions of this book may not be available in other formats.

Library of Congress Cataloging-in-Publication Data

Names: Dean, John R., author.
Title: Extraction techniques for environmental analysis / John R. Dean, Northumbria University, Newcastle, UK.
Description: First edition. | Hoboken, NJ, USA : Wiley, 2022. | Includes bibliographical references and index.
Identifiers: LCCN 2021050906 (print) | LCCN 2021050907 (ebook) | ISBN 9781119719045 (cloth) | ISBN 9781119719052 (adobe pdf) | ISBN 9781119719038 (epub)
Subjects: LCSH: Extraction (Chemistry) | Environmental chemistry.
Classification: LCC TP156.E8 D4349 2022 (print) | LCC TP156.E8 (ebook) | DDC 660/.28424–dc23/eng/20211109
LC record available at https://lccn.loc.gov/2021050906
LC ebook record available at https://lccn.loc.gov/2021050907

Cover Design: Wiley
Cover Image: © John Dean

Set in 9.5/12.5pt STIXTwoText by Stravie, Pondicherry, India
Printed and bound by CPI Group (UK) Ltd, Croydon, CR0 4YY

C9781119719045_150222

This book has been written during the global SARS-CoV-2 (COVID-19) pandemic from 2020 to 2021. So, as well as working from home, with occasional visits to the laboratory on-campus, most of my time has been sat in front of a PC in the dining room (I suspect like many of you).

Significant changes have also taken place during this time in the Dean family.

My wife Lynne was ordained deacon in the Church of England on Saturday 3 July 2021, so is now known as the Reverend Lynne.

My children, though no longer at home, continue in their respective careers. Naomi, a chemistry teacher in the North Lake District, and Sam, now a part-time outdoor instructor and full-time driver for a well-known manufacturer of shortbread in North-East Scotland.

And just pre-pandemic, we welcomed Harris (a border terrier) into our family.

Contents

Preface

This book provides a comprehensive overview of the approaches required to obtain relevant and validated data in an environmental context. A comprehensive range of extraction techniques for the recovery of organic compounds from environmental matrices are reviewed in their function and application with illustrated case studies. In addition, other areas within the whole environmental process are included to allow the reader a fuller understanding. Each chapter, as well as being illustrated with tables and figures, also contains example Case Studies that allow a more in-depth view. The book has been designed to be user-friendly allowing the reader to investigate one topic at a time or to get an overview. The book is arranged into eight sections (and eighteen chapters) that allow the entire breadth of the environmental context to be highlighted.

Section A (Chapter 1) focuses on the initial considerations necessary to undertake environmental analysis. Specifically, it introduces the most important organic compounds of concern in the environment. The essentials of practical work are considered alongside health and safety. Then, the considerations necessary on data presentation and the importance in understanding the most appropriate units, and the recording of the appropriate number of digits (the use of significant figures). Quantitative analysis is covered in terms of preparation of standards, calibration data, the use of an internal standard, the calculation of limits of detection and quantitation, as well as considerations for sample extracts in terms of dilution and concentration factors. Finally, quality assurance aspects are considered, as well as the role and importance of certified reference materials.

Section B (Chapter 2) considers aspects of sampling and sample storage across the different matrices of aqueous, solid, and gaseous samples. Initial considerations are given to the design of an appropriate sampling strategy, followed by the different types of aqueous and solid matrices that might be encountered in an environmental context. The key physicochemical properties of water and soil matrices are then considered and discussed. A summary of the terminology and inter-relationships between sampling and analytical operations are outlined. Finally, an overview, and then detailed consideration of the sample storage options and preservation techniques available for liquid, solid, and gaseous samples.

Section C (Chapters 3–8) consider the options available for the extraction of aqueous samples. Each individual chapter, generally, considers the theoretical basis of the selected technique, its method of application and practice, as well as an applications section. The following techniques are covered in this collection of chapters: classical approach for aqueous extraction (Chapter 3), solid-phase extraction (Chapter 4), solid-phase microextraction (Chapter 5), in-tube extraction (Chapter 6), stir-bar sorptive extraction (Chapter 7) and membrane extraction (Chapter 8).

Section D (Chapters 9–13) consider the options available for the extraction of solid samples. Each individual chapter, generally, considers the theoretical basis of the selected technique, its method of application and practice, as well as an applications section. The following techniques are covered in this collection of chapters: classical approach for extraction of solid samples (Chapter 9), pressurized liquid extraction (Chapter 10), microwave-assisted extraction (Chapter 11), matrix solid-phase dispersion (Chapter 12) and supercritical fluid extraction (Chapter 13).

Section E (Chapter 14) considers air sampling with respect to gaseous samples and particulate matter. Consideration is provided of the available techniques, methods of recovery of collected compounds, as well as the legal aspects of workplace exposure limits.

Section F (Chapters 15–16) considers the pre-concentration, and associated sample extraction procedures necessary for action on sample extracts are considered in Chapter 15. A broad range of solvent evaporation techniques are considered, as well as clean-up procedures. Finally, the process of chemical derivatization for gas chromatography is outlined. While Chapter 16 provides extensive coverage of the key analytical chromatographic techniques for environmental analysis. From first principles, through the theory of separation and detailed instrumentation for both gas chromatography and high-performance liquid chromatography. Finally, a brief section on other analytical techniques for organic compound analysis is provided.

Section G (Chapter 17) considers the post-analysis decision-making processes by using selected environmental problem-solving case studies. The first case study considers the initial necessary planning in attempting an environmental problem, the second case study considers the whole concept of environmental analysis and finally, the third case study uses a novel environmental chemistry Escape Room as a problem-solving tool.

Section H (Chapter 18) considers the historical development of a range of extraction and chromatographic techniques from its earliest stage to the modern day.

Finally, the **Appendices** provide a useful approach to reframe and consider the key extraction techniques for aqueous and solid samples, as well as the instrumental techniques using crossword puzzles (with available solutions).

John R. Dean
Summer 2021

About the Author

John R. Dean
DSc, PhD, DIC, MSc, BSc, FRSC,
CChem, CSci, PFHEA

Since 2004 John R. Dean has been Professor of Analytical and Environmental Sciences at Northumbria University, where he is also currently Head of Subject in Analytical Sciences, which covers all Chemistry and Forensic Science Programmes. His research is both diverse and informed covering such topics as the development of novel methods to investigate the influence and risk of metals and persistent organic compounds in environmental and biological matrices, to development of new chromatographic methods for environmental and biological samples using gas chromatography and ion mobility spectrometry, the development of novel approaches for pathogenic bacterial detection/identification and most recently the use of an unmanned aerial vehicle for precision agriculture applications. Much of the work is directly supported by industry and other external sponsors.

He has published extensively (over 225 papers, book chapters and books) in analytical and environmental science. He has also supervised over 40 PhD students.

John remains an active member of the Royal Society of Chemistry (RSC) and serves on several of its committees, including Committee for Accreditation and Validation of Chemistry Degrees and the Research Mobility Grant committee. His dedication to the society over 40 years was acknowledged in 2021 with the award for Exceptional Service.

After a first degree in Chemistry at the University of Manchester Institute of Science and Technology (UMIST), this was followed by an MSc in Analytical Chemistry and Instrumentation at Loughborough University of Technology, and finally a PhD and DIC in Physical Chemistry at the Imperial College of Science and Technology (University of London). He then spent two years as a postdoctoral research fellow at the Government Food Laboratory in Norwich. In 1988, he was appointed to a lectureship in Inorganic/Analytical Chemistry at Newcastle Polytechnic (now Northumbria University) where he has remained ever since.

John is also active in paddlesport; he holds performance (UKCC level 3) coach awards in open canoe and white water kayak. In 2012, his involvement in a local club was acknowledged by the award of an 'outstanding contribution' by the British Canoe Union.

Acknowledgements

Several colleagues and students (undergraduate and postgraduate (past and present) have provided data that has been used in case studies within the chapters. Unless mentioned specifically, all colleagues and students were located at Northumbria University. In addition, I also acknowledge research sponsors who have provided the funding that has allowed the science behind the case studies to have been done. My appreciation is expressed to all of them for their dedication to the science.

Specifically, the following are acknowledged. In Chapter 1, Dr Graeme Turnbull for the original design of the template on the Control of Substances Hazardous to Health (COSHH), and Brooke Duffield and Samantha Bowerbank for the data used in Case Study A. In Chapter 3, Dr Wanda Scott and Edwin Ludkin for the data in Case Study A, as well as the funder: Engineering and Physical Sciences Research Council for the award of an Industrial case award in collaboration with LGC limited, London, with support from the Department of Trade and Industry under the National Measurement System Valid Analytical Measurement (VAM) Programme. In Chapter 4, for the data used in Case Study B, Dr Paul Bassarab, Professor Justin Perry and Edwin Ludkin, as well as the funder: The Royal Commission for the Exhibition of 1851 for the award of an industrial fellowship. For the high-resolution mass spectra (HRMS), the EPSRC UK National Mass Spectrometry Service Centre, Swansea, UK. For the data used in Case Study C, (the late) Dr Robert Downs, Professor Justin Perry and Edwin Ludkin, as well as the funder: The Engineering and Physical Sciences Research Council (EPSRC) and AkzoNobel Ltd. For the high-resolution mass spectra (HRMS), the EPSRC UK National Mass Spectrometry Service Centre, Swansea, UK. In Chapter 5, Dr Emma Reed (néeTait), Professor Stephen Stanforth and Professor John Perry (Freeman Hospital, Newcastle), as well as the funder: bioMérieux S.A. In Chapter 6, Samantha Bowerbank along with assistance from Daniela Cavagnino, ThermoFisher, as well as the funder: Northumbria University. In Chapter 9, for the data used in Case Study A, Dr Francesc A.E. Turrillas, University of Valencia, and the funder: V Segles, University of Valencia, and Dr Wanda Scott and Edwin Ludkin, and the funder: Engineering and Physical Sciences Research Council for the award of an Industrial case award, in collaboration with LGC Limited, London, with support from the Department of Trade and Industry under the National Measurement System Valid Analytical Measurement (VAM) Programme. In Chapter 10, for Case Study A, Dr Wanda Scott and Edwin Ludkin, and the funder: Engineering and Physical Sciences Research Council for the award of an Industrial case award, in collaboration with LGC Limited, London, with support from the

Department of Trade and Industry under the National Measurement System Valid Analytical Measurement (VAM) Programme. For Case Study B, Dr Damian Lorenzi, and the funder: Northumbria University in collaboration with British Geological Survey, Keyworth. For Case Study C, Joel Sánchez-Piñero and Samantha Bowerbank, and the funder: Xunta de Galicia and the European Union (European Social Fund - ESF) for a pre-doctoral grant, as well as Northumbria University. In Chapter 11, Case Study A, Dr Ian Barnabas and Edwin Ludkin, and the funder: Northumbria University and Analytical and Environmental Services Ltd., Northumbria Water plc. For Case Study B, Dr Claire Costley and Edwin Ludkin, and the funder: ICI Research and Technology Centre, Wilton, Middlesbrough and Northumbria University. In Chapter 12, Dr Carolyn Heslop and Edwin Ludkin, and the funder: Unilever, Port Sunlight. In Chapter 13, Dr Ian Barnabas and Edwin Ludkin, and the funder: Northumbria University and Analytical and Environmental Services Ltd., Northumbria Water plc. In Chapter 15, Dr Damien Lorenzi, and the funder: Northumbria University in collaboration with British Geological Survey, Keyworth. In Chapter 16, for Case Study A, Dr Damien Lorenzi, and the funder: Northumbria University in collaboration with British Geological Survey, Keyworth. In Chapter 17, Case Study B, Joel Sánchez-Piñero and Samantha Bowerbank, and the funder: Xunta de Galicia and the European Union (European Social Fund - ESF) for a predoctoral grant as well as Northumbria University. For Case Study C, the funder: Northumbria University.

Section A

Initial Considerations

1

The Analytical Approach

LEARNING OBJECTIVES

After completing this chapter, students should be able to:

- Contextualize an environmental problem.
- Comprehend the implications of persistent organic pollutants in the environment.
- Undertake a COSHH assessment.
- Develop a strategy for effective practical work.

1.1 Introduction

Environmental analysis does not start in the laboratory but outside (e.g. in a field, river, lake, urban environment or industrial atmosphere). Therefore, environmental analysis requires more than just knowledge of the analytical technique to be used (e.g. chromatography). It requires a consideration of the vast array of extraction techniques that are available pre-analysis. The focus of these pre-analysis extraction techniques (covered in Chapters 3–15) is to recover the organic compounds of interest from a matrix. The matrices can be diverse in their form but generically can be considered as solid, liquid or gas. The purpose of the extraction techniques is to, therefore, recover the organic compounds from the matrices and allow pre-concentration/matrix clean-up to take place.

In addition, it is important to place the extraction and subsequent analysis in its context. Important aspects, therefore, are as follows:

- Consideration of the appropriate health and safety aspects in the laboratory (and the external environment).
- What do you know already about the site to be investigated?
- What are the expectations about the results?
- What type of sampling regime is planned?
- How might the collected samples be stored and preserved?
- What type of sample preparation methodologies are appropriate to the sample?
- How might the analysis be done?
- What are the quality control procedures to be used (including calibration strategies and the use of certified reference materials)?

Extraction Techniques for Environmental Analysis, First Edition. John R. Dean.
© 2022 John Wiley & Sons Ltd. Published 2022 by John Wiley & Sons Ltd.

- How will the knowledge of the results and their interpretation, contextualization and subsequent action be considered?

While all of these are covered to some extent in this book, the reader should also consult other resources, e.g. books, scientific journals and the web.

1.2 Environmental Organic Compounds of Concern

The range of potential organic compounds to be identified and quantified is vast. Their sources are equally diverse and varied. The Stockholm Convention is a global initiative, established in 2001 from former international collaborators, by the United Nations Environmental Programme (UNEP) and requires its signatories to take measures to eliminate or reduce the release of persistent organic pollutants (POPs) into the environment to protect human health where exposure is often via the food chain. UNEP identified 12 (initial) POPs that cause adverse effects on humans and the ecosystem (Table 1.1). There chemical structures, molecular formulae and molecular weights are shown in Figure 1.1.

Aldrin: A pesticide applied to soils to kill termites, grasshoppers, corn rootworm and other insect pests; it can also kill birds, fish and humans. In humans, the fatal dose for an adult male is estimated to be about 5 g. Human exposure is mostly through dairy products and animal meats.

Chlordane: It is used to control termites and as a broad-spectrum insecticide on a range of agricultural crops. It can remain in the soil for an extended time and has a reported half-life of one year. Chlordane may affect the human immune system and is therefore classified as a possible human carcinogen.

Table 1.1 Stockholm Convention: The 12 initial persistent organic pollutants (POPs).

Category	Chemical[a]
Pesticides	Aldrin
	Chlordane
	DDT
	Dieldrin
	Endrin
	Heptachlor
	Hexachlorobenzene
	Mirex
	Toxaphene
Industrial	Hexachlorobenzene
	Polychlorinated biphenyls (PCBs)
Byproducts	Hexachlorobenzene
	Polychlorinated dibenzo-p-dioxins (PCDD)
	Polychlorinated dibenzofurans (PCDF)

[a] Note: Hexachlorobenzene appears under all three categories.

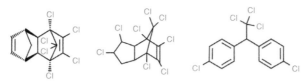

Name	Aldrin	Chlordane	Dichlorodiphenyl trichloroethane (DDT)
Molecular formula	$C_{12}H_8Cl_6$	$C_{10}H_6Cl_8$	$C_{14}H_9Cl_5$
Molecular weight	364.91	409.78	354.49

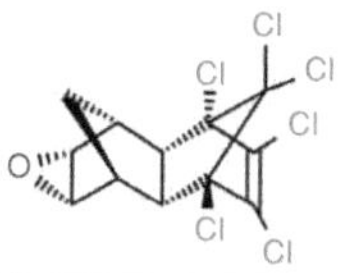

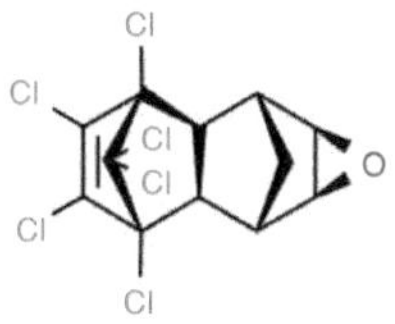

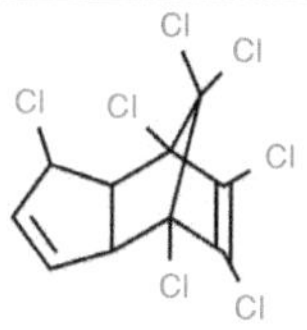

Name	Dieldrin	Endrin	Heptachlor
Molecular formula	$C_{12}H_8Cl_6O$	$C_{12}H_8Cl_6O$	$C_{10}H_5Cl_7$
Molecular weight	380.91	380.91	373.32

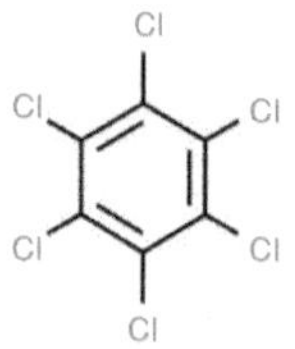

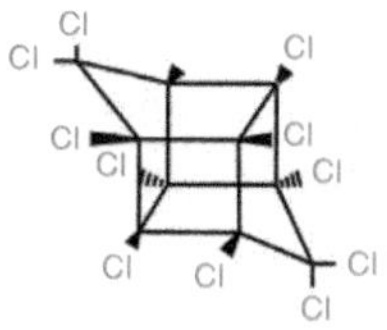

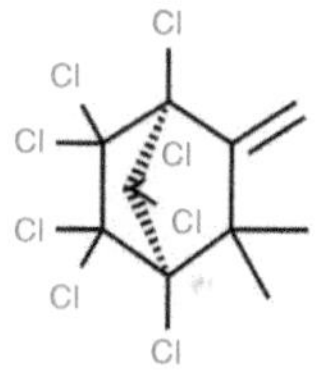

Name	Hexachlorobenzene	Mirex	Toxaphene
Molecular formula	C_6Cl_6	$C_{10}Cl_{12}$	$C_{10}H_5Cl_7$
Molecular weight	284.78	545.54	373.32

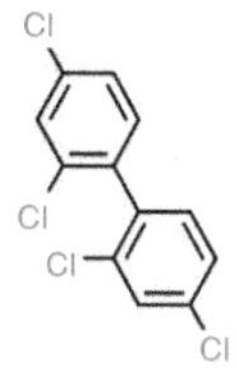

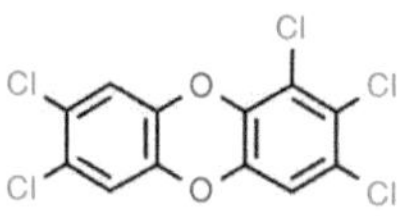

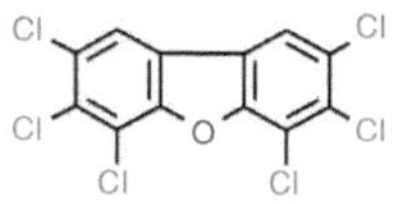

Name	Example, 2, 2′, 4, 4′ - PCB	Example, 1, 2, 3, 7, 8- Pentachlorodibenzo-p-dioxin	Example, 2, 3, 4, 6, 7, 8- Hexachlorodibenzofuran
Molecular formula	$C_{12}H_6Cl_4$	$C_{12}H_3Cl_5O_2$	$C_{12}H_2Cl_6O$
Molecular weight	291.99	356.42	374.86

Figure 1.1 Chemical structures of the 12 initial persistent organic pollutants.

Dichlorodiphenyltrichloroethane (DDT): It was widely used during World War II to protect soldiers and civilians from malaria, typhus and other diseases spread by insects. Subsequently, it has continued to be used to control disease in crops (e.g. cotton) and insects (e.g. mosquitoes). DDT continues to be applied against mosquitoes in developing countries to control malaria. DDT has long-term soil persistence (10–15 years) after application. It has been used extensively, and so its residues can be found everywhere.

Dieldrin: It has been used mainly to control termites and textile pests, as well as to control insect-borne diseases and insects living in agricultural soils. It has a half-life in soil of approximately five years. Aldrin (see earlier) can rapidly convert to dieldrin, so higher concentrations of dieldrin than expected can be found in the environment. Dieldrin is highly toxic to fish and other aquatic animals (e.g. frogs). Residues of dieldrin can be found in air, water, soil, fish, birds and mammals, including humans.

Endrin: It is sprayed on the leaves of crops (e.g. cotton and grains) to protect from insects. It can also be used to control rodents (e.g. mice and voles). It has a long half-life (persisting up to 12 years in soils). In addition, endrin is highly toxic to fish.

Heptachlor: It is used to kill soil insects and termites, as well as cotton insects, grasshoppers, other crop pests and malaria-carrying mosquitoes. Heptachlor is classified as a possible human carcinogen.

Hexachlorobenzene (HCB): It is used to kill fungi that affect food crops (e.g. to control wheat bunt). It is also a byproduct from the manufacture of industrial chemicals and can occur as an impurity in several pesticide formulations. In high doses, HCB is lethal to some animals and, at lower levels, adversely affects their reproductive success.

Mirex: It is used to control ants and termites. In addition, it has also been used as a fire retardant in plastics, rubber and electrical goods. Direct exposure to Mirex does not appear to cause injury to humans; however, the results of animal studies have led it to be classified as a possible human carcinogen. It has a half-life in soil of up to 10 years.

Toxaphene: It is used to protect cotton, cereal grains, fruits, nuts and vegetables from insects. It has also been used to control ticks and mites in livestock. It has a half-life, in soil, of up to 12 years. It has been listed as a possible human carcinogen due to its effects on laboratory animals.

Polychlorinated biphenyls (PCBs): These compounds (209 different types of which 13 exhibit a dioxin-like toxicity) are used in industry as heat exchange fluids, in electric transformers and capacitors, and as additives in paint, carbonless copy paper and plastics. Their persistence in the environment corresponds to the degree of chlorination, and half-lives can vary from 10 days to 1.5 years.

Polychlorinated dibenzo-p-dioxins (PCDDs): These compounds (75 different types of which 7 are of concern) are produced unintentionally due to incomplete combustion, as well as during the manufacture of pesticides and other chlorinated substances. They are emitted mostly from the burning of hospital waste, municipal waste and hazardous waste, as well as automobile emissions, peat, coal and wood. They can have a half-life in soil of up to 10–12 years. They are associated with a number of adverse effects in humans, including immune and enzyme disorders and chloracne, and they are classified as possible human carcinogens.

Polychlorinated dibenzofurans (PCDFs): These compounds (135 different types) are produced unintentionally from many of the same processes that produce dioxins, as well as during the production of PCBs. They have been detected in emissions from waste incinerators and automobiles. Furans are structurally like dioxins and share many of their toxic effects. They are persistent in the environment for long periods and are classified as possible human carcinogens.

In addition, in subsequent revisions of the original Stockholm Convention, another 16 POPs have been added to the listings (Table 1.2).

α-Hexachlorocyclohexane and β-hexachlorocyclohexane: The technical mixture of hexachlorocyclohexane (HCH) contains mainly five forms of isomers, namely α-, β-, γ-, δ- and ε-HCH. Lindane is the common name for the γ isomer of HCH. The α- and β-HCH are highly persistent in water in colder regions and may bioaccumulate and biomagnify in biota and arctic food webs. They are subject to long-range transport, are classified as potentially carcinogenic to humans and adversely affect wildlife and human health in contaminated regions. The use of α- and β-HCH as insecticides has been phased out but are produced as byproducts of lindane. For each ton of lindane produced, around 6–10 tons of α- and β-HCH are also produced. This has led to large stockpiles, which can cause site contamination.

Chlordecone: It is chemically related to Mirex (Table 1.1). It is highly persistent in the environment, has a high potential for bioaccumulation and biomagnification and based on physico-chemical properties and modelling data, chlordecone can be transported for long distances. It is classified as a possible human carcinogen and is very toxic to aquatic organisms. Chlordecone is a synthetic chlorinated organic compound, which was mainly used as an agricultural pesticide. While it was commercially introduced in 1958, it has now been banned for sale and use in many countries.

Table 1.2 Stockholm convention: the additional 16 persistent organic pollutants (POPs).

The 16 new additional POPs
α-Hexachlorocyclohexane
β-Hexachlorocyclohexane
Chlordecone
Hexabromobiphenyl
Hexabromocyclododecane
Hexabromodiphenyl ether and heptabromodiphenyl ether (commercial octabromodiphenyl ether)
Hexachlorobutadiene
Lindane
Pentachlorobenzene
Pentachlorophenol and its salts and esters
Perfluorooctane sulphonic acid (PFOS), its salts and perfluorooctane sulphonyl fluoride (PFOSF)
Polychlorinated naphthalenes
Technical endosulphan and its related isomers
Tetrabromodiphenyl ether and pentabromodiphenyl ether (commercial pentabromodiphenyl ether)
Decabromodiphenyl ether (commercial mixture, cDecaBDE)
Short-chain chlorinated paraffins (SCCPs)

Hexabromobiphenyl: It belongs to the group of polybrominated biphenyls (i.e. brominated hydrocarbons formed by substituting hydrogen with bromine in biphenyl). It is highly persistent in the environment, highly bioaccumulative and has a strong potential for long-range environmental transport. It is classified as a possible human carcinogen and has other chronic toxic effects. It has historically been used as a flame retardant. It is no longer produced or used in most countries due to restrictions under national and international regulations.

Hexabromocyclododecane (HBCD): It has a strong potential to bioaccumulate and biomagnify. It is persistent in the environment and has a potential for long range environmental transport. It is very toxic to aquatic organisms. It is particularly harmful to humans as a neuroendocrine carcinogen. It was used as a flame-retardant additive on polystyrene materials (in the 1980s) and as part of safety regulation for articles, vehicles and buildings.

Hexabromodiphenyl ether and heptabromodiphenyl ether (commercial octabromodiphenyl ether): These are the main components of commercial octabromodiphenyl ether. The commercial mixture of octaBDE is highly persistent, has a high potential for bioaccumulation and food-web biomagnification, as well as for long-range transport. The only degradation pathway is through debromination and producing other bromodiphenyl ethers.

Hexachlorobutadiene: It is a halogenated aliphatic compound, mainly created as a byproduct in the manufacture of chlorinated aliphatic compounds. It is persistent, bioaccumulative and very toxic to aquatic organisms and birds. It can be long-range transported leading to significant adverse human health and environmental effects, and it is classified as a possible human carcinogen. It is mainly used as a solvent for other chlorine-containing compounds. It occurs as a byproduct during the chlorinolysis of butane derivatives in the large-scale production of both carbon tetrachloride and tetrachloroethene.

Lindane: It is persistent, bioaccumulates easily in the food chain and bioconcentrates rapidly. There is evidence for long-range transport and toxic effects (immunotoxic, reproductive and developmental effects) in laboratory animals and aquatic organisms. It has been used as a broad-spectrum insecticide for seed and soil treatment, foliar applications, tree and wood treatment and against ectoparasites in both veterinary and human applications. Its production has decreased.

Its production has decreased rapidly in the last few years due to the introduction of regulations in several countries.

Pentachlorobenzene (PeCB): It belongs to a group of chlorobenzenes that are characterized by a benzene ring in which the hydrogen atoms are substituted by one or more chlorines. It is persistent in the environment, highly bioaccumulative and has a potential for long-range environmental transport. It is moderately toxic to humans and very toxic to aquatic organisms. Previously, it was used in PCB products, in dyestuff carriers, as a fungicide and a flame retardant. It is produced unintentionally during combustion, thermal and industrial processes, and can occur in the form of impurities in solvents or pesticides.

Pentachlorophenol (PCP) and its salts and esters: It can be found in two forms: PCP itself or as its sodium salt (which dissolves easily in water). It is detected in the blood, urine, seminal fluid, breast milk and adipose tissue of humans. It is likely, because of its long-range environmental transport, to lead to significant adverse human health and/or environmental effects. It has been used as a herbicide, insecticide, fungicide, algaecide, disinfectant and as an ingredient in antifouling paint. Its use has significantly declined due to the high toxicity of PCP and its slow biodegradation; its main contaminants include other polychlorinated phenols, polychlorinated dibenzo-pdioxins and polychlorinated dibenzo furans.

Perfluorooctane sulphonic acid (PFOS), its salts and perfluorooctane sulphonyl fluoride (PFOSF): PFOS is a fully fluorinated anion, which is commonly used as a salt or incorporated into larger polymers. It is extremely persistent and has substantial bioaccumulations and biomagnifying properties; however, it does not partition into fatty tissues but instead binds to proteins in the blood and the liver. It has a capacity to undergo long-range transport. PFOS is both intentionally produced (for use in electric and electronic parts, firefighting foam, photo imaging, hydraulic fluids and textiles), and an unintended degradation product of related anthropogenic chemicals.

Polychlorinated naphthalenes (PCNs): They are mixtures (up to 75 chlorinated naphthalene congeners plus byproducts) often described by the total fraction of chlorine. While some PCNs can be broken down by sunlight and, at slow rates, by certain microorganisms, many PCNs persist in the environment. Bioaccumulation has been confirmed for tetra- to heptaCNs. Chronic exposure can lead to increased risk of liver disease. PCNs make effective insulating coatings for electrical wires; they are also used as wood preservatives, as rubber and plastic additives, for capacitor dielectrics and in lubricants. Intentional production of PCN is assumed to have ended; however, they can be formed during high-temperature industrial processes in the presence of chlorine.

Technical endosulphan and its related isomers: It occurs as two isomers: α- and β-endosulphan. They are both biologically active. Technical endosulphan (CAS No: 115-29-7) is a mixture of the two isomers along with small amounts of impurities. It is persistent in the atmosphere, sediments and water. Endosulphan bioaccumulates and has the potential for long-range transport. It is toxic to humans and has been shown to have adverse effects on a wide range of aquatic and terrestrial organisms. The use of endosulphan is banned or will be phased out in 60 countries that, together, account for 45% of current global use. It has been used as an insecticide to control crop pests, tsetse flies and ectoparasites of cattle and as a wood preservative.

Tetrabromodiphenyl ether and pentabromodiphenyl ether (commercial pentabromodiphenyl ether): Tetrabromodiphenyl ether and pentabromodiphenyl ether are the main components of commercial pentabromodiphenyl ether. They belong to a group of chemicals known as 'polybromodiphenyl ethers' (PBDEs). The commercial mixture of penta-BDE is highly persistent in the environment, bioaccumulative and has a potential for long-range environmental transport (it has been detected in humans throughout all regions). There is evidence of its toxic effects in wildlife, including mammals. Polybromodiphenyl ethers including tetra-, penta-, hexa- and hepta-BDEs inhibit or suppress combustion in organic materials and therefore are used as additive flame retardants. The production of tetra- and penta-BDEs has ceased in certain regions of the world, while no production of hexa- and hepta-BDEs is reported.

Decabromodiphenyl ether (commercial mixture, cDecaBDE): The commercial mixture consists primarily of the fully brominated decaBDE congener in a concentration range of 77.4–98%, and smaller amounts of the congeners of nona-BDE (0.3–21.8%) and octa-BDE (0–0.04%). The deca-BDE is highly persistent, has a high potential for bioaccumulation and food-web biomagnification, as well as for long-range transport. Adverse effects are reported for soil organisms, birds, fish, frog, rat, mice and humans. Deca-BDE is used as an additive flame retardant and has a variety of applications including plastics/polymers/composites, textiles, adhesives, sealants, coatings and inks. Deca-BDE-containing plastics are used in housings of computers and TVs, wires and cables, pipes and carpets. Commercially available deca-BDE consumption peaked in the early 2000s, but c-deca-BDE is still extensively used worldwide.

Short-chain chlorinated paraffins (SCCPs): Chlorinated paraffins (CPs) are complex mixtures of certain organic compounds. Their degree of chlorination can vary between 30 and 70 wt%. They are sufficiently persistent in air for long-range transport to occur and appear to be hydrolytically stable. Many SCCPs can accumulate in biota. They are likely, because of their long-range environmental transport, to lead to significant adverse environmental and human health effects. They are used as a plasticizer in rubber, paints, adhesives, flame retardants for plastics, as well as an extreme pressure lubricant in metal working fluids. They are produced by chlorination of straight-chained paraffin fractions. The carbon chain length of commercial chlorinated paraffins is usually between 10 and 30 carbon atoms; however, the short-chained chlorinated paraffins vary between C10 and C13. The production of SCCPs has decreased globally as jurisdictions have established control measures.

In addition, a whole range of other organic pollutants are investigated in the environment including some classes of compounds, e.g. volatile organic compounds (e.g. BTEX: benzene, toluene, ethylbenzene and xylenes); solvents (e.g. carbon tetrachloride, chloroform) and polycyclic aromatic hydrocarbons (e.g. naphthalene, acenaphthylene, acenaphthene, fluorene, phenanthrene, anthracene, fluoranthene, pyrene, benzo(a)anthracene, chrysene, benzo(b) fluoranthene, benzo(k)fluoranthene, benzo(a)pyrene, benzo(g,h,i)perylene, indeno(1,2,3-c,d)pyrene and dibenzo(a,h)anthracene). In addition, a whole range of emerging pollutants (EPs) are now of concern (Table 1.3). The term emerging is used as a descriptor not because these compounds are new, but because their environmental impact is raising cause for concern. While EPs are an increasing list of organic compounds with diverse sources and include antibiotics, analgesics, anti-inflammatory drugs, psychiatric drugs, steroids and hormones, contraceptives, fragrances, sunscreen agents, insect repellents, microbeads, microplastics, antiseptics, pesticides, herbicides, surfactants and surfactant metabolites, flame retardants, industrial additives and chemicals, plasticizers and gasoline additives, among others, and often enter the environment via waste water discharge into rivers, lakes etc. Within this identifier of EPs, a sub-group of organic pollutants has been identified and are often named pharmaceuticals and personal care products (PPCPs). Often the exposure to these EPs is at low concentration, so concern is focused on chronic low-level exposure and mixtures of these compounds with additive or unexpected effects.

Table 1.3 Emerging pollutants in the environment.

Emerging pollutants	Compounds
Perfluorinated compounds (PFCs)	Perfluorooctanesulphonate (PFOS), perfluorooctanoic acid (PFOA) (see also Table 1.2), and their salts are the most essential representative PFCs and are widely used in fire-fighting foams, lubricants, metal spray plating and detergent products, inks, varnishes, coating formulations (for walls, furniture, carpeting and food packaging), waxes and water and oil repellents for leather, paper and textiles
Water disinfection byproducts	Disinfection chemicals used in swimming pool and drinking water purification. Disinfection byproducts (DBPs), particularly chlorinated DBPs (CDBPs) in purified water, and nearly all humans are exposed to these chemicals in developed regions through swimming pools and drinking water. More than six hundred DBPs have been discovered, including iodinated trihalomethanes (THMs), aldehydes, ketones, halomethanes, hydroxy acids, carboxylic acids, alcohols, keto acids, esters and even nitrosamines
Gasoline (petrol) additives	Gasoline encompasses more than five hundred components, such as the known or suspected carcinogenic substances benzene, 1,3-butadiene and methyl tert-butyl ether (MTBE) (an unleaded petrol additive)
Manufactured nanomaterials	Manufactured nanomaterials (with a particle size of approximately 1–100 nm) include amorphous silicon dioxide (SiO_2), carbon nanotubes (CNTs) and titanium dioxide (TiO_2), which due to their large surface area can adsorb organic compounds, e.g. PAHs
Human and veterinary pharmaceuticals	Pharmaceuticals are emerging contaminants in the environment because of their increasing applications in humans and animals. Approximately three thousand different chemicals involved in human medicine, including lipid regulators, anti-inflammatory drugs, analgesics, contraceptives, neuroactive medicine, antibiotics and beta-blockers, exist. The main pathway through which pharmaceuticals enter the surface water is human intake, followed by subsequent excretion in municipal wastewater, hospitals, pharmaceutical waste and landfills. Veterinary Antibiotics (VAs) are being increasingly used in many regions to protect the health of animals and treat diseases to improve the feed efficiency of livestock, poultry, pets, aquatic animals, silkworms, bees etc. The VAs are mainly divided into several pharmacological types: antimicrobial, anthelmintic, steroidal and nonsteroidal, anti-inflammatory, antiparasitic, astringent, estrus synchronization, nutritional supplement and as growth promoters
UV-filters	Sunscreens/ultraviolet filters (UV-filters) are mainly used in personal care products, such as lipsticks, perfumes, hairsprays, hair dye and moisturizers, skin care products, shampoos and makeup, as well as in non-cosmetic products, including furniture, plastics, carpets and washing powder. Organic sunscreens absorb photons of UV and include 3-(4-methylbenzylidene) camphor (4-MBC), benzophenone-3 (BP-3), 2-ethylhexyl 4-methoxycinnamate (OMC), 2-ethylhexyl 4-dimethylaminobenzoate (OD-PABA), 3-benzylidene camphor (3- BC), homosalate (HMS) and 4-aminobenzoic acid (PABA). These compound types enter the aquatic environment by bathing, washing clothes and swimming
Personal care products	Personal care products encompass many different substances that are in regular use, such as fragrances and cosmetic ingredients. They are often released in raw sewage water, then often inefficiently treated in sewage treatment plants and finally discharged in rivers, streams or lakes. They can also be directly released into water during swimming and bathing in lakes or rivers

1.3 Essentials of Practical Work

It is important to develop a good understanding of the underlying principles of good laboratory practice (GLP) and apply them from the start to the end of the process. The effective recording of all relevant data and information at the time of obtaining the scientific information is essential. Therefore, a systematic and appropriate method of recording all information accurately is essential.

All experimental observations, and data, should be recorded in either an A4 notebook or electronically (and backed up on an external hard drive or USB stick). Example templates are shown for sample collection (Table 1.4), sample treatment (Table 1.5) and sample preparation (Table 1.6). Some key reminders include the recording of the sampling geographical location, the date the sample was collected, the total weight or volume of each individual sample, sample storage conditions, sample treatment and sample preparation. In addition, the quality assurance used to ensure that the data is fit for purpose, and the actual recording of the results and their initial interpretation. It is important to remember to record all information at the point of collection; it is easy to forget it later if not written down.

Important factors to remember when recording information:

- Record data correctly in tabular format and legibly if by hand, even you may not be able to read your own writing later.
- Include the date and title of individual experiments and/or areas of investigation.

Table 1.4 Example template for use electronically or in notebook form: sample collection.

Template: Sample collection
• Date of sample collection: day / month / year
• Location of sampling site:
Address: . . . .
Grid reference: . . . .
• Whom was permission obtained to obtain samples:
Name: . . . . Tel. no. / mobile: . . . .
Email: . . . .
• Weather (and source of information) when samples collected: e.g. dry, sunny, overcast. Temperature, wind direction etc.
. . . .
Temperature: . . . . °C
Wind Direction: . . . .
Precipitation: . . . . mm
Source of information: . . . .
• Method of obtaining samples: . . . .
• Number of samples obtained: . . . .
Unique sample code added to each container Yes / No
Was the date added Yes / No

Table 1.5 Example template for use electronically or in notebook form: sample treatment.

Template: Sample Treatment (e.g. solids)

- Grinding and sieving
Grinder used (model/type) . . . .
Particle size (sieve mesh size) . . . .

- Mixing of the sample
Manual shaking yes / no
Mechanical shaking yes / no . . . . rpm
Other (specify) . . . .

- Sample storage
Fridge yes / no Temperature . . . . °C
Frieezer yes / no Temperature . . . . °C
Other (specify) . . . .

- Other comments
pH . . . .

Table 1.6 Example template for use electronically or in notebook form: sample preparation.

- Sample weight(s)
(record to 4 decimal places)
sample 1 . . . . .g sample 2. . . . . .g

- Soxhlet extraction method
Drying agent added and weight (specify) . . . .
Type of solvent(s) used . . . .
Volume of solvent(s) used. . . . . .ml
Any other details. . . . .

- Other method of sample extraction e.g. SFE, PLE, MAE, and operating conditions (specify)
. . . .
. . . .
. . . .
- Sample clean-up yes / no
Specify . . . .

. . . .
- Pre-concentration of the sample yes / no
Method of solvent reduction . . . .
Final volume of extract. . . . .

(Continued)

Table 1.6 (Continued)

- Sample derivatization

Specify . . . .

. . . .

Reagent concentration . . . . .mol l^{-1}

Reagent volume used . . . . .ml

Heat required (specify) . . . .

- Sample dilution

Specify with appropriate units the dilution factor involved.

. . . .

. . . .

- Addition of an internal standard

Specify . . . .

Added before extraction yes / no

Added after extraction yes / no

- Sample and reagent blanks

Specify . . . .

. . . .

- Recovery

Specify . . . .

Added before extraction yes / no

Added after extraction yes / no

- Briefly outline the purpose of the investigation/experiment, i.e. what you hope to achieve by the end.
- Identify and record the hazards and risks associated with the chemicals/equipment being used. It is a requirement to complete a Control of Substances Hazardous to Health (COSHH) form for each chemical and a risk assessment (see Section 1.4).
- Refer to the method/procedure being used (undergraduate laboratory) or write a full description of the method/procedure and its origins (postgraduate research).
- Record your observations (and note your interpretation at this stage), e.g. accurate weights, volumes, how standards and calibration solutions were prepared and instrumentation settings (and the actual operating parameters).
- Record data with the correct units, e.g. mg, µg g^{-1}, mmol l^{-1} and to an appropriate number of significant figures, e.g. 26.3 mg and 0.48 µg g^{-1} (and not 26.3423 mg and 0.4837 µg g^{-1}).
- Interpret data in the form of tables, graphs (including calibration graphs) and spectra.
- Record initial conclusions.
- Identify any actions for future work.

1.4   Health and Safety

It is a legal requirement for institutions to provide a working environment that is both safe and without risk to health. In the United Kingdom, the Health and Safety at Work Act 1974 provides the legal framework for health and safety. The introduction of the COSHH regulations in 2002 imposed specific legal requirements for risk assessment when hazardous chemicals (or biological agents) are used. In the European Union (EU), the system for controlling chemicals is the Registration, Evaluation, Authorization and restriction of Chemicals (REACH). While in the United States, the Environmental Protection Agency (EPA) is responsible for chemical safety relating to human health and the environment. Adherence to health and safety requirements is often evidenced by the provision of training and information on safe working practices in the laboratory (and external environment). For the student, this is often done by attending an appropriate safety briefing, the reading (and subsequent signing) of a safety guide or booklet acknowledging an understanding of safety and their role in protecting themselves and other students, as well as receiving appropriate training in the use of scientific equipment.

In all cases, however, it is important to understand the definitions applied to hazard and risk.

- A hazardous substance is one that has the ability to cause harm.
- A risk is about the likelihood that the substance may cause harm.

Prior to undertaking any laboratory work, a risk assessment must be undertaken by an appropriately identified person (e.g. supervisor, academic or technical staff). The purpose of the risk assessment is to identify laboratory activities that could cause injury to people and then to provide control measures to ensure that the risk is reduced. Important considerations are:

- substance hazards;
- how the substance is to be used;
- how it can be controlled;
- who is exposed;
- how much exposure and its duration.

It is important to distinguish between the hazard of a substance and its risk from exposure. This can be done by doing a Risk Matrix Analysis (RMA). The RMA allows a prioritization of the likelihood and severity, to the individual, from the hazard identified. All manufacturers of hazardous chemicals are required to provide a Material Safety Data Sheet (MSDS) for the stated chemical. The MSDS will contain information about the chemical including:

- manufacturer;
- name of chemical;
- chemical components;
- hazards associated with the product (including a Hazard Statement and a Precautionary Statement);
- first aid measures;
- firefighting measures;
- handling and storage;
- accidental release procedures;

- exposure control and personal protection;
- physical and chemical properties;
- stability and reactivity;
- toxicological and ecological information;
- disposal practices;
- other miscellaneous information.

With this information, the user must then complete a COSHH form (an example is shown in Table 1.7). As part of the COSHH process, specific details of the Hazard Statement

Table 1.7 An example of Control of Substances Hazard to Health (COSHH) form.

Section 1: Overview

Names of chemicals to be used:	*Enter the name of each hazardous chemical to be used*
Title of activity:	*Enter the title of the activity*
Brief description of theactivity:	*Briefly describe the activity to be undertaken*
Responsible person:	*Enter name of the member of staff responsible for your work e.g. supervisor*
Faculty / Department	*Enter the name of your Faculty / Department*
Date of assessment	*Enter the date*	Date of Re-assessment	*Enter the date one year from now*
Location of work:	*Enter the name of Building / Laboratory in which the work will be carried out.*

Section 2: Emergency Contacts (e.g. project supervisor).

Name	Position	Contact Telephone Number
Enter the name	*Enter their position*	*Enter their telephone number*

Section 3: Hazard Identification

3.1 For hazardous substances in this activity, click all that apply.

	☐ Toxic		☐ Severe Health Hazards		☐ Health Hazards
	☐ Explosive		☐ Flammable		☐ Oxidising
	☐ Corrosive		☐ Gases Under Pressure		☐ Environmental

3.2 Select the hazard phrases (H-phrases) for each hazardous substance.

1.	*Select a Hazard phrase*	e.g.	H302-Harmful if swallowed
e.g.	H226-Flammable liquid and vapour	e.g.	EUH014-Reacts violently with water

Section 4: Hazard Properties

Name of substance	Physical form	Quantity	Frequency	Route of exposure
Enter substance name	*Enter physical form*	*Enter the quantity*	*Enter the frequency*	*Select route*
e.g. *Chemical name*	solid dust	1 g	weekly	Ingestion

Section 5: Identifying Those at Risk

5.1 Who might be at risk? Select all that apply.

☐ Staff/PGRs	☐ Taught Students	☐ Young persons (under 18 years old)
☐ New or expectant mothers	☐ Others:	

Table 1.7 (Continued)

5.2 Assessment of risk to human health before control measures are in place.
Select the likelihood and severity of harm in the presence of the identified hazards **before** the control measures outlined above are implemented. Calculate the risk rating and act accordingly.

Likelihood of harm	Severity	Risk Rating and Outcome (likelihood x Severity)
e.g. 2. Unlikely	*2. Minor Injury*	*5. Good lab practice required.*

Section 6: Control Measures (Specify control procedures to each hazardous substance identified in section 4.)

6.1.Physical or engineering controls				
☐ Laboratory	☐ Controlled area	☐ Total containment	☐ Glove Box	☐ Fume cupboard
☐ Microbial safety cabinet	☐ Local exhaust ventilation	☐ Access control	☐ Other: *Enter details*	

You must alsospecify below at which point in the work activity they are to be used.
Specify at which point the control measures should be implemented

6.2 Administrative controls.
Describe administrative controls

6.3 Personal protective equipment (PPE).	
☐ Eyewear protection (Minimum standard CE EN166)	☐ Disposable lab coat
☐ Lab coat	☐ Chemical suit
☐ Specialised footwear. (Minimum standard EN ISO 20345) State type: *Enter details here*	☐ Hearing protection (Minimum standard EN352-1) State type: *Enter details here*
☐ Gloves State minimum standard: ☐ BS EN455 – single use for chemical hazards. ☐ BS EN374 – single use for chemicals hazards and microorganisms. State type used: *Enter details here*	☐ Respirator State type: ☐ Disposable P3 (Minimum standard EN149) ☐ Replaceable filter (Minimum standard EN140) ☐ Powered respirator. State type used: *Enter details here*
☐ Full-face visor	☐ Other State: *Enter details here*

6.4 Storage requirements.
Describe storage conditions

6.5 Transport of hazardous substances.
Describe how you will transport the hazardous substances

6.6 Disposal of waste. If specialised waste is to be generated, you must discuss this with a member of technical staff and consult the university waste policy.

(Continued)

Table 1.7 (Continued)

Waste type	Waste subtype	Detail method of disposal
Select waste type e.g. liquid	*Select a waste sub-type e.g. Inorganic waste*	*Describe the method of disposal e.g. down the sink with plenty of water*

6.7 Emergency procedures.

Minor spillage (for less than 250 mL / 250 g of materials with a low-medium risk rating).	**Major spillage (for greater than 250 mL / 250 g of materials with a low-medium riskrating, or <u>any</u> high risk materials).**
☐ Secure the spill area.	☐ Evacuate and secure the laboratory/area.
☐ Inform a competent person (e.g. a member of technical staff or your supervisor).	☐ Inform a competent person (e.g. a member of technical staff or your supervisor).
☐ Other *Describe other emergency procedures*	☐ Evacuate the building using the fire alarm.

In the event of fire, assuming you are trained in the handling of extinguishers <u>and</u> it is safe to do so, specify which types of fire control may be used:

☐ Carbon dioxide	☐ Water	☐ Dry powder	☐ Foam	☐ Fire blanket	☐ Automatic fire suppression

☐ Other *Describe other fire control measures here, if applicable*

In the event of an accident requiring first aid, seek assistance as soon as possible.
Detail below any specific considerations, which must be made for the hazardous substances in use.

☐ If hazardous material comes into contact with skin, remove any affected clothing and wash the area with copious amounts of water. ☐ For large areas rinse the skin using the emergency shower.	☐ If hazardous material comes into contact with the eyes, rinse the eyes using an eye wash station. ☐ For serious eye burns, use diphoterine station.
☐ For phenol burns, wash with copious amounts of water and apply polyethylene (PEG) 300 to the area.	☐ For hydrofluoric acid burns, wash with copious amounts of water and apply calcium gluconate gel to the area.
☐ If cyanide has been inhaled, move the victim to fresh air.	☐ Other. Please state: *Enter details here*

6.8 <u>Assessment of risk</u> to human health once control measures are in place.
Select the likelihood and severity of harm in the presence of the identified hazards **after** the control measures outlined above are implemented. Calculate the risk rating and act accordingly. Guidance may be found in the appendix by clicking <u>here</u>.

Likelihood of harm	Severity	Risk Rating and Outcome (Likelihood x Severity)
Enter likelihood e.g. 2. Unlikely	*Enter severity e.g. 1. Delay only*	*Enter risk rating e.g. 2. Good laboratory practice only required*

6.9 Instruction, training and supervision.
In consultation with the approver, specify the level of training and supervision required to safely carry out the work described. **Select all that apply.**

☐ Special instructions are required to safely carry out the work.	
☐ Special training is required to safely carry out the work.	
☐ Work may be carried without direct supervision.	☐ Work may be carried without indirect supervision.

Table 1.7   (Continued)

Work may not be started without the advice and approval of the approver.	Work may not be carried out without close supervision.

Section 7: Approval
I hereby confirm that the above is a suitable and sufficient risk assessment for the work activity described.

7.1 The assessor.		
Name	Signature	Date

7.2 The approver (if required).		
Name	Signature	Date

and Precautionary Statement, for each chemical, must be included (Table 1.8). Then, an assessment of the likelihood of harm coming to pass given the amount/nature of the chemical to be used and the environment/manner it is to be used in; at this stage, the likelihood is assessed on the basis that no specific control measures are being taken. The likelihood therefore assesses the highest risk. After assessing the likelihood, the next stage is to consider the severity of the risk. This is done by considering the substance-specific risk (rather than the activity specific risk). Again, like the likelihood, this considers the highest severity. By then performing the RMA (Risk = Likelihood × Severity) (Table 1.9), you arrive at the risk for using the chemical.

The individual working in the laboratory is also a major source of contamination. Therefore, as well as the normal laboratory safety practices of wearing a laboratory coat

Table 1.8   Examples of (a) Hazard[a] and (b) Precautionary[b] Statements.

(a)

Letter	Type of hazard	Intrinsic properties of the substance	Example
H	2 = physical	e.g. explosive properties for codes 200–210; flammability for codes 220–230; etc.	H302 harmful if swallowed.
H	3 = health		
H	4 = environmental		

[a] There are 72 individual and 17 combined Hazard statements.

(b)

Letter	Type of precaution	Examples
P	1 = general precaution	P102 Keep out of the reach of children
P	2 = prevention precaution	P281 Use personal protective equipment as required
P	3 = response precaution	P301 If swallowed:
P	4 = storage precaution	P404 Store in a closed container
P	5 = disposal precaution	P501 Dispose of contents/container to . . .

[b] There are 116 individual and 33 combined Precautionary statements.

Table 1.9 Risk matrix analysis[a].

			Severity					
			6	5	4	3	2	1
			multiple fatalities	single fatality	major injury	lost time injury	minor injury	delay only
Likelihood	6	certain	36	30	24	18	12	6
	5	very likely	30	25	20	15	10	5
	4	likely	24	20	16	12	8	4
	3	may occur	18	15	12	9	6	3
	2	unlikely	12	10	8	6	4	2
	1	remote	6	5	4	3	2	1

[a] **Note: Low risk:** numerical score 1–10. Good laboratory practice (including Personal Protective Equipment of a laboratory coat and safety glasses) required.
High risk: numerical score 12–18. Specific identified control measures must be used.
Very high risk: numerical score 20+. Trained personnel only.

and safety glasses, it may be necessary to take additional steps such as the wearing of 'contaminant-free' gloves, a close-fitting hat or mask (COVID-19) as well as working in a fume cupboard.

The basic generic rules for laboratory work (and as appropriate for associated work outside the laboratory using chemicals) are as follows:

- Always wear appropriate protective clothing, typically, this involves a clean laboratory coat fastened up, eye protection in the form of safety glasses or goggles, appropriate footwear (open toed sandals or similar are inappropriate) and ensure long hair is tied back. In some circumstances, it may be necessary to put on gloves, e.g. when using concentrated acids.
- Never eat or drink in the laboratory.[1]
- Never work alone in a laboratory.[2]
- Make yourself familiar with fire regulations in your laboratory and building.[3]
- Be aware of accident/emergency procedures in your laboratory and building.[3]
- Use appropriate devices for transferring liquids, e.g. pipette, syringe, Gilson.
- Only use/take the minimum quantity of chemical required for your work. This can prevent cross-contamination, as well as reducing the amount to be disposed.
- Use a fume cupboard for hazardous chemicals, e.g. volatile organic compounds. Check that the fume cupboard is functioning properly (i.e. has an air flow that takes fumes away from the worker) before starting your work.

---

1 Smoking is banned in public buildings in the United Kingdom.
2 This is strictly enforced with undergraduate students; however, postgraduate researchers often work in the proximity of others to ensure some safety cover is available. Universities will have procedures in place to allow such work to take place, and it will always involve notifying others of your name and location. In the case of postgraduate researchers, the proximity of a (mobile) telephone is additionally beneficial to alert others.
3 This might involve additional training.

- Clear up spillages and breakages as they occur; for example in the undergraduate laboratory, notify the demonstrator/technician immediately to ensure that appropriate disposal takes place, e.g. broken glass in the glass bin.
- Always work in a logical and systematic manner; it saves time and can prevent a waste of resources, e.g. only weighing out the amount of chemical required when it is required.
- Always think ahead and plan your work; accordingly, this involves reading the laboratory script before you enter the laboratory, as well as checking that you are following the script while undertaking the experiment.

1.5 Considerations for Data Presentation

1.5.1 Useful Tips on Presenting Data in Tables

Tables are a useful method for recording numerical data in a readily understandable form. Tables provide the opportunity to summarize data and to allow comparisons between methods. Typically, the data is shown in columns (running vertically) and rows (running horizontally). Columns may contain details of the sample (a sample code identifier), concentration (with units), names of compounds, as well as the properties measured, while rows contain the written or numerical information for the columns. An example is shown in Table 1.12.

1.5.2 Useful Tips on Presenting Data in Graphical Form

Graphs are used, normally, to represent a relationship between two variables, x and y. It is normal practice to identify the x-axis as the horizontal axis (absicca axis) and to use this for the independent variable e.g. concentration ($\mu g \, ml^{-1}$). The vertical or ordinate axis (y-axis) is used to plot the dependent variable, e.g. signal response (mV). An example is shown in Figure 1.2a.

1.6 Use and Determination of Significant Figures

A common issue when recording data from practical work is the reporting of significant figures. The issue is important as it conveys, to the reader, an understanding of the underlying practical work. A few examples will illustrate the issues and how they can be interpreted.

Example 1.1   When asked to *accurately* weigh out approximately 0.5 g of sample, how many decimal places should be reported?

Response 1.1   In this situation, it would be expected that a four decimal place analytical balance would be used to accurately weigh out the sample. On that basis, the sample would be recorded as, for example 0.5026 g. In practice, the sample would have been weighed by difference, i.e. a sample container would be first weighed, then the sample placed inside the container and the weight again recorded, and finally, the sample transferred to an extraction vessel, and the sample container re-weighed. By taking the weights of the container with/without the sample allows an accurate recording of the weight of sample transferred to the container.

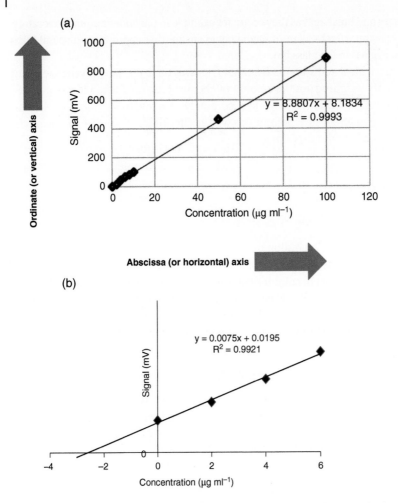

Figure 1.2   Calibration graphs. (a) A direct calibration graph and (b) a standard additions method calibration graph.

Example 1.2   Is it appropriate to round up/down numbers?
Response 1.2   Yes, for example if you have a numerical value, representing a weight or concentration, of 276.643, it would be reasonable to represent this as 276.6 or even 277. If the value was 0.828, then it may be reasonable to round up to 0.83. Whereas for a value of 12 763, it would be reasonable to report as 12 763 or, in some circumstances 12 760.

Example 1.3   In calculating a result in a spreadsheet, for the concentration (mg kg^{-1}) of an organic compound in a solid sample, a numerical value of 25.21 345 678 is obtained. Is this correct?
Response 1.3   No, this is a totally unrealistic representation in terms of the number of decimal places, in terms of the actual determination of the concentration, its interpretation as a concentration and demonstrates a lack of understanding of the data. A more appropriate, and realistic, reporting of the concentration would be 25.2 mg kg^{-1}.

In general terms, the following guidance is provided:

- When rounding up numbers, add one to the last figure if the number is greater than 5, e.g. 0.54667 would become 0.5467.
- When rounding down numbers, remove one to the last figure if the number is less than 5, e.g. 0.54662 would become 0.5466.
- For a number 5, round to the nearest even number, e.g. 0.955 would become 0.96 (to two significant figures) OR if the value before 5 is even, it is left unchanged, e.g. 0.945 would become 0.94 (to two significant figures) OR if the value before 5 is odd, its value is increased by one, e.g. 0.955 would become 0.96 (to two significant figures).
- Zero is not a significant figure when it is the first figure in a number, e.g. 0.0067 (this has two significant figures **6** and **7**). In this situation, it is best to use scientific notation, e.g. 6.7×10^{-3}.

1.7 Units

The Systeme International d'Unites (SI) is the internationally recognized system for measurement (Table 1.10). The most used SI derived units are shown in Table 1.11. It is also common practice to use prefixes (Table 1.12) to denote multiples of 10^3. This allows numbers to be kept between 0.1 and 1000. For example, 1000 ppm (parts per million) can also be expressed as $1000 \mu g \, ml^{-1}$ or $1000 \, mg \, l^{-1}$ or $1000 \, ng \, \mu l^{-1}$.

Table 1.10 Some commonly used base SI Units.

Measured quantity	Name of SI unit	Symbol
Length	Metre	m
Mass	Kilogram	kg
Amount of substance	Mole	mol
Time	Second	s
Thermodynamic temperature	Kelvin	K

Table 1.11 Some commonly used derived SI units.

Measured quantity	Name of unit	Symbol	Definition in base units	Alternative in derived units
Electric charge	Coulomb	C	$A \, s$	$J \, V^{-1}$
Energy	Joule	J	$m^2 \, kg \, s^{-2}$	$N \, m$
Force	Newton	N	$m \, kg \, s^{-2}$	$J \, m^{-1}$
Frequency	Hertz	Hz	s^{-1}	—
Pressure	Pascal	Pa	$kg \, m^{-1} \, s^{-2}$	$N \, m^{-2}$
Power	Watt	W	$m^2 \, kg \, s^{-3}$	$J \, s^{-1}$

Table 1.12   Commonly used prefixes.

Multiple	Prefix	Symbol
10^{15}	peta	P
10^{12}	tera	T
10^{9}	giga	G
10^{6}	mega	M
10^{3}	kilo	k
10^{-3}	milli	m
10^{-6}	micro	μ
10^{-9}	nano	n
10^{-12}	pico	p
10^{-15}	femto	f

1.8   Calibration and Quantitative Analysis

Quantitative analysis is the cornerstone of environmental analysis with all analyses requiring some form of calibration associated with the determination of contaminants. To assess the level of contamination requires the determination of a concentration; this is achieved using a calibration graph. To perform a calibration graph requires the preparation of calibration solutions from a stock solution, as well as the practical skills inherent in weighing, dilutions and quantitative transfer of solids and liquids. Figure 1.3 shows the manual dexterity required to quantitatively remove a known volume of stock solution from a volumetric flask. All these skills require knowledge of balances, pipettes and volumetric flasks, as well as the correct choice of standards (e.g. pentachlorophenol, benzo(a)pyrene) and their associated grades (i.e. purity), including the use of solvents (including grades of acetone, dichloromethane).

1.9   Terminology in Quantitative Analysis

In doing any quantitative analyses, it is important to be consistent in the terminology used. A concise description of some of the key terms is shown below:

Accuracy: The closeness of agreement between a test result (i.e. measured in the laboratory) and the accepted reference value (i.e. from a certified reference material; see Section 1.16).

Error (of measurement): The result of a measurement minus the true value of the measurand.

   Random error *(of a result):* A component of the error which, in the course of a number of test results for the same characteristic, varies in an unpredictable way; it is not possible to correct for random error.

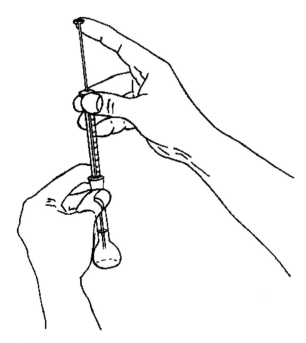

Figure 1.3 Quantitative transfer of a known volume of stock solution for preparation of calibration solutions, using a microsyringe. Note the insertion of the syringe containing the organic solvent/ extract in the receiving solution.

Systematic error: A component of the error which, in the course of a number of test results for the same characteristic, remains constant or varies in a predictable way; systematic errors and their causes may be known or unknown.

Precision: The closeness of agreement between independent test results obtained under stipulated conditions.

Repeatability: Precision under repeatability conditions, i.e. conditions where independent test results are obtained with the same method on identical test items in the same laboratory, by the same operator, using the same equipment within short intervals of time.

Reproducibility: Precision under reproducibility conditions, i.e. conditions where test results are obtained with the same method on identical test items in different laboratories, with different operators, using different equipment.

Uncertainty: Parameter, associated with the result of a measurement, that characterizes the dispersion of the values that could reasonably be attributed to the compound.

A pictorial representation of the terms accuracy and precision is given in Figure 1.4, with explanatory notes.

1.10 Preparing Solutions for Quantitative Work

The basis of any quantitative work is that you start with a known concentration of the substance (e.g. pentachlorophenol, benzo(a)pyrene) as a stock solution. Serial dilutions are then required to produce a set of calibration solutions, which are then run through a specific

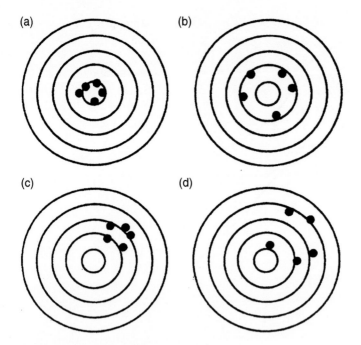

The centre of the bullseye represents the 'true' value.

NOTES: (a) the data points would be classed as accurate and precise, (b) the data points would be classed as accurate but imprecise, (c) the data points would be classed as inaccurate but precise and (d) the data points would be classed as inaccurate and imprecise.

Figure 1.4   A pictorial representation of the term's accuracy and precision.

analytical instrument and the responses recorded. The generated data is then used to produce a calibration graph (see, Section 1.11) against which other unknown samples can then be compared.

Solutions are usually prepared in terms of their molar concentrations, e.g. mol l^{-1}, or mass concentrations, e.g. $\mu g \ ml^{-1}$. Both units refer to an amount per unit volume, i.e. concentration = amount/volume. It is important to use the highest (purity) grade of chemicals (liquids or solids) for the preparation of solutions for quantitative analysis, e.g. TraceCERT® or a specified purity, e.g. 97%.

Example 1.4   Prepare a 1000 $\mu g \ ml^{-1}$ solution of benzo(a)pyrene (BaP) in a 10 ml volumetric flask. The CAS number for benzo(a)pyrene is 50-32-8.

Response 1.4   For a 1000 $\mu g \ ml^{-1}$ solution BaP, dissolve 10.00 mg of BaP in acetone (or another solvent, e.g. dichloromethane) and dilute to 10.0 ml in acetone. This will give you a 1000 $\mu g \ ml^{-1}$ solution of BaP.

Example 1.5   Prepare a 0.1 mol l^{-1} solution of benzo(a)pyrene (BaP) in a 10 ml volumetric flask. The CAS number for benzo(a)pyrene is 50-32-8. The molecular weight of benzo(a)pyrene ($C_{20}H_{12}$) is 252.31.

Response 1.5 For a 0.1 mol l^{-1} solution of BaP, dissolve 252 mg of BaP in acetone and dilute to 10 ml in acetone. This will give you a 0.1 mol l^{-1} solution of BaP.

Example 1.6 Prepare a 1000 µg ml^{-1} solution of pentachlorophenol (C_6HCl_5O) (PCP), from the pure compound, in a 10 ml volumetric flask. The CAS number of PCP is 87-86-5.
Response 1.6 For a 1000 µg ml^{-1} solution of PCP, dissolve 10.00 mg of PCP in methanol and dilute to 10.0 ml in methanol. This will give you a 1000 µg ml^{-1} solution of PCP.

Example 1.7 Prepare a 0.1 mol l^{-1} solution of pentachlorophenol (C_6HCl_5O) (PCP), from the pure compound, in a 10 ml volumetric flask. The CAS number of PCP is 87-86-5, and its molecular weight is 266.34.
Response 1.7 For a 0.1 mol l^{-1} solution of PCP: Dissolve 266 mg of PCP in methanol and dilute to 10 ml in methanol. This will give you a 0.1 mol l^{-1} solution of PCP.

1.11 Calibration Graphs

Calibration graphs can be done in two different formats, typically a 'direct' plot or a standard additions method plot (Figure 1.2). In the normal 'direct' calibration graph, the most common type of calibration graph, a plot of signal response (Y on the ordinate axis) versus increasing concentration (x on the abscissa axis) of the compound is made (Figure 1.2a). It is then possible to estimate the concentration of a compound in an unknown sample by interpolation, either graphically or by regression (using, for example Microsoft excel) (see, further reading section at the end of this chapter). Assuming a linear response allows a plot of the line of regression of y on x to be made:

$$y = m.x + c \qquad (1.1)$$

where y is the signal response, e.g. absorbance, signal (mV); x is the concentration of the calibration solution (in appropriate units, e.g. µg ml^{-1} or ppm); m is the slope of the graph and c is the intercept on the x-axis.

By simple re-arrangement allows the determination of the unknown sample concentration (x):

$$\left(y - c\right) / m = x \qquad (1.2)$$

Alternatively, the method of standard additions can be used; this approach is useful if the sample is known to contain a potentially interfering matrix. In this approach, a known (and fixed) volume of the sample is added to each of the calibration solutions. The volume of sample to be added to the calibration solutions needs to be estimated; this can be done by first running the unknown sample and interpolating from the direct plot. By again plotting the signal response against concentration of compound (as above), a different format of graph is obtained (Figure 1.2b). The graph no longer passes through zero on either axis; by extending the graph toward the x-axis (extrapolation) until it intercepts, it allows the concentration of the analyte in the unknown sample to be estimated. It is essential that the standard additions plot is linear over its entire length, otherwise considerable error will be introduced; it may be therefore necessary to either add a smaller volume of sample to the calibration solutions or alter the concentration range used. The term *linearity* is used to define the ability of the method to obtain test results proportional to the concentration of the analyte, whereas the *linear dynamic range* is the concentration range over which the analytical working calibration curve remains linear.

Example 1.8   What is the linear dynamic range of the calibration plot shown in Figure 1.5a?
Response 1.8   The linear dynamic range extends from 0 to 160 mg l^{-1} (Figure 1.5b).

1.12   The Internal Standard

Sometimes a chemical compound is added in equal amounts to all samples, and this is referred to as the internal standard. One consequence of adding an internal standard is that it influences the linearity of measurement with an analytical technique. In these situations,

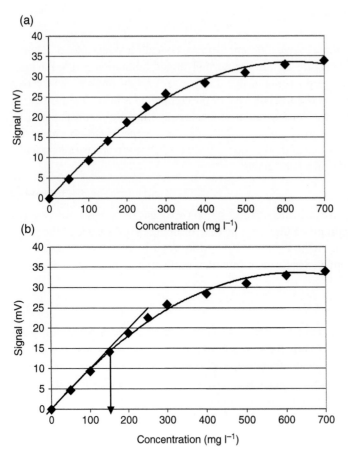

Figure 1.5   An investigation of linear dynamic range. (a) Calibration graph, and (b) interpretation of linear dynamic range.

the y-axis (response) is plotted as the ratio of the compound to the internal standard signal versus the compound concentration. The benefits of adding an internal standard are multi-fold, as it allows correction for the following:

- Compound loss during sample preparation
- Non-quantitative transfer of samples between containers
- Compound adsorption loss on storage
- Evaporation loss during pre-concentration (Chapter 15)
- Variation in injection volume (e.g. in gas chromatography)
- Variation in mass spectrometer response due to ion suppression or enhancement (e.g. HPLC-MS).

However, selection of an appropriate compound to act as an internal standard requires consideration. Ideally, it should have similar physicochemical properties and display similar behaviour to the compound or compounds for which it is acting as the internal standard. The two generic types of internal standard that are used are either structural analogues or stable isotope-labelled compounds. In the case of stable isotope-labelled compounds, the label is often ^{13}C, ^{15}N or ^{2}H (D) (Figure 1.6). In the case of selecting a structural analogue compound as an internal standard, the proposed compound ideally has similar structure and functionalities, e.g. COOH, -CHO and -Cl etc. with differences only being with C–H moieties (Figure 1.7).

1.13 Limits of Detection/Quantitation

The limit of detection (LOD) of an analytical procedure is the lowest amount of compound in an unknown sample which can be detected but not necessarily quantified, i.e. recorded as an exact value. Various definitions exist as to the method of determining the LOD. When quoting concentrations as LOD, it is appropriate to indicate the exact method of determination. The limit of detection, expressed as a concentration (in appropriate units), is derived

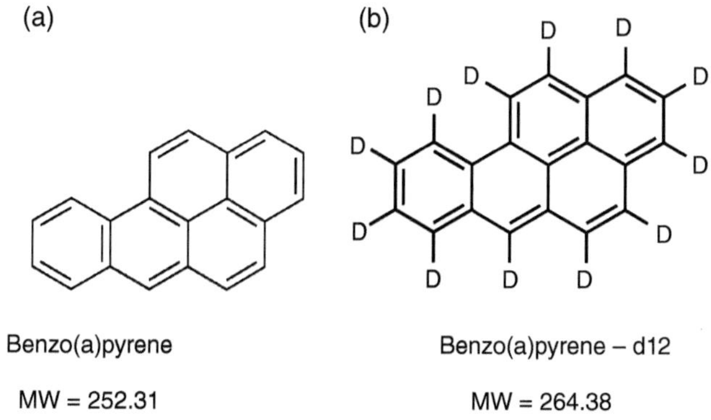

Figure 1.6 An example of a stable isotope-labelled compound as an internal standard. (a) Analyte compound and (b) labelled compound.

(a)   (b)

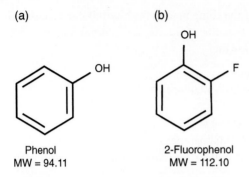

Phenol   2-Fluorophenol
MW = 94.11   MW = 112.10

Figure 1.7 An example of a structural analogue compound as an internal standard. (a) Analyte compound and (b) analogue compound.

from the smallest measure, X, that can be detected with reasonable certainty for a given procedure. One approach to determine the LOD is to measure the signal of a known concentration at or near the lowest concentration that is observable (normally at least seven times). The value X is given by the equation:

$$X = X_{LCS} + K.SD_{LCS} \tag{1.3}$$

where X_{LCS} is the mean of the low concentration standard, SD_{LCS} is the standard deviation of the low concentration standard and K is a numerical factor chosen according to the confidence level required (typically 2 or 3).

An alternate approach, useful in chromatography, to calculate the LOD is by determining the concentration of compound providing a signal-to-noise ratio (S/N) of three. In this approach, the signal is measured from decreasing standard solutions until a signal is found whose height is three times taller than the maximum height of the baseline (measured at both sides of the chromatographic peak). The concentration corresponding to that peak is taken as the LOD.

Another approach to calculate the LOD is as follows:

$$LOD = 3.3.\sigma / m \tag{1.4}$$

where σ is the standard deviation of the response and m is the slope of the calibration curve (see Eq. 1.1).

An estimate for the value of σ can be made by either (i) measurement of the magnitude of the analytical background response (i.e. analyse a blank sample (a minimum of seven times) and calculate the standard deviation), or (ii) use the standard deviation of the y-intercept (c term in Eq. 1.2) of multiple regression lines.

As LODs are often not practically measurable, a more realistic value is to use the limit of quantitation (LOQ) of an analytical procedure. The LOQ is the lowest amount of an analyte in a sample, which can be quantitatively determined with suitable uncertainty; the LOQ can be taken as 10× 'the signal-to-noise ratio' or K = 10 in Eq. (1.3), or substitute 10 (instead of 3.3) in Eq. (1.4).

1.14 Dilution or Concentration Factors

Once the concentration of the compound has been determined from the calibration graph, it is necessary to report its actual concentration in the original starting material, e.g. soil or river water sample. This requires the use of either a dilution or concentration factor. The following examples illustrate the use of these factors.

Example 1.9 Use of the dilution factor. Based on the following information, calculate the concentration of a compound (in units of mg kg^{-1}) in the original soil sample.
An accurately weighed soil sample (2.1189 g) was extracted with dichloromethane. The extract was quantitatively transferred to a 25 ml volumetric flask and made up to the mark in dichloromethane. This solution was then serially diluted by taking 1 ml of the solution and transferring it to a further 10 ml volumetric flask where it is made up to the mark with dichloromethane.
What is the dilution factor?

Response 1.9 The dilution factor is calculated as follows:

$$\left(25\,\mathrm{ml}/2.1189\,\mathrm{g}\right)\times\left(10\,\mathrm{ml}/1\,\mathrm{ml}\right)=119.0\,\mathrm{ml\,g^{-1}} \tag{1.5}$$

If the solution were then analysed and found to be within the linear portion of the graph (see Figure 1.5b), the value for the dilution factor would then be multiplied by the concentration obtained from the graph. So, if the concentration from the graph was determined to be 15.1 mg l^{-1} (or 15.5 µg ml^{-1}), it would produce a final value, representative of the compound under investigation, i.e. 1797 µg g^{-1}. It is important to consider the number of significant figures quoted; in this case, 1797 µg g^{-1} (or 1797 mg kg^{-1}). The amount of the compound in the original soil sample is therefore 1797 mg kg^{-1}.

Example 1.10 Use of concentration factor. Based on the following information, calculate the concentration of pentachlorophenol (in units of µg l^{-1}) in the original wastewater sample.
A wastewater sample (1000 ml) was extracted into dichloromethane (3×5 ml) using liquid–liquid extraction. The extract was then quantitatively transferred to a 25 ml volumetric flask and made up to the mark in dichloromethane. What is the concentration factor?

Response 1.10 The concentration factor is calculated as follows:

$$\left(25\,\mathrm{ml}/1000\,\mathrm{ml}\right)=0.025 \tag{1.6}$$

If the solution was then analysed and found to be within the linear portion of the graph (see Figure 1.5b), the value for the concentration factor would then be multiplied by the concentration from the graph. So, if the concentration from the graph was determined to be 58.8 ng ml^{-1}, it would produce a final value, representative of the compound under investigation, i.e. 1.47 ng g^{-1}. It is important to consider the number of significant figures quoted; in this case 1.5 ng ml^{-1} is appropriate; also, be careful with prefixes on units (see, Table 1.12)]. The amount of pentachlorophenol in the original wastewater sample is therefore 1.5 µg l^{-1}.

1.15   Quality Assurance

Quality assurance is all about getting the correct results that are representative of the original sample, i.e. the contaminated land site from which the soil sample was obtained or the river water from which the sample was obtained. In practice, in environmental analyses, this is extremely challenging as it involves multiple steps: sampling, sample collection, sample storage, sample preparation, analytical determination and data interpretation and action. Considerations can be taken to control and inform the sampling, sample collection and storage (see Chapter 2). Subsequently, the samples are prepared and analysed in a laboratory. It is possible to ensure that the laboratory is functioning appropriately by adopting a good quality assurance scheme. The main objectives of a quality assurance scheme are:

- selection and validation of an appropriate method of sample preparation;
- selection and validation of an appropriate method of analysis;
- regular maintenance (and upgrading) of analytical instruments;
- ensure appropriate record of methods and results are maintained;
- ensure that high quality data is produced;
- overall to ensure that a high quality of laboratory performance is maintained.

Examples of important aspects of establishing and maintaining such a QA scheme are as follows:

- Individual performing the analyses
  - o Has the individual been trained in the use of the instrumentation and/or procedures? If so by whom (where they trained or experienced)?
  - o Was the training formal (formal qualification or certificate of competency obtained) or done in-house?
  - o Can the individual use the instrumentation alone or do they require oversight?

- Laboratory procedures and practices
  - o Do the procedures use certified reference materials to assess the accuracy of the method? (see Section 1.16)
  - o Do the procedures use spiked samples to assess recoveries? The samples are spiked with a known concentration of the analyte under investigation and their recoveries noted; this allows an estimate of analyte matrix effects.
  - o Do the procedures include analysis of reagent blanks? Analysing reagents whenever the batch is changed or a new reagent introduced allows reagent purity to be assessed and, if necessary controlled, and also acts to assess the overall procedural blank; typically introduce a minimum number of reagent blanks, i.e. 5% of the sample load).
  - o Do the procedures use standards to calibrate instruments? A minimum number of standards should be used to generate the analytical curve, e.g. minimum of 5. Daily verification of the calibration plot should be done using one or more standards within the linear working range.
  - o Do the procedures include the analysis of duplicate samples? Analysis of duplicates or triplicates allows the precision of the method to be determined and reported.

o Do the procedures include known standards within the sample run? A known standard should be run after every 10 samples to assess instrument stability; this also verifies the use of a daily calibration plot.

1.16 Use of Certified Reference Materials

A certified reference material (CRM) is a substance for which one or more analytes have certified values, produced by a technically valid procedure, accompanied with a traceable certificate (Table 1.13) and issued by a certifying body. Examples of certifying bodies include National Institute of Science and Technology NIST), USA and LGC, UK.

Table 1.13 An example of a certificate for a certified reference material[d]

National Institute of Science and Technology Certificate of Analysis
Standard Reference Material 1234: Sandy loam soil Certified Values

Constituent	Certified value[a,b] ($\mu g\ kg^{-1}$)	Uncertainty[a,b,c] ($\mu g\ kg^{-1}$)	Weight of sample (g)[d]
PCB101	57	6	5
PCB118	120	4	5

Certified Values

Constituent	Certified value[a,b] ($mg\ kg^{-1}$)	Uncertainty[a,b,c] ($mg\ kg^{-1}$)	Weight of sample (g)[d]
Phenanthrene	192	6	5
Fluoranthene	320	7	5
Benzo[a]pyrene	31	1	5

Indicative Values

Constituent	Indicative value ($mg\ kg^{-1}$)[a,e]	Number of laboratories
Anthracene	4	8
Chrysene	12	7
Fluorene	8	20

[a] Values expressed on a dry weight basis
[b] The certified values were obtained using procedures involving pressurised liquid extraction.
[c] The uncertainty interval calculated provides a level of confidence of 95%.
[d] Weight of sample taken for homogeneity assessment
[e] Interlaboratory mean of means of the final data set.

It can be seen on the certificate (Table 1.13) that some of the concentration values are 'certified' while others are 'indicative'. The use of the term certified means that the concentrations stated are reliable, whereas the term indicative means that the concentrations stated have some uncertainty (or an insufficient number of methods have been used in their characterization). In addition, other information would be contained on the certificate relating to the actual material supplied, specifically, details of the minimum amount of material that is representative of the whole. If a smaller sample size is taken than recommended on the certificate, then the certified value and its uncertainty are not guaranteed, the expiry date or shelf-life. This is the last date that the material should be used and remains within its certified value and its uncertainty, and moisture correction. Often the material will report its certified value and its uncertainty based on its dry mass. In these situations, correction for the moisture content can be made provided that the dry mass is determined on a separate sub-sample. An extensive range of CRMs are available for organic compounds in different matrices. Typical compounds include dioxins and furans; hydrocarbons and petrochemicals; PCBs and related compounds; pesticides and metabolites, pharmaceutical and veterinary compounds and metabolites; phenols and aromatic compounds; polycyclic aromatic hydrocarbons; and volatile organic compounds, while matrices include waters (e.g. river, waste); sediments (e.g. lake); soils (e.g. loam); sewage sludges (e.g. domestic); and ash, particulate and dusts (e.g. fine).

1.17 Applications

Case Study A   Calculation of the Limit of Detection and Limit of Quantitation

Background: The compound atropine (Figure 1.8) is a prescription medicine used to treat the symptoms of low heart rate (bradycardia), reduce salivation and bronchial secretions before surgery or as an antidote for overdose of cholinergic drugs or mushroom poisoning. It is common practice within the determination of the key figures of merit of an analytical technique is to estimate the limit of detection and limit of quantitation. An approach uses the following equations:

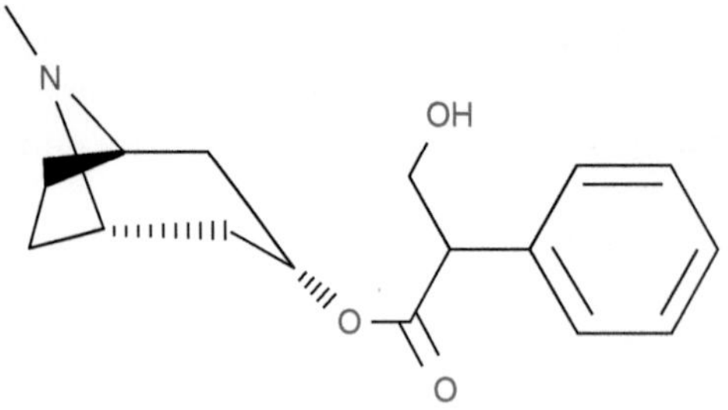

Figure 1.8   Structure of atropine.

$$LOD = 3.3.\sigma / m \tag{1.4}$$

$$LOQ = 10.\sigma / m \tag{1.7}$$

where σ is the standard deviation of the y-intercepts (c term in Eq. 1.2) of multiple regression lines and m is the (average) slope of the calibration curve (see Eq. 1.1).

Activity: An organic compound (atropine) was analysed using HPLC-UV. To determine the LOD and LOQ, a calibration was produced over the concentration range 0–100 μg ml^{-1} (using 11 data points, 0, 10, 20, 30, 40, 50, 60, 70, 80, 90 and 100 μg ml^{-1}). The calibration was run six times to obtain six multiple calibration graphs. Using the data obtained (Table 1.14), calculate the LOD and LOQ.

$$LOD = (3.3 \times 6.533) / 11.848$$
$$LOD = 1.8 \, \mu gml^{-1} \tag{1.8}$$

and

$$LOQ = (10 \times 6.533) / 11.848$$
$$LOQ = 5.5 \, \mu gml^{-1} \tag{1.9}$$

Table 1.14 Data for calculation of LOD and LOQ.

Data set	Calibration equation (y = mx+c)	Slope (m)	Intercept (c)
1	y = 11.854x − 28.082	11.854	28.082
2	y = 11.942x − 33.706	11.942	33.706
3	y = 11.640x − 26.002	11.640	26.002
4	y = 11.902x − 34.188	11.902	34.188
5	y = 11.765x − 36.253	11.765	36.253
6	y = 11.987x − 44.464	11.987	44.464
Average		**11.848**	
Standard deviation			**6.533**

Case Study B   Some Numerical Worked Examples

Background: By a series of worked examples, the numerical and graph plotting aspects of environmental analysis are considered. The essential stages to the calculations are as follows:

- Determine the concentration of the working solutions.
- Plot the calibration graph (concentration versus signal response). Graph plotting can be done using either a suitable spreadsheet, e.g. Microsoft Excel, or on graph paper.

- Determine the best fit for the calibration data. If we assume that a straight-line graph is obtained, then the following applies. If using a suitable spreadsheet, e.g. Microsoft Excel, then this can be done by selecting 'add trendline' followed by 'display equation on chart' and 'display r squared value on chart'. If using graph paper manually plot the data points, then by using a ruler or flexi curve establish the best fit of the data points to each other. Determine the intercept of the fitted line on the x-axis and calculate the slope of the line. In either case, you should now have the formula for a straight-line equation, i.e. $y = mx + c$, where y is the signal response, m is the slope of the graph, x is the concentration (in appropriate units) and c is the intercept of the line of best fit on the x-axis].
- Calculate, using the equation $y = mx + c$, the concentration (in appropriate units) of the sample based on its generated signal response. This can be done by re-arranging the equation $y = mx + c$ such that the concentration of the sample, x, can be determined as follows: $x = (y - c)/m$].
- Then, establish the dilution/concentration factor associated with the sample preparation (see Section 1.14).
- Calculate the concentration (in appropriate units) based on the dilution/concentration factor (in appropriate units) multiplied by the sample concentration (in appropriate units) as determined from the calibration graph.
- Finally, check that the reported concentration in the original sample is in appropriate units.

Activity 1: An aqueous sample was analysed by GC-FID for pentachlorophenol. The sample was extracted by placing 5 ml of the aqueous sample into a separating funnel with 2 × 5 ml of dichloromethane. The extract was quantitatively transferred to a volumetric flask (25 ml) and made up to volume with dichloromethane (including the addition of internal standard).

A calibration plot was generated by diluting a 500 µg ml^{-1} stock solution of pentachlorophenol. Then, 1 ml of the stock solution was placed in a 10 ml volumetric flask and made up to the mark with acetone (working solution). The working solution is further diluted (Table 1.15) to make the standard solutions.

Table 1.15   Data for worked example 1.

Flask	Pentachlorophenol working solution (ml)	final volume (ml)	GC-FID response (signal)
1	0.00	10	0
2	0.50	10	1523
3	1.50	10	3567
4	3.00	10	6235
5	6.00	10	13 563
Diluted sample			8563

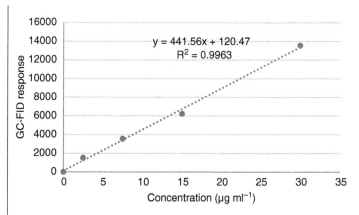

Figure 1.9 Calibration graph for worked example 1, the determination of pentachlorophenol by GC-FID.

By plotting a fully annotated calibration graph of signal response (y-axis) versus concentration ($\mu g \ ml^{-1}$) (x-axis), determine the concentration of pentachlorophenol, in units of $\mu g \ ml^{-1}$, as determined from the graph. Using the dilution/concentration factor and units of the sample extract, determine the concentration of pentachlorophenol, in units of $mg \ l^{-1}$, in the original aqueous sample.

The calibration graph is shown in Figure 1.9. The regression coefficient (R^2) indicates the linearity of the graph (i.e. close to a perfect straight-line of 1.0000). Using the equation $y = mx+c$, the concentration (in appropriate units) was determined.

$$x = (y-c)/m$$
$$x = (8563-120.47)/441.56$$
$$x = 19.1 \mu gml^{-1}$$

Then, using the dilution/concentration factor, determine the concentration of PCP in the original sample.

$$\text{Dilution / concentration factor is} (25ml)/(5ml) = 5$$

The concentration in the original sample was:

$$19.1 \mu gml^{-1} \times 5 = 95.5 \mu gml^{-1}$$

The concentration of pentachlorophenol in the aqueous sample is $95.5 \mu g \ ml^{-1}$, which is equivalent to $95.5 \ mg \ l^{-1}$.

Activity 2: A soil sample was analysed for benzo(a)pyrene as follows. An accurately weighed sample of 2.1351 g was extracted. The extract was quantitatively transferred to a volumetric flask (25 ml) and made up to volume with solvent. A calibration plot was generated by diluting a $1000 \mu g \ ml^{-1}$ stock solution of benzo(a)pyrene. Then, 1 ml of the stock solution was placed in a 10 ml volumetric flask and made up to the mark with solvent (working solution). This solution (Table 1.16) was diluted to make the following standard solutions:

Table 1.16   Data for worked example 2.

Flask	Benzo(a)pyrene working solution (ml)	Final volume (ml)	GC-MS signal
1	0	10	0
2	0.1	10	150
3	0.2	10	290
4	0.3	10	435
5	0.5	10	730
Extracted sample		25	490

Plot a fully annotated calibration graph of signal response (y-axis) versus concentration (μg ml^{-1}) (x-axis). Then, determine the concentration of benzo(a)pyrene, in units of μg ml^{-1}, as determined from the graph. Calculate the dilution/concentration factor and units of the sample extract. Finally, determine the concentration of benzo(a)pyrene, in units of mg kg^{-1}, in the original sample.

The calibration graph is shown in Figure 1.10. The regression coefficient (R^2) indicates the linearity of the graph (i.e. close to a perfect straight-line of 1.0000). Using the equation $y = mx + c$, the concentration (in appropriate units) was determined.

$$x = (y - c) / m$$
$$x = (490 - 0.8108) / 145.5$$
$$x = 3.36 \, \mu g \, ml^{-1}$$

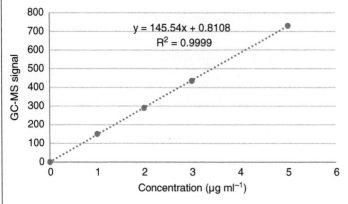

Figure 1.10   Calibration graph for worked example 2, the determination of benzo(a)pyrene by GC-MS.

Then, using the dilution/concentration factor, determine the concentration of PCP in the original sample.

Dilution / concentration factor is $(25 \text{ ml}) / 2.1351 \text{ g} = 11.71 \text{ ml g}^{-1}$

The concentration in the original sample was:

$3.36 \text{ } \mu\text{g ml}^{-1} \times 11.71 \text{ ml g}^{-1} = 39.3 \text{ } \mu\text{g g}^{-1}$

The concentration of pentachlorophenol in the aqueous sample is $39.3 \text{ } \mu\text{g g}^{-1}$, which is equivalent to 39.3 mg kg^{-1}.

Further Reading

Dean, J.R., Jones, A.M., Holmes, D. et al. (2017). *Practical Skills in Chemistry*, 3e. Harlow, UK: Pearson.

Section B

Sampling

2

Sampling and Storage

LEARNING OBJECTIVES

After completing this chapter, students should be able to:

- Understand the underlying principles of environmental sampling of solid, liquid and gaseous samples.
- Be aware of the key physicochemical properties of water and soil samples.
- Comprehend the interrelationships and terminology of the sampling and analytical operations.
- Understand the principle and limitations of sample storage with respect to organic compounds.

2.1 Introduction

Sampling involves taking a representative sample (i.e. a sub-sample) from the inhomogeneous whole (sample). Inhomogeneity can be both temporal (i.e. vary in time) and spatial (i.e. vary in space), and the inter-sample variability can be large (Figure 2.1). Therefore, sampling in its simplest form involves taking a sample from an agricultural field, a water sample from a river or an indoor air sample from the home. The development of a sampling strategy involves consideration of some key aspects including:

- Identification of sampling site location, i.e. grid reference or latitude/longitude on map.
- Identification of spatial sampling site data, e.g. altitude or depth from surface or sea level.
- Confirm the date/time and number of sampling visits planned.
- Identification of the method of sampling, including a description of the sampling equipment used and type of samples.
- Confirm the size (mass/volume) of samples, i.e. the amount of sample to be removed from the bulk. The larger the sample the more homogeneous it is likely to be.
- Consider the additional information required, e.g. climatic conditions (i.e. precipitation, wind direction and speed); topography of site; tidal information (neaps/springs); anthropogenic activities adjacent to the site (e.g. land use, agricultural practice, industry etc.).

Extraction Techniques for Environmental Analysis, First Edition. John R. Dean.
© 2022 John Wiley & Sons Ltd. Published 2022 by John Wiley & Sons Ltd.

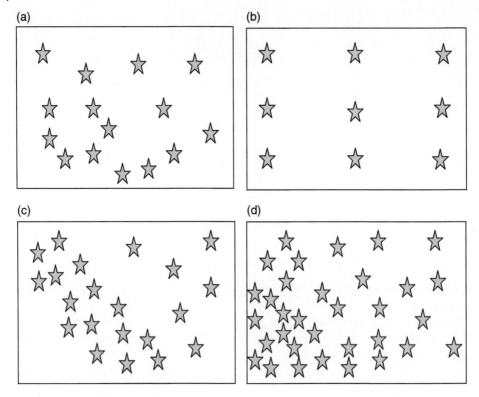

Figure 2.1 Potential contaminant distribution across an environmental site. (a) Random. (b) Uniform. (c) Stratified. (d) Gradient.

2.2 Sampling Strategy

In developing a sampling strategy for a field site (sampling site), the approaches for sampling can be considered as random, systematic or in a stratified random manner (Figure 2.2).

- In random sampling (Figure 2.2a), a two-dimensional co-ordinate grid is superimposed on the area to be investigated. The selection of samples is completely 'down to the luck of the draw', or random, without regard to the variation of the contaminant. The entire sample area is not sampled, but every site on the grid has an equal chance of being selected for sampling. This type of sampling is ideal if the contaminant is homogeneous within the site.
- In systematic sampling (Figure 2.2b), a two-dimensional co-ordinate grid is superimposed on the area to be investigated and the position of the first sample is selected at random. Then, further samples are taken at fixed distances/directions (e.g. at intervals of 5 m). This type of sampling has the potential to provide more accurate results than simple random sampling as the whole site will be sampled.
- In stratified sampling (Figure 2.2c and d) is commonly used in locations that are known to have contaminants heterogeneously distributed. This is therefore the most common approach to sampling. In this type of sampling, the site is sub-divided into smaller areas each of which is fairly homogeneous. Each sub-area is then randomly sampled. The sub-dividing of the site can be done into either equal areas (Figure 2.2c) or related to known features within the site (Figure 2.2d).

(a)

(b)

(c)

(d)

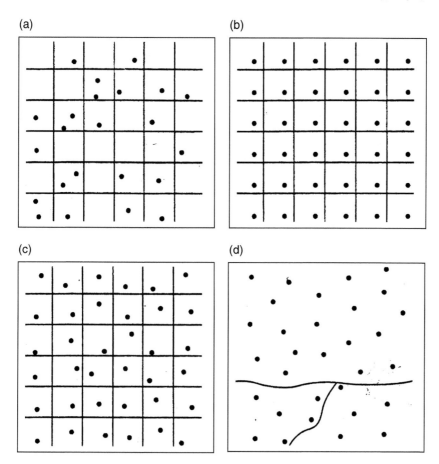

Figure 2.2 Some example approaches for sampling. (a) Random. (b) Systematic. (c) Stratified random based on sub-dividing site into equal areas. (d) Stratified random based on Known features within the site.

2.3 Types of Aqueous Matrices

Water can be divided into six types: rainwater, freshwater, estuarine water, seawater, drinking water and wastewater.

- **Rainwater:** It is generally unbuffered and contains low levels of dissolved salts. The initial washout, from the ground, after a dry period may contain a high proportion of particulates, but the overall bulk of the precipitation is relatively free from colloidal or suspended material.
- **Freshwater:** The quality and content of freshwater is dominated by its source and location of the sampling point in the river catchment and the flow of the system at the time. Soft water is low in ionic strength and frequently characterized by the presence of humic substances predominantly in high flow conditions. Hard water can have a varying load of calcium and magnesium salts, which will adsorb trace organic contaminants.
- **Estuarine water:** Estuarine waters have a high load of suspended solids from the river inputs. Changes in salinity and pH alone cause coagulation, flocculation and precipitation of metal oxides onto the suspended particulates creating a dynamic heterogeneous

exchange between each state. These waters ultimately receive much of the run-off from agricultural inputs and the discharges from industrial waste.

- *Seawater:* This is relatively more homogeneous over a wider area, although variation can occur due to a lack of mixing around estuarine and coastal regions. The high salt content and the relatively low concentrations of trace organic compounds requires different analytical methodologies to those used for fresh water supplies. High-volume sampling is often required to obtain a representative sample. Emulsions are easily formed when extracting seawater with organic solvents. Chlorinated solvents, e.g. dichloromethane, are used to increase extraction efficiency.
- *Drinking water:* This is provided to a common minimum standard under UK legislation and will have similar characteristics to its source. Although the suspended soils are removed from surface water supplies and the bacterial activity minimized by disinfection, it can still contain quantities of colloidal humic substances and other dissolved components. Drinking water reflects the characteristics of the geology and land use of the catchment and the purification treatment.
- *Wastewater:* The quality and content of wastewater is extremely varied. The parameters that generally cover discharge contents include pH, chemical oxygen demand (COD), biological oxygen demand (BOD), suspended solids and the main anions (NO_3^-, SO_4^{2-}, Cl^- and PO_4^{3-}) and cations (Na^+, K^+, Ca^{2+} and Mg^{2+}). The varied concentration of dissolved organic matter can also produce an emulsion with liquid–liquid extraction (LLE) techniques and reduces extraction efficiency both for LLE and solid-phase extraction (SPE) by lowering the value of the octanol–water partition coefficient (K_{ow}). The suspended solids themselves are usually highly organic in nature and will absorb most lipophilic trace organic contaminants.

The characteristics of these type of waters are summarized in Table 2.1.

2.4 Types of Soil Matrices

Soil is formed through the gradual breakdown of rock, by several mechanisms, including weathering and erosion where it is gradually ground down to smaller components. Soil can best be defined as a composition of the following components: clay minerals, organic matter, water and a living component, all of which will vary according to the soil type and their location.

- *Clay minerals:* The effect of clay and organic matter is considered as having significant effects on the extraction of organic compounds. Clay minerals, for instance, have been associated with difficulties in extracting planar (or nearly planar) organic molecules from soil. Clay minerals are based primarily on silicates and oxides, and typically have a particle size of <2 μm. An important set of crystalline silicate minerals, that form layer structures, are kaolinite (1:1 layer structure) and montmorillonite (2:1 layer structure) (Figure 2.3). When water is introduced to either of these minerals, the polar molecules (i.e. water and organic compounds) get retained in between the layers causing swelling. As the clay dries, the layers return to their original interplanar distance, trapping within

Table 2.1 The characteristics of natural waters [1].

Characteristic	Rainwater	Freshwater		Seawater		Drinking water	Wastewater
		Low flow	High flow	Estuarine	Open ocean		
pH	3.5–6.8	5.0–8.5	4.2–8.5	7.5–8.5	7.5–8.5	5.0–8.0	3–10
Buffer capacity	Low	Medium	Low	Very high	Very high	Low	Medium–high
salts	Low $(100\,\mu g\,l^{-1})$	Medium $(1–100\,\mu g\,l^{-1})$	Low	Very high $(3.5\,g\,l^{-1})$	Very high $(3.5\,g\,l^{-1})$	Low–medium	Low–high $(10–100\,mg\,l^{-1})$
Total organic carbon (TOC)	Low $(<0.5\,mg\,l^{-1})$	Low	Medium–high	Medium $(0.5–1\,mg\,l^{-1})$	Low $(0.5–1\,mg\,l^{-1})$	Medium–high $(<0.5\,mg\,l^{-1})$	Low–high $(10–100\,mg\,l^{-1})$
Natural organics	Low $(1:10^{-12})$	Low	Medium–high $(1:10^{-6})$	Medium–high $(1:10^{-6})$	Medium–high $(1:10^{-6})$	Low	Medium–high $(1:10^{-6})$
Suspended solids	Low	Low $(0.1–1\,mg\,l^{-1})$	Medium–high $(1–100\,mg\,l^{-1})$	Medium–high	Very low $(0.01–1\,mg\,l^{-1})$	Low	Medium–high
Sample volume	10–100l	1–10l	10–50l	1–10l	10–100l	1–5l	0.1–1l

(a) (b)

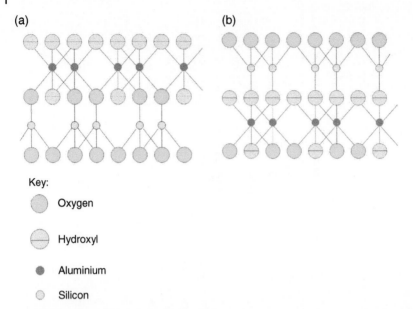

Key:

⬤ Oxygen

⬤ Hydroxyl

⬤ Aluminium

⬤ Silicon

Figure 2.3 The important crystalline silicate mineral layer structures of (a) kaolinite and (b) montmorillonite. **(a)** The 1:1 layer structure is based on a repeating pattern of tetrahedrally (four coordinated) silicon and octahedrally (six coordinated) aluminium atoms. The presence of oxygen atom and hydroxyl groups allows hydrogen bonding and consequently helps to stabilize the crystal structure. **(b)** The 2:1 layer is based on an octahedrally coordinated aluminium atom sandwiched between two layers of tetrahedrally coordinated silicon atoms.

the matrix the organic compounds. This can have implications for the recovery of organic compounds from clay minerals within soils.

- **Organic matter:** The organic matter content of any soil varies according to their location. For example, a soil in an area with high vegetation will have a larger amount of organic matter compared to a soil with little flora around. Organic matter can be sub-classified into three distinct components: humic acid, fulvic acid and humin. Humin is the decomposing remains of both plants and animals, while humic and fulvic acids are classed as a mixture of acids with similar properties. A summary of how the components can be identified is outlined in Figure 2.4.
- **Water:** Water plays a principal role in the soil environment. Not only does it affect plant growth, but also it can help to create or destroy soil structure. The relationship between soil and water is complex, it affects a lot of the physical properties of a soil, for example the expansion of the clay fraction (see earlier) and the transport of nutrients through the soil profile. The water content of a soil varies immensely from being totally saturated to completely dry.
- **Living portion:** The living portion of the soil is composed of micro-organisms such as fungi and bacteria. They are responsible for breaking down dead and decaying matter, and they also help to release nutrient into the soil through decomposition. The living population can make up around 5% of the soil volume. The presence of micro-organisms in soil can influence the stability of the organic compounds, e.g. degradation of the parent organic compound into its metabolites.

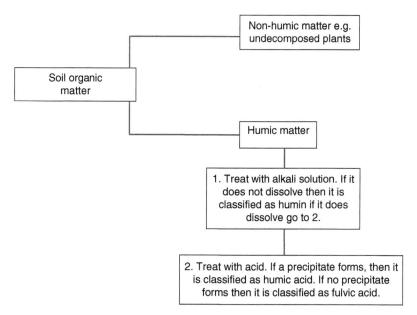

Figure 2.4 An approach to classify soil organic matter.

2.5 Physicochemical Properties of Water and Solid Environmental Matrices

It is important to know the key physicochemical properties of aqueous (water) and soil environmental matrices. Firstly, as a physicochemical property can influence the recovery of the compounds from its matrix, and secondly, these properties can vary both in space and time.

2.5.1 Aqueous (Water) Samples

The key physicochemical properties are:

- **Temperature:** The solubility and stability of organic compounds are influenced by temperature. Temperature measurements, in the range −5 to 50 °C, can be determined using a calibrated mercury thermometer or an electronic thermometer based on a sensor.
- **Density:** The density of natural waters can vary between pure water and water soured from discharges, due to the presence of dissolved and suspended matter; differences will also exist between river, estuarine and sea waters. Density can be determined by accurate weighing and is normally reported against a specific temperature, e.g. 20 °C.
- **Colouring:** Pure water is colourless, but the presence of dissolved impurities, e.g. iron compounds, dissolved humic and plant material, will impart coloration. Quantitative analysis of colour can be done using spectrophotometry or by qualitative analysis by description, e.g. brownish.
- **Turbidity:** This is the scattering of light (in the water) by the presence of suspended or colloidal dissolved compounds (or inorganic substances). Turbidity can be measured instrumentally by using nephelometry, whereas qualitative analysis by description, e.g. opaque.

- **pH:** This is defined as the negative decadic logarithm of the proton (i.e. hydronium ion) activity and is an important parameter of natural waters. Determination of pH, using a meter, is important as it strongly influences the stability, reactivity and mobility of organic compounds (and elements) in natural waters. The pH of natural waters can vary between 4 and 9.
- **Redox potential:** The potential for reduction and oxidation (red-ox) reactions in natural waters influences the behaviour of compounds (and elements), along with pH and the concentration of the compound (or element). The redox potential (E_h), defined by the Nernst equation, is measured by a meter. Natural waters can be determined as anaerobic (based on a positive E_h value) or anaerobic (based on a negative E_h value).
- **Conductivity:** It is the ability of the natural water to conduct electric current, reported as electrical conductivity (e.g. mS cm^{-1}), and measured using a meter. This ability of the natural water to conduct an electric current depends on the presence of ions, their concentration, mobility and the temperature of the matrix. The background conductivity of water, due to the dissociation of water, is 0.06 µS cm^{-1} at 25 °C.
- **Surface tension:** The intermolecular attractive forces, in the aqueous state, leads to the property of surface tension. In pure water, the surface tension is 72.8 mN m^{-1} at 20 °C. The presence of any impurities (e.g. detergents in natural waters) greatly affects the surface tension.
- **Total/dissolved organic carbon (TOC/DOC):** The determination of the presence of the detritus from plant and animal tissues in natural waters is termed total dissolved organic carbon. Using filtration, of the natural water, allows determination of the dissolved organic carbon (by removal of the particulates). Natural and potable waters have a TOC between 1 and 2 mg l^{-1}; polluted waters considerably higher.
- **Biochemical oxygen demand (BOD):** The BOD represents the amount of oxygen consumed by bacteria and other microorganisms while they decompose organic matter under aerobic (when oxygen is present) conditions at a specified temperature.
- **Chemical oxygen demand (COD):** The COD represents the amount of oxygen consumed by reactions in solution. A COD test can be used to assess the amount of oxidizable pollutants (e.g. organic compounds) in natural waters.

2.5.2 Solid (Soil) Samples

The key physicochemical properties are:

- **Temperature:** Solid samples are typically heterogeneous in composition and structure and contain water. It is possible to determine temperature, in the range −5 to 50 °C, using a calibrated mercury thermometer or an electronic thermometer based on a sensor; care is needed in their operation in a solid matrix.
- **Water content:** Water is ubiquitous in all solid matrices. It is important to know, and measure, the water content of solids to assess their accurate weight (for subsequent sample preparation/analysis). Often the method for the determination of the water content of a solid sample involves its oven drying at 105 °C, and using accurate weighing pre- and post-drying, to assess the mass of water lost.
- **Density:** The complex nature of a solid matrix means a simple calculation cannot be done by the sample mass and volume.

- *Cation exchange capacity (CEC):* The presence within a solid environmental matrix of clay and amorphous minerals, and organic matter means that soil has a negatively charged surface. The cation exchange capacity of a soil is an important property as it strongly influences the retention of charged organic compounds (and metal ions).
- *pH:* (as per the water description on pH) The pH measurement of a solid environmental matrix is done in suspension (e.g. using a solid:liquid ratio of 1:2.5) and determination using a meter. Often for pH measurement of soils, a neutral salt solution (e.g. 0.1 M KCl or 0.01 M CaCl$_2$) is used, instead of water, for suspending the solid matrix. This is done to ensure a constant, and hence comparable, ion activity in the soil solution. The pH of soils can vary between 2 and 9.
- *Redox potential:* (as per the redox potential description in water).
- *Particle size distribution:* This parameter can greatly influence recovery of compounds by extraction. Soils can be considered in different size fractions: coarse particles (2 μm and above); fine particles (2–0.5 μm); and ultrafine particles (<0.5 μm). Alternate classifications are also used based on the sand (<1 cm to 63 μm), silt (<63 to 2 μm) and clay (<2 μm) composition of soil (Figure 2.5). So, for example, a soil with a 30% clay content, 70% silt content and 40% sand content would be described as a silty clay loam soil. Sieving is the most common method of determining particle size, though laser-based technology can be used.

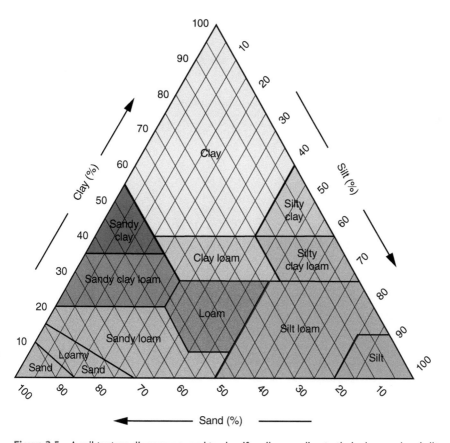

Figure 2.5 A soil texture diagram as used to classify soils according to their clay, sand and silt content.

- *Porosity and surface area:* The surface area of a solid environmental matrix can be determined using a gas adsorption method based on the BET (Brunauer, Emmett and Teller) theory. While the external surface area of a soil sample may be obvious, solids often have a considerable inner surface area. The internal surface can consist of a network of pores, hence porosity is an important consideration.
- *Total organic carbon (TOC):* Plant debris and animal degradation material form the organic matter of soils and sediments. The TOC, in soils, can vary between 0.5 and 5% (on a weight basis), but can vary above this range in, for example, composts. Instruments are available, based on combustion, to determine TOC.

2.6 Sampling Soil (and/or Sediment)

Soil is a heterogeneous material with significant variations possible within a single sampling site due to different topography, farming procedures, soil type (e.g. clay content), drainage and the underlying geology. It is additionally important (see Section 2.1) to also record the following information:

- The soil horizon samples. A typical soil profile (Figure 2.6) may consist of the following:
  - L layer: the litter layer, composed of the debris from plants and animals.
  - A horizon (top soil): the uppermost horizon, contains mineral matter and some organic matter from the L layer.
  - B horizon (sub-soil): lies below the A horizon and has a lower organic matter content.
  - C horizon: it is part of the underlying parent rock from which the soil above was derived.

Figure 2.6 A typical nomenclature for assessing a soil horizon.

Colour can also be used to identify the different soil horizons within the profile. Standardized descriptions of colour can be obtained by use of the Munsell Soil Colour Chart System (Figure 2.7). This is a master atlas of colour that contains almost 1600 colour comparison chips. The colours are prepared according to an international standard. There are 40 pages, each is 2.5 hue steps apart. On each page, the colour chips are arranged by Munsell value and chroma. The standard way to describe a colour using Munsell notation is to write the numeric designation for the Munsell hue (H) and the numeric designation for value (V) and chroma (C) in the form H V/C. In the example in Figure 2.7, the soil sample would be identified using the Munsell notation as: 7.5YR/4/4.

- Soil texture identified either by particle size range (e.g. <2 mm) or clay, sand, silt content (e.g. sandy loam) (see Figure 2.5)).
- pH.
- Organic carbon content (organic matter).
- Water content.
- Other common parameters include cation-exchange capacity, redox conditions, type of clay mineral etc.

Figure 2.7 An example illustrating the use of the Munsell Soil Colour Chart.

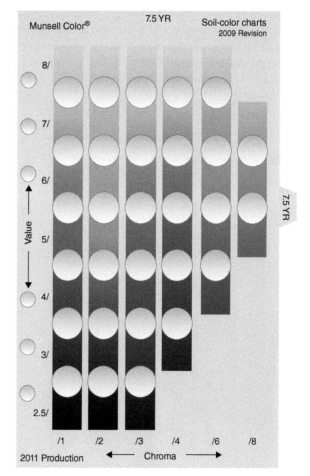

In addition, the land use (i.e. arable, pasture, forest, industrial, brownfield etc.,) should be identified.

Soil sampling can be done, for example, by using an auger (Figure 2.8), spade or trowel. A hand auger (e.g. corkscrew type) allows a sample to be acquired from a reasonable depth (e.g. up to 2 m), whereas a trowel is more appropriate for surface material. As all three devices are made of stainless steel, the risk of contamination is reduced; however, great care needs to be taken to avoid cross-contamination from one sampling position to another. Special care is needed to decontaminate (clean) the sampling device between each sample.

Once the sample has been obtained, it should be placed inside a suitable container (e.g. a geological soil bag, Kraft®), sealed and clearly labelled with a permanent marker pen. Record an abbreviated sample location and/or number, date of sampling, depth sample taken from and name of person collecting the sample. After obtaining the soil sample, replace any unwanted soil and cover with a grass sod, if appropriate. The sample should then be transported back to the laboratory for pre-treatment.

In the laboratory, the soil sample (in its sample bag) would be dried by either air drying (left in a contamination-risk free area) or drying cabinet. If information is likely to be required on the original fresh weight of sample and/or moisture content, then details (e.g. weight) need to be obtained in relation to the sample prior to drying. The duration of drying and temperature are variable but typically air drying at <20 °C may require seven days, whereas in a drying cabinet at 40 °C may be 48 hours. When the soil sample no longer loses weight, it is assumed to be dry and is ready for extraction/analysis. This process is known as 'drying to constant weight'.

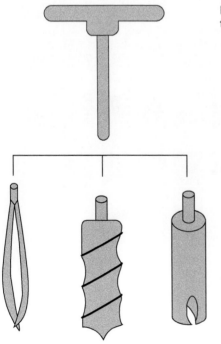

Figure 2.8 A hand-held auger (with options for three different sampling tools).

Figure 2.9 A phase diagram for water.

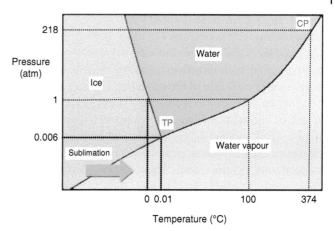

An alternative to the use of a drying cabinet is to freeze-dry the soil sample. In this case, the soil is first frozen and then dried in the frozen state under vacuum. This is a particularly useful approach for compounds that are heat-sensitive or where loss of volatiles is likely, i.e. the potential loss of organic contaminants due to the use of raised temperature (e.g. naphthalene, when extracting PAHs).

Freeze-drying, or lyophilization, is a low-temperature dehydration process that occurs by freezing the sample, lowering the pressure, then removing the water content, as ice, by sublimation. Sublimation is the physical process in which a solid substance (ice) becomes a gas (water vapour) (Figure 2.9). The components of a freeze-dryer (Figure 2.10) are as follows:

- **Drying chamber:** Sample is placed on temperature-controlled shelves within the vacuum tight drying chamber where heating/cooling takes place.
- **Condenser:** The condenser collects the water vapour released by the sample. As the water vapour interacts with the condensing surface, it turns into ice crystals that are removed by the system.
- **Vacuum pump:** This is used to create the vacuum within the drying chamber.
- **Heat source:** This provides the heat that allows the sublimation process to occur.

Figure 2.10 A schematic diagram of a freeze-dryer.

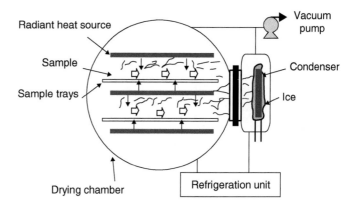

After drying, the sample should be sieved through a stainless-steel sieve (Figure 2.11). Prior to sieving, it is necessary to physically remove stones and large roots. Typically, the soil samples would be sieved to <2 mm particle size. Depending on the information likely to be required, with respect to the sample result and its implication, it may be necessary to reduce the particle size of the sample still further, e.g. to <125 μm.

It may be necessary to reduce the overall quantity of the sample required for the subsequent sample treatment/analysis while still retaining the sample homogeneity. This may be done using a process called 'coning and quartering' (Figure 2.12). The process involves decanting the soil sample on to an inert and contamination-free surface, e.g. a clean sheet of polythene, to form a cone. The cone is then manually divided in to four quarters, using for example, a stainless steel trowel. Then, two opposite quarters of the cone are removed and re-formed in to a new, but smaller cone. By repeating the process as many times as necessary, a suitable-sized sub-sample (e.g. 5 g) can be obtained. The representative (of the whole) sub-sample is now ready for sample extraction.

Figure 2.11 Grinding and sieving of a soil sample (a) grinding of dried sample in a mortar and pestle, (b) coarse ground soil sample, (c) sieving of coarse ground soil sample through a stainless steel sieve (<2 mm), (d) soil sample with non-sieved waste and (e) final <2 mm sieved soil sample.

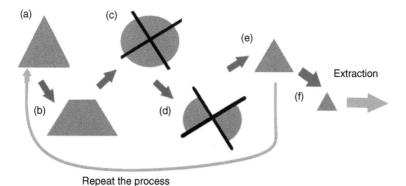

Figure 2.12   Soil sub-sampling: coning and quartering. (a) soil sample arranged in the shape of a cone, (b) top of the cone is flattened, (c) whole sample split into four approximately equal portions, (d) two quarters, from opposite sides, are made into a new cone, (e) with new cone process is repeated or (f) sample is then ready for extraction.

2.7   Sampling Water

Water is a major constituent of the earth's surface (approx. 70%) and is available in a variety of different types or classifications. A classification system might identify the following:

- Water resulting from atmospheric precipitation and condensation (e.g. rain, snow, fog and dew)
- Surface, ground or spring water (e.g. river, lake and runoff)
- Saline water, estuarine water and brine
- Potable (drinking) water
- Wastewater (e.g. mine drainage, landfill leachate and industrial effluent).

It may appear that water is homogeneous, but in fact, in most cases, it is not. Spatial and temporal variations in water make it heterogeneous. An illustration (Figure 2.13) shows how the main flow of a stream is influenced by the influx from a minor tributary (of the stream). This can make it difficult to obtain a representative sample. Spatial variation, for example, can occur within a lake due to changes in the flow (i.e. a lake will often have at least one inlet source via a stream and an outlet flow via a river), differences in chemical composition (i.e. due to the different underlying geology of the lake bed) as well as temperature variation (i.e. a deep lake will be cooler than a shallow lake due to the sun's thermal heating). In addition, temporal variation can occur due to heavy precipitation (e.g. snow and rainfall) as well as seasonal changes resulting in low lake water levels (e.g. leading to a concentration of contaminants) and vice versa. It is additionally important (see Section 2.1) to also record the following information when sampling water:

- Depth water sampled from (i.e. distance from the surface)
- pH and temperature (of the water)
- Sampling method (e.g. suction, pressure etc.)
- Actual and average flow rate (e.g. of river water)
- Other common parameters including dissolved organic carbon, conductivity, redox conditions, colour, odour, turbidity, dissolved oxygen, ionic strength etc.

Main flow of stream

'Boundary' between
the two flows

Influx from a minor tributary

#Stanley Burn, Northumberland

Figure 2.13   Spatial and temporal variation in a flowing stream.#

Water samples can be collected using the sampling device shown in Figure 2.14, as a grab or spot sample. It is essentially an open tube with a closure mechanism at either end; the tube is made of either stainless steel or PVC. Between 1 and 30 l of sample can be collected. The sampling device is lowered into the water to the desired depth using a distance calibrated line. Consideration needs to be given to the potential risk of contamination from the individual (and any associated peripherals) doing the sampling. For

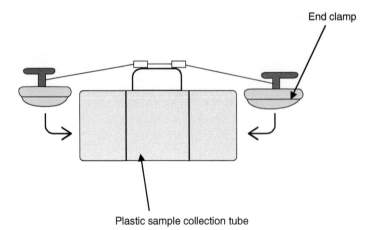

End clamp

Plastic sample collection tube

Figure 2.14   Schematic diagram of a spring-loaded water sampling device.

example, if the sample is to be taken from a lake, then the person sampling may be in a boat. In this scenario, every care should be taken such that the composition of the boat does not influence the sample being taken. Then, both ends of the device are mechanically, and remotely, opened for a short time. After closing both ends, the sampler is brought back to the surface and the sample transferred into a suitable container. Amber glass is the preferred sample container for water samples; the use of plastic containers is discouraged for water samples as they have the potential to leach organic contaminants (and metals) into the acquired sample. Typically, the sample should be stored at <4 °C and analysed as soon as possible. The representative (of the whole) sub-sample is now ready for sample extraction/pre-concentration.

2.8 Sampling Air

Contaminants in the air result from both anthropogenic and natural sources. Anthropogenic sources, i.e. those derived from human activity, largely occur as a result of burning fossil fuels and include emissions from power plants, motor vehicles, controlled burning practices (e.g. agriculture and forest management), fumes from sprays (e.g. paint) and municipal waste incineration and gas (methane) generation. Whereas natural sources include volcanic activity, wind generated dust from exposed land and smoke from wildfires. It is additionally important (see Section 2.1) to also record the following information when sampling air:

- Air temperature, barometric pressure, relative humidity, wind speed and direction
- Precipitation during sampling period (and immediately before e.g. a day)
- Sampling device (container, flow-through, adsorbents or absorbents) including dimensions and technical details
- For particulate matter (cut-off range of sampling device, filter or impactor type)
- Specification of sampling efficiency and capacity
- Pre-sampling processes (e.g. air passage through filter or drier)
- Duration and rate of sampling
- Other parameters include duration of daylight hours, local point sources, air flow obstacles etc.

Air sampling can be classified into two different sample types: the sampling of air particulates on filters (passive sampling); and gaseous vapour sampling on a sorbent (non-passive sampling). While the sampling approaches are different, both types seek to determine the presence of either naturally volatile air-borne contaminants or those contaminants that become air-borne because of other activities, e.g. wind-generated.

In passive sampling, air-borne material diffuses on to fibre glass or cellulose fibre filters where the material is collected (Figure 2.15). The collected material is then extracted from the filter prior to analysis. In non-passive sampling, air-borne material is actively pumped through a sorbent (e.g. ion-exchange resin or polymeric substrate) and collected (Figure 2.16). By sampling a known quantity of air (10–500 m³), quantitative sampling is possible. After collection, the sample containing sorbent tube is sealed and transported back to the laboratory for analysis. Often the analysis is for volatile organic compounds

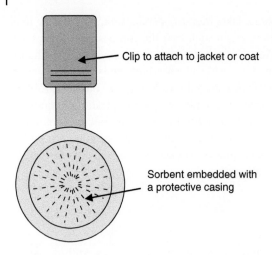

Figure 2.15 Air sampling using a passive sampler.

Clip to attach to jacket or coat

Sorbent embedded with a protective casing

(VOCs). If this is the case, desorption of the VOCs takes place either using organic solvent (solvent extraction, see Chapter 9) or heat (thermal desorption, see Chapter 14) followed by analysis using gas chromatography (see Section 16.4).

2.9 Sampling and Analytical Operations Interrelationships and Terminology

It is possible to create two workflow diagrams that explain the key aspects of sampling operations (Figure 2.17) and their analytical operations (Figure 2.18). Each of these operational approaches will now be considered with definitions of the associated terminology.

2.9.1 Sampling Operations

- *Sampling site:* This is defined as a well delimited area, where sampling operations take place.
- *Sampling strategy:* This is defined as the result of the selection of the sampling points within a sampling site.
- *Increment/single sample:* This is defined as the individual portion of material collected by a single operation of a sampling device.
- *Primary sample:* This is defined as the collection of one or more increments initially taken from a population.
- *Composite/aggregate sample:* This is defined when two or more increments are mixed in appropriate proportions from which the average value of a desired characteristic may be obtained.
- *Sub-sample:* This is defined as the sample obtained by a procedure in which the items of interest are randomly distributed in parts of equal or in equal size.
- *Laboratory sample:* This is defined as the sample or sub-sample sent to or received by the laboratory.

(a)

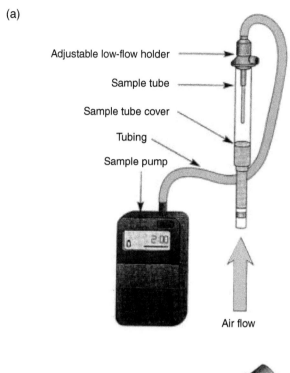

Adjustable low-flow holder

Sample tube

Sample tube cover

Tubing

Sample pump

Air flow

(b)

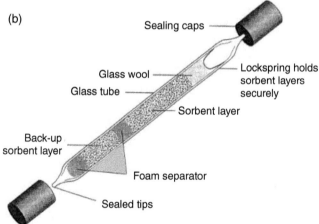

Sealing caps

Glass wool

Glass tube

Lockspring holds
sorbent layers
securely

Sorbent layer

Back-up
sorbent layer

Foam separator

Sealed tips

Figure 2.16 Air sampling using (a) sorbent tube sampling system and (b) a typical sorbent tube.

- **Test sample:** This is defined as the sample, prepared from the laboratory sample, from which the test portions are removed for testing and for analysis.

2.9.2 Analytical Operations

- **Test portion:** This is defined as the quantity of material, of proper size for measurement of the concentration (or other property of interest), removed from the test sample.

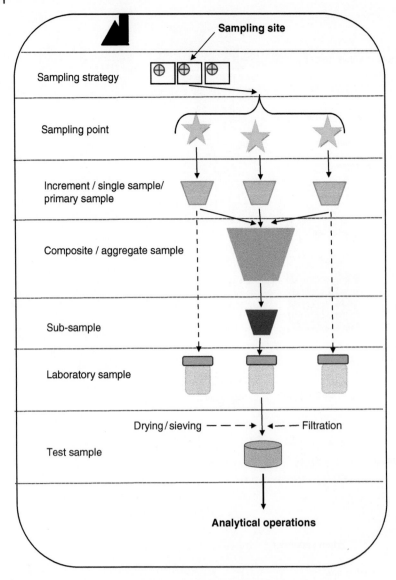

Figure 2.17 Sampling operations interrelationships and terminology.

- **Test solution:** This is defined as the solution prepared from the test portion.
- **Treated solution:** This is defined as the test solution that has been subjected to reaction or separation procedures prior to measurement of some property.
- **Aliquot:** This is defined as the known amount of a homogeneous material, assumed to be taken with negligible sampling error.
- **Measurement:** This is defined as the description of a property of a system by means of a set of specified rules, which maps the property onto a scale of specified values, by direct or 'mathematical' comparison with specified reference(s).

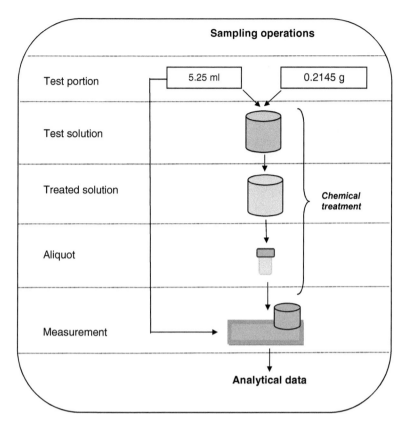

Figure 2.18 Analytical operations interrelationships and terminology.

2.10 Storage of Samples

The storage of samples is a necessary necessity between sample collection, sample preparation and the analysis. Issues that can be controlled, and hence addressed, include the choice of storage vessel and its location, as well as preservation techniques and the time duration prior to analysis.

The major factors affecting sample storage are as follows:

- Chemical, e.g. contamination of the sample from its container; oxidation of compounds; and photochemical decomposition of compounds
- Physical, e.g. sorption of metals on to container wall
- Biological, e.g. decomposition of compounds due to microorganisms

It would be impossible to eliminate these factors, so the normal procedure is to reduce them as far as possible.

2.10.1 Choice of Storage Container for Liquid Samples

No container is inert. Ultimately, the container can influence the sample integrity in two ways: sorption of the analyte on to its surface or leaching of contaminants into the sample.

Therefore, the choice of the container composition is important (as well as its preparation prior to sample contact). Sample containers (beakers, volumetric flasks) are made of glass (e.g. borosilicate glass) or a type of plastic (e.g. low/high density polyethylene (PE), polypropylene (PP), polyvinyl chloride (PVC), polytetrafluoroethylene (PTFE), polycarbonate (PC), perfluoroalkoxy fluorocarbons (PFA)).

While borosilicate glass may appear to be inert it is not; its composition, primarily of, SiO_2, Al_2O_3, Na_2O, K_2O, B_2O_3, CaO and BaO, as well as elemental impurities inherent within the constituents, can pose problems in the laboratory. In addition, glass also presents a chemically highly reactive surface in the form of Si-OH groups that, depending on the pH of the stored solution, can act as an effective ion exchange medium. Under alkali conditions, the Si-OH surface will become Si-O$^-$; allowing positively charged compounds to be retained unless the pH of the solution is controlled. While plastics may appear to be inert, they are not; while their composition does vary their formation, often using metal catalysts (e.g. Al, Cr, Ti, V and Zn), can result in contamination of the sample solution. Also, the leaching of plastics by, for example, an organic solvent can lead to the presence of phthalates in the sample solution.

As the main vessels used for quantitative work (e.g. preparation of a series of calibration standards) are volumetric flasks and vials (e.g. sample container for an autosampler), it is also necessary to consider the importance of the stopper (glass/plastic) and cap for volumetric flasks and vials, respectively. Contact between the sample solution and the stopper is often minimal, i.e. sample container is stored vertically, except when you shake the contents. It is normal practice before sub-sampling the volumetric flask to invert a few times to ensure that homogeneity within the vessel is restored; it may not be visually apparent but some 'settling' may have taken place within the storage container. Stoppers for volumetric flasks come in a variety of materials including glass and plastic; often with a glass volumetric flask, you might have a plastic stopper. Always select a stopper that is colourless, i.e. transparent; a coloured stopper, by definition, has a potential contamination risk. Vial caps are available in a variety of forms including caps with/without septa; care should be taken in the choice of the cap liner (i.e. the inner part that is exposed to the sample solution). Select a vial cap that is (potentially) the most inert, e.g. PTFE; however, some trial and error may be required to identify a low-contamination risk material.

2.10.2 Cleaning of Storage Container for Liquid Samples

Pre-cleaning of the storage container is also especially important to reduce contamination prior to sample contact. The following generic procedure is recommended for glass or plastic containers (e.g. volumetric flask and beaker):

1) Wash the container in detergent to remove any previous sample residue.
2) Soak the container overnight in an acid bath (i.e. 10% v/v nitric acid). A 10% v/v nitric acid solution is made up from 10 ml of concentrated nitric acid in every 100 ml volume of distilled water).
3) Rinse with ultrapure water. Ultrapure water should be available in the laboratory based on a combination of de-ionized, distilled water.
4) Repeat the rinse step at least twice more with ultrapure water. It is not necessary to fill the container to the top with ultrapure water; simply add approximately 10–20% water

(as a percentage of the total volume of the container), agitate the water in the container, ensuring that the water is in contact with all internal surfaces; then discard.

5) Add your sample solution in either ultrapure water (for HPLC) or organic solvent (for GC). Prior to addition of organic solvent, ensure the container is either fully dried by placing the container in a drying oven overnight or by displacing the water with some organic solvent.

2.11 Preservation Techniques for Liquid Samples

The available methods of preservation are limited and are essentially fulfilling a limited number of functions that include: controlling pH (e.g. to minimize Si-O$^-$ interactions for charged compounds), temperature reduction (e.g. to reduce microorganism activity) and removal of light (e.g. to reduce the possibility of photochemical decomposition).

To reduce the potential for photochemical and/or microorganism degradation of organic compounds, samples are normally stored in the refrigerator at 4 °C (short-term storage) or freezer (−18 °C) for long-term storage. Storage under these conditions reduces most enzymatic and oxidative reactions; in addition, samples can be stored in amber/brown vessels. This is additionally beneficial for light sensitive compounds when they are not being stored in the refrigerator/freezer. The potential risk of sample oxidation can also be minimized by the reduction of air (and hence oxygen) in the sample vessel. A reduction of air in the sample storage vessel can be achieved by ensuring the correct sized vessel is used for the sample and that the sample fills the vessel. Some examples of preservation methods are shown in Table 2.2.

Table 2.2   Selected examples of preservation techniques[a] for water samples.

Organic compound	Container	Preservation	Maximum holding time
Pesticides (organochlorine)	Glass	1 ml of a 10 mg ml^{-1} HgCl$_2$ or adding of extraction solvent (500 ml of water).	7 d, 40 d after extraction
Pesticides (organophosphorus)	Glass	1 ml of a 10 mg ml^{-1} HgCl$_2$ or adding of extraction solvent (500 ml of water).	14 d, 28 d after extraction
Pesticides (chlorinated herbicides)	Glass	Cool to 4 °C, seal, add HCl to pH <2 (500 ml of water).	14 d
Pesticides (polar)	Glass	1 ml of a 10 mg ml^{-1} HgCl$_2$ (500 ml of water).	28 d
Phenolic compounds	Glass	Cool to 4 °C, add H$_2$SO$_4$ to pH <2 (500 ml of water).	28 d
Biological oxygen demand (BOD)	Plastic or glass	Cool to 4 °C (1000 ml of water).	48 h
Chemical Oxygen Demand (COD)	Plastic or glass	Cool to 4 °C, add H$_2$SO$_4$ to pH <2 (50 ml of water).	28 d

[a] As recommended by different agencies, i.e. Environmental Protection Agency (EPA) and the International Organization for Standardization (ISO).

2.12   Preservation Techniques for Solid Samples

A solid by its physical appearance and texture will have less contact points with the storage container than a liquid. Nevertheless, it needs to be selected on its relative inertness. However, often some pre-treatment needs to take place prior to sample storage. In the case of a soil sample, for example, it is important to dry the soil sample in an oven at a temperature of 35 °C for at least 48 hours to allow moisture to be removed. The soil sample having previously been stored in a Kraft™ paper bag; this allows moisture to evaporate with time. Once dried the sample can then be stored in either a fridge (at 4 °C) for a few weeks or freezer (at −18 °C) for an extended period in a suitable container, e.g. glass container. Alternatively, the sample could be freeze-dried and then stored in the fridge or freezer.

2.13   Preservation Techniques for Gaseous Samples

In the case of gaseous samples, it is normal practice to analyse them as soon as possible. As the concentration of gaseous samples often involves a form of trapping on a sorbent (see Chapter 14), it could be envisaged that the sorbent trap could be stored for a limited period of time in the fridge at 4 °C (and dark).

2.14   Applications

Case Study A   Designing a Sample Strategy

Background: Consider a site that has been identified for chemical analysis, due to its inherent former industrial activity. The site (Figure 2.19a) is an urban environment with public open spaces, as well as residential and commercial buildings.

Activity: Design a suitable sampling strategy to obtain an in-depth coverage of the site and consider the consequences in an urban environment.

In this situation, a systematic sampling strategy (Figure 2.2b) was selected. This approach was selected as it involves the following:

- selection of the sample points,
- the size of the sample area,
- the shape of the sample area, and
- the number of sampling units in each sample.

This sampling approach is advantageous as it allows a consideration of any distribution of contaminants (Figure 2.1). Using a systematic sampling strategy, a two-dimensional co-ordinate grid is first overlaid on a map of the site (Figure 2.19b). Then, a sampling plan is produced (Figure 2.19c) for which 68 sampling points are identified. However, the identified sampling points are hindered by the presence of buildings, footpaths and other infrastructure obstacles (e.g. stanchions for bridges) (Figure 2.19a). So, the revised (and hence actual) systematic sampling needs to be re-defined with the resultant reduction of sampling points to 28 (Figure 2.19d). To

(a)

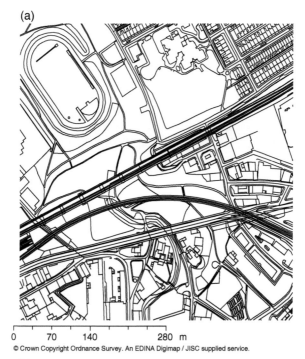

| 0 | 70 | 140 | 280 m |

© Crown Copyright Ordnance Survey. An EDINA Digimap / JISC supplied service.

(b)

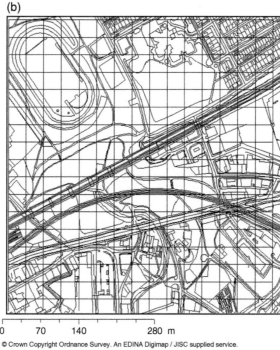

| 0 | 70 | 140 | 280 m |

© Crown Copyright Ordnance Survey. An EDINA Digimap / JISC supplied service.

Figure 2.19 An example of a potential contaminated land site for investigation. (a) The whole site to be investigated, (b) overlay of a two-dimensional sampling grid, (c) systematic sampling plan, (d) actual systematic sampling plan and (e) photographs of the whole site.

(c)

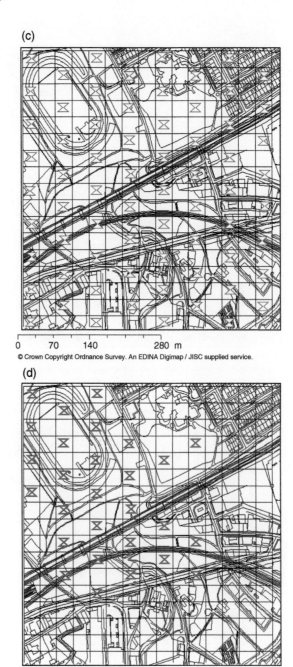

0   70   140   280 m
© Crown Copyright Ordnance Survey. An EDINA Digimap / JISC supplied service.

(d)

0   70   140   280 m
© Crown Copyright Ordnance Survey. An EDINA Digimap / JISC supplied service.

Figure 2.19 (Continued)

(e)

Northern edge of sampling site; close to playing field

Northern edge of sampling site; close to school and residential area

Slope connecting northern edge and lower sampling site

Lower sampling site with stream, metro line bridge and main railway bridge

Lower sampling site with stream, road fly-over and metro line bridge

Lower sampling site with stream, metro line bridge and main railway bridge

Figure 2.19 (Continued)

visualize the sampling points would infer that an actual site visit had taken place (Figure 2.19e), which would allow visualization of the possibilities of taking an incremental sample. The photographs of the northern edge of the sampling site show the playing field, as well as the school/residential area. Connecting the northern edge and the lower sampling site is a steep slope, through which a path intertwines, surrounded by dense low-growing bushes and trees. The lower sampling site has the emergence of a small stream (from an underground culvert), as well as the presence of footpaths, buildings and several bridge stanchions, all of which prevent sampling points to be used and hence the resultant sampling points identified in Figure 2.19d.

Case Study B   Sampling: How Many Samples is Enough?

Background: Obtaining samples that are representative of the whole site is often a major consideration. One approach is to consider the acceptable error that is tolerable in the final data. However, economic considerations often restrict both the quantity of material removed and the number of samples to be taken. To determine the tolerable error, the scientist needs to decide the magnitude of the error (E), by considering the following:

$$E = \pm t(V)^{0.5} \tag{2.1}$$

$$V = S^2 / n \tag{2.2}$$

where t = test value, V = variance, S^2 = sum of squares and n = number of samples. The sum of squares, S^2, is calculated using the following equation:

$$S = \Sigma(x_i - x)^2 / (n-1) \tag{2.3}$$

The number of samples to be taken can then be calculated using the following equation:

$$n = t^2 S^2 / E^2 \tag{2.4}$$

Activity 1: A series of 10 samples were taken at random from a site. The following results were obtained by analysis of the samples for a pesticide, P. Using the above equations, calculate the magnitude of error that can be tolerated.

Sample	1	2	3	4	5	6	7	8	9	10
P (x_i) µg ml^{-1}	91	95	104	82	95	103	97	89	85	89
X (mean)	93	93	93	93	93	93	93	93	93	93
$(x_i - x)$	−2	2	11	−11	2	10	4	−4	−8	−4
$(x_i - x)^2$	4	4	121	121	4	100	16	16	64	16

$$\Sigma(x_i - x)^2 = 466$$

$$S^2 = 466/9 = 51.8$$

$$V(\text{variance}) = 51.8/10 = 5.18$$

$$E(\text{at the 95\% confidence interval}) = \pm 2.262(5.18)0.5 = \pm 5.15 \, \mu g \, ml^{-1}$$

Note: The value of 2.262 was obtained from tables (Table 2.3) of critical values of Students' t statistic at the 95% confidence interval at n−1.

This infers that by taking 10 samples, an error of 5.15 μg ml^{-1} was tolerated, and that the concentration of the pesticide, P, in the sample should be expressed as $93 \pm 5.15 \, \mu g \, ml^{-1}$.

Activity 2: If the level of precision obtained in Activity 1 is not acceptable and a maximum error of 2 μg ml^{-1} is only allowed, how many samples would now need to be taken?

The number of samples required can be calculated using Eq. (2.4):

$$(1.960)^2 (51.8)/2^2 = 49.7$$

Therefore, 50 samples are required.

Note: The value of 1.960 was obtained from tables (Tables 2.3) of critical values of Students' t statistic at the 95% confidence interval at n = α.

Table 2.3   Critical values, two-sided, of Students' t-statistics at various confidence intervals.

Number of degrees of freedom	Confidence interval (%)						
	90	95	98	99	99.5	99.8	99.9
1	6.314	12.71	31.82	63.66	127.3	318.3	636.6
2	2.920	4.303	6.965	9.925	14.09	22.33	31.60
3	2.353	3.182	4.541	5.841	7.453	10.21	12.92
4	2.132	2.776	3.747	4.604	5.598	7.173	8.610
5	2.015	2.571	3.365	4.032	4.773	5.893	6.869
6	1.943	2.447	3.143	3.707	4.317	5.208	5.959
7	1.895	2.365	2.998	3.499	4.029	4.785	5.408
8	1.860	2.306	2.896	3.355	3.833	4.501	5.041
9	1.833	2.262	2.821	3.250	3.690	4.297	4.781
10	1.812	2.228	2.764	3.169	3.581	4.144	4.587

(Continued)

Table 2.3 (Continued)

Number of degrees of freedom	Confidence interval (%)						
	90	95	98	99	99.5	99.8	99.9
11	1.796	2.201	2.718	3.106	3.497	4.025	4.437
12	1.782	2.179	2.681	3.055	3.428	3.930	4.318
∝	1.645	1.960	2.326	2.576	2.807	3.090	3.291

Reference

1 Wells, D.E. (1998). *Analyst* 123: 983–989.

Section C

Extraction of Aqueous Samples

3

Classical Approaches for Aqueous Extraction

LEARNING OBJECTIVES
After completing this chapter, students should be able to: • Comprehend the role and function of liquid-liquid extraction. • Understand the guiding principles behind solvent selection. • Be aware of the issues likely to be encountered when performing liquid–liquid extraction. • Be aware of alternate approaches, their function and issues, for aqueous extraction.

3.1 Introduction

The recovery of organic compounds from aqueous samples (e.g. rainwater, freshwater, estuarine, seawater, drinking water and wastewater) is necessary to evaluate the concentration of soluble pollutants. This might be because the organic compound is present in low concentration (e.g. so preconcentration is required) or to ensure the organic compounds are in an appropriate solvent for analysis (e.g. analysis of volatile organic compounds (VOCs) by gas chromatography). A range of options are possible, which involves transferring organic compounds from the aqueous phase into an organic phase; the organic phase may be a solvent or a sorbent or membrane. If a sorbent is used, the process for retention maybe either adsorption (i.e. the sorbent retains the compounds as a thin film) or absorption (i.e. the compounds enter the sorbent; they are retained within its bulk). The most common approach for the extraction of compounds from aqueous samples is liquid–liquid extraction (LLE). In addition, alternate approaches for the recovery of VOCs from aqueous samples are described.

3.2 Liquid–Liquid Extraction

The principal of LLE is that a sample is distributed, or partitioned, between two immiscible liquids or phases in which the compound and matrix have different solubilities. Normally, one phase is aqueous (often the denser or heavier phase) and the other phase is an organic

Extraction Techniques for Environmental Analysis, First Edition. John R. Dean.
© 2022 John Wiley & Sons Ltd. Published 2022 by John Wiley & Sons Ltd.

solvent (the lighter phase). The basis of the extraction process is that the more polar hydrophilic compounds prefer the aqueous (polar) phase and the more non-polar hydrophobic compounds prefer the organic solvent.

3.2.1 Theory of LLE

The term used to describe the distribution of a compound between two immiscible solvents is the distribution coefficient. The distribution coefficient is an equilibrium constant that describes the distribution of a compound, X, between two immiscible solvents, e.g. an aqueous phase and an organic phase. For example, an equilibrium can be obtained by shaking the aqueous phase containing the compound, X, with an organic phase, such as hexane. This process can be written as an equation:

$$X(aq) \Leftrightarrow X(org) \tag{3.1}$$

where (aq) and (org) are the aqueous and organic phases, respectively. The ratio of the activities of X in the two solvents is constant and can be represented by:

$$K_d = \{X\}_{org} / \{X\}_{aq} \tag{3.2}$$

where K_d is the distribution coefficient. While the numerical value of K_d provides a useful constant value, at a particular temperature, the activity coefficients are neither known or easily measured. A more useful expression is the fraction of compound extracted (E), often expressed as a percentage:

$$E = C_o V_o / (C_o V_o + C_{aq} V_{aq}) \tag{3.3}$$

or

$$E = K_d V / (1 + K_d V) \tag{3.4}$$

where C_o and C_{aq} are the concentrations of the compound in the organic phase and aqueous phases, respectively; V_o and V_{aq} are the volumes of the organic and aqueous phases, respectively; and V is the phase ratio V_o/V_{aq}.

For a one-step LLE, K_d must be large, i.e. >10, for quantitative recovery (>99%) of the compound in one of the phases, e.g. the organic solvent. This is a consequence of the phase ratio, V, which must be maintained within a practical range of values: $0.1 < V < 10$ (Eq. 3.4). Typically, two or three repeat extractions are required with fresh organic solvent to achieve quantitative recoveries. Equation (3.5) is used to determine the amount of compound extracted after successive multiple extractions:

$$E = 1 - \left[1 / (1 + K_d V) \right]^n \tag{3.5}$$

where n = number of extractions.

3.2.2 Selection of Solvents

The selectivity and efficiency of LLE is critically governed by the choice of the two immiscible solvents (Table 3.1). Often the organic solvent for LLE is chosen because of its:

- Low solubility in the aqueous phase (typically < 10%)
- High volatility for solvent evaporation in the concentration stage (see Chapter 15)
- High purity (any solvent impurities would also be pre-concentrated during the solvent evaporation process)
- Compatibility with the choice of chromatographic analysis. For example, do not use chlorinated solvents, such as dichloromethane, if the method of analysis is GC-ECD (a detector for GC that is sensitive and selective for halogenated compounds) or strongly UV-absorbing solvents if using HPLC-UV (the choice of solvent purity, at least with respect to its UV-contaminants is important)
- Polarity and hydrogen-bonding properties that can enhance compound recovery in the organic phase, i.e. increase the value of K_d (Eq. 2.2).

The equilibrium process (K_d) can be influenced by several factors that include adjustment of pH to prevent ionization of acids or bases, by formation of ion-pairs with ionizable compounds, by formation of hydrophobic complexes with metal ions or by adding neutral salts to the aqueous phase to reduce the solubility of the compound (salting out). Examples of the choice of solvents for LLE are shown in Table 3.2.

Table 3.1 A solvent miscibility table.

Solvent	Polarity index	Refractive index @20°C	UV(nm) cutoff @1AU	Boiling point(°C)	Viscosity (cPoise)	Solubility in water (%w/w)
Acetic acid	6.2	1.372	230	118	1.26	100
Acetone	5.1	1.359	330	56	0.32	100
Acetonitrile	5.8	1.344	190	82	0.37	100
Benzene	2.7	1.501	280	80	0.65	0.18
n-Butanol	4.0	1.394	254	125	0.73	0.43
Butyl acetate	3.9	1.399	215	118	2.98	7.81
Carbon tetrachloride	1.6	1.466	263	77	0.97	0.08
Chloroform	4.1	1.446	245	61	0.57	0.815
Cyclohexane	0.2	1.426	200	81	1.00	0.01
1,2-Dichloroethane[a]	3.5	1.444	225	84	0.79	0.81
Dichloromethane[b]	3.1	1.424	235	41	0.44	1.6
Dimethylformamide	6.4	1.431	268	155	0.92	100
Dimethyl sulphoxide[c]	7.2	1.478	268	189	2.00	100
Dioxane	4.8	1.422	215	101	1.54	100
Ethanol	5.2	1.360	210	78	1.20	100
Ethyl acetate	4.4	1.372	260	77	0.45	8.7
Di-ethyl ether	2.8	1.353	220	35	0.32	6.89
Heptane	0.0	1.387	200	98	0.39	0.0003
Hexane	0.0	1.375	200	69	0.33	0.001
Methanol	5.1	1.329	205	65	0.60	100
Methyl-t-butyl ether[d]	2.5	1.369	210	55	0.27	4.8
Methyl ethyl ketone[e]	4.7	1.379	329	80	0.45	24
Pentane	0.0	1.358	200	36	0.23	0.004
n-Propanol	4.0	1.384	210	97	2.27	100
Iso-Propanol[f]	3.9	1.377	210	82	2.30	100
Di-Iso-propyl ether	2.2	1.368	220	68	0.37	
Tetrahydrofuran	4.0	1.407	215	65	0.55	100
Toluene	2.4	1.496	285	111	0.59	0.051
Tichloroethylene	1.0	1.477	273	87	0.57	0.11
Water	9.0	1.333	200	100	1.00	100
Xylene	2.5	1.500	290	139	0.61	0.018

Immiscible / Miscible

Immiscible means that in some proportions two phases will be produced

Synonym table
[a] Ethylene chloride.
[b] Methylene chloride.
[c] Methyl sulphoxide.
[d] Tert-butyl methyl ether.
[e] 2-Butanone.
[f] 2-Propanol.

Column labels (diagonal): Xylene, Water, Tichloroethylene, Toluene, Tetrahydrofuran, Di-iso-propyl ether, Iso-Propanol, n-Propanol, Pentane, Methyl-t-butyl ether, Methanol, Methyl ethyl ketone, Hexane, Heptane, Di-ethyl ether, Ethyl acetate, Ethanol, Dioxane, Dimethyl sulphoxide, Dimethylformamide, Dichloromethane, 1,2-Dichloroethane, Cyclohexane, Chloroform, Carbon tetrachloride, Butyl acetate, n-Butanol, Benzene, Acetonitrile, Acetone, Acetic acid

https://openwetware.org/wiki/File:Solvent_miscibility_chart.jpg.

Table 3.2 Choice of organic solvent for liquid–liquid extraction.

Aqueous solvents	Water-immiscible organic solvents
Water	Hexane, isooctane, petroleum ether (or other aliphatic hydrocarbons)
Acidic solution	Diethylether
Basic solution	Dichloromethane (methylene chloride)
High salt (salting out effect)	Chloroform
Complexing agents (ion pairing, chelating and chiral)	Ethyl acetate
Any two (or more) of the above	Aliphatic ketones (C6 and above)
	Aliphatic alcohols (C6 and above)
	Toluene, xylenes (UV absorbance)
	Any two (or more) of the above

3.2.3 Solvent Extraction

Two distinct approaches for LLE are possible: discontinuous LLE, where equilibrium is established between two immiscible phases, or continuous LLE, where equilibrium may not be reached.

In discontinuous extraction, the most common approach uses a separating funnel (Figure 3.1). The aqueous sample (1 l, at a specified pH) is introduced into a large separating funnel (2 l capacity with Teflon stopcock) and 60 ml of a suitable organic solvent, e.g. dichloromethane, is added. A stopper is then placed into the top of the separating funnel and the separating funnel is then shaken manually. By placing the stoppered end of the separating funnel into the palm of the hand, an inversion of the funnel can take place. In between each inversion, and while the stopper is in the palm of the hand, the stopcock is opened to release any gases that may build-up with the funnel. However, remember to close the stopcock before inverting the funnel again! This process is repeated for approximately 1–2 minutes (inverting the separating funnel approximately 5–6 times).

The process can also be automated by using a mechanical bed-shaker. The shaking process allows thorough interspersion between the two immiscible solvents, thereby maximizing the contact between the two solvent phases and hence assisting mass transfer and allowing efficient partitioning to occur. After a suitable resting period (approximately 5 minutes), the organic solvent is collected by opening the stopcock and carefully running out the lower phase (assuming this is the organic phase) and quantitatively transferring it to a volumetric flask. Fresh organic solvent is then added to the separating funnel and the process repeated. This should be done at least three times in total. The three organic extracts should be combined, ready for concentration (see Chapter 15).

In some cases, the kinetics of the extraction process are slow, such that the equilibrium of the compound between the aqueous and organic phases is poor, i.e. K_d is very small, then continuous LLE can be used. This approach can also be used for large volumes of aqueous

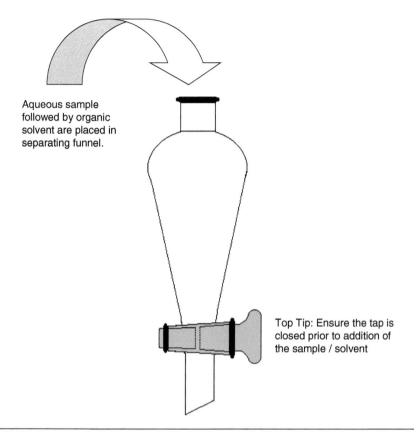

Aqueous sample followed by organic solvent are placed in separating funnel.

Top Tip: Ensure the tap is closed prior to addition of the sample / solvent

Procedure for discontinuous LLE:

- Place aqueous sample in separating funnel.
- Add extraction solvent, e.g. DCM (and salt, if required).
- Shake vigorously for 1–2 minutes (release excess pressure periodically by inverting an opening the tap).
- Allow organic layer to separate (10 minutes), then remove this fraction and retain.
- Repeat the extraction twice more using fresh organic solvent.
- Combine the extracts (i.e. three separate repeat extractions of the same sample).
- Pre-concentration of the combined extracts maybe necessary for subsequent analysis.

Figure 3.1 A separating funnel for discontinuous liquid–liquid extraction.

sample. In this situation, fresh organic solvent is boiled, condensed and allocated to percolate repetitively through the compound-containing aqueous sample. Two common versions of continuous liquid extractors are available, using either heavier-than water (Figure 3.2a) or lighter-than (Figure 3.2b) organic solvents. Extractions usually take several hours but do provide concentration factors of up to x10^5. Obviously, several systems can be operated unattended and in series, allowing multiple samples to be extracted. Typically, a 1l sample, pH adjusted if necessary, is added to the continuous extractor. Then organic solvent, e.g. dichloromethane (in the case of a system in which the solvent has a greater density than the sample), of volume 300–500 ml is added to the distilling flask together

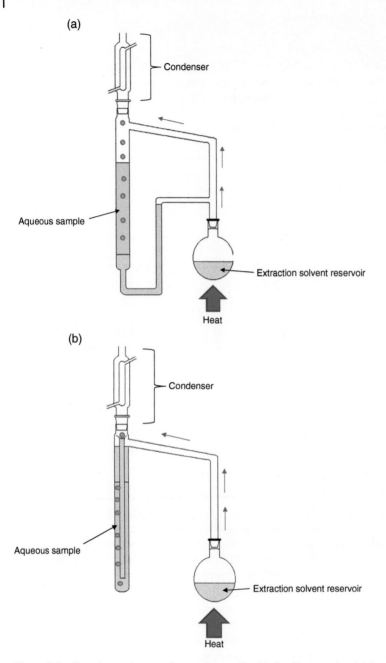

Figure 3.2 Experimental set-up for continuous liquid–liquid extraction (a) for use with an organic solvent heavier than water and (b) for use with an organic solvent lighter than water.

with several boiling chips. The solvent is then boiled, using a water bath, and the extraction process continues for 18–24 hours. After completion of the extraction process, and allowing for sufficient cooling time, the boiling flask is detached, and solvent evaporation can then occur (see Chapter 15).

3.2.4 Problems with the LLE process and Their Remedies

Practical problems with LLE can occur and include emulsion formation. An emulsion appears as a 'milky white' colouration within the separating funnel with no distinct boundary between the aqueous and organic phases. Emulsion formation can occur particularly for samples that contain surfactants or fatty materials. The remedy is to disrupt or 'brake-up' the emulsion by:

- Centrifugation of the mixture
- Filtration through a glass wool plug or phase separation paper
- Heating (e.g. place in oven) or cooling (e.g. place in refrigerator) the separating funnel
- Salting out by addition of sodium chloride salt to the aqueous phase
- Addition of a small amount of a different organic solvent.

3.3 Liquid Microextraction Techniques

3.3.1 Single-Drop Microextraction (SDME)

Single-drop microextraction (liquid-phase microextraction, solvent microextraction or liquid–liquid microextraction) is a miniaturized version of LLE. In SDME, a standard GC syringe is used to suspend a microdroplet of organic solvent, e.g. 1 µl of toluene. Toluene is a useful solvent for SDME as it has a low water solubility. The microdroplet can be either suspended above the aqueous sample (i.e. for headspace extraction) or inserted in the aqueous sample (i.e. direct immersion) (Figure 3.3). In the case of direct immersion, an improvement in extraction efficiency can be made by agitating the aqueous sample using a magnetic stir bar and/or the addition of salt, i.e. 'salting out'. In addition, in headspace extraction, the sample can be heated to increase the extraction efficiency. After a selected period of time

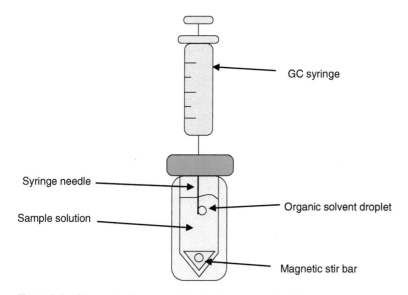

GC syringe

Syringe needle

Organic solvent droplet

Sample solution

Magnetic stir bar

Figure 3.3 Schematic diagram of single drop microextraction procedure.

(e.g. 30 minutes), the micro-droplet is drawn back into the syringe and then injected into the injection port of a GC, and the analysis run.

The main issues to address in use of SDME are as follows:

- Selection of extraction solvent. In addition to toluene, other potential solvents are n-octyl acetate, isoamyl alcohol, undecane, octane, nonane and ethylene glycol. The selected extraction solvent should be capable of dissolving the compound.
- Chromatographic interference. The choice of organic solvent needs to be considered such that it does not co-elute in the GC run time near the compounds of interest; this may be an issue when using GC detectors other than MS.
- Ideal droplet size is typically between 1 and 2 μl.
- The sampling temperature. While room temperature is appropriate, an increase in temperature will lead to enhanced recoveries, but it can also lead to drop instability.
- The equilibration and extraction time need to be optimized so as to allow enough time for compounds to partition into the organic solvent droplet.
- While agitation, by stirring of the sample solution, will increase the partitioning rate, it can also lead to droplet instability.

In addition, additional consideration is required when using SDME for headspace extraction.

- Extraction solvent volatility. If the solvent is too volatile, it will evaporate during the extraction process thereby negating any possibility of extraction.
- To maximize extraction efficiency a larger droplet is best; consideration, therefore, is required of the headspace volume to droplet ratio.
- The sample solution temperature is important; a higher temperature will need to more availability of the organic compounds in the headspace.
- The sample solution pH and/or ionic strength can influence the headspace extraction recovery.

3.3.2 Dispersive Liquid–Liquid Microextraction (DLLME)

Dispersive liquid–liquid microextraction (DLLME) is based on a three-component solvent system to recover organic compounds from an aqueous sample (Figure 3.4). The DLLME normally takes place at a micro-scale, e.g. in a centrifuge tube. An extraction solvent (e.g. 20 μl) is mixed with a dispersive solvent (e.g. 1 ml), and then rapidly injected into the aqueous sample (e.g. 5 ml). This rapid injection of the extraction and dispersive solvent causes a cloudy suspension (emulsion) to form; the emulsion is a suspension of microdroplets of the extraction solvent. The formation of this emulsion allows the rapid partitioning of the organic compounds from the aqueous sample into the extraction solvent; this enhanced separation is possible due to the creation of the large surface area by the microdroplets. The emulsion (cloudy suspension) is then centrifuged creating a two-phase system. The centrifugation allows the recovery of the extraction solvent containing organic compounds effectively. The sedimented phase, because of the centrifugation process, can be either removed to access the extraction solvent or left in situ and the extraction solvent removed.

The extraction solvent must be immiscible with water and miscible with the dispersive solvent, as well as have an affinity for the organic compounds to be extracted. In addition,

(a) (b) (c) (d) (e)

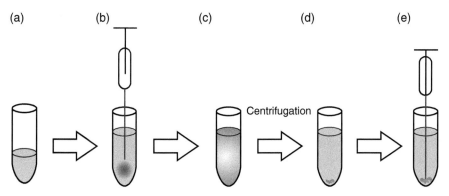

Centrifugation

Figure 3.4 Procedure for dispersive liquid–liquid microextraction (DLLME). A = Aqueous sample solution in centrifuge tube containing compounds to be extracted. B = Rapid injection of a mixture of the extraction solvent and dispersive solvent into sample solution. C = A cloudy dispersion forms that facilitates fast extraction of organic compounds from the aqueous samples into the extraction solvent. D = After centrifugation, a precipitate (or sedimented phase) forms. E = Either removal of the sediment phase is done with the syringe (leaving the extraction solvent containing the organic compounds) or the syringe is used to remove the extraction solvent containing the organic compounds.

and ideally to prevent further evaporation of the extract and reconstitution in an appropriate solvent, the extraction solvent needs to be compatible with the selected analytical technique, e.g. GC or HPLC. Conversely, the dispersive solvent must be miscible with both the aqueous sample and extraction solvent. The selection of both the extraction and dispersing solvents are important. Extraction solvents that have been used, which are denser than the aqueous sample, include tetrachloroethylene, carbon tetrachloride, chlorobenzene and carbon disulphide; this process is sometimes referred to as the traditional DLLME. Alternatively, extraction solvents have been used that are lighter than the aqueous sample, and these include toluene, xylene, hexane and heptane; this process is referred to as low-density solvent-based DLLME (LDS-LLDME). In this case, the solvent floats on the surface of the aqueous sample after phase separation (Figure 3.4, part D). The choice of dispersing solvents has included acetone, methanol and acetonitrile. The efficiency of the DLLME approach is calculated by using an enrichment factor (EF) and high relative recovery (RR). The enrichment factor is calculated as follows:

$$EF = C_{sed} / C_o \tag{3.6}$$

where C_{sed} is the concentration of the organic compounds in the sedimented extraction solvent and C_o is the concentration of the organic compounds in the original aqueous sample.

The relative recovery is calculated as follows:

$$RR = \left(\left(C_{found} - C_{real} \right) / C_{add} \right) \times 100 \tag{3.7}$$

where C_{found} is the total amount of organic compound found after addition of a known concentration standard, C_{real} is the original concentration of organic compound in the aqueous sample, and C_{add} is the concentration of the organic compound standard that was spiked into the original aqueous sample.

The main issues to address in use of DLLME are as follows:

- Selection of extraction solvent and its volume. The selected extraction solvent should be capable of dissolving the compound.
- Selection of dispersing solvent and its volume.
- The pH of the aqueous sample.
- The extraction time.
- The extraction temperature. While room temperature is appropriate, an increase in temperature may enhance recoveries.
- The centrifugation rate and time (e.g. 6000 rpm for 2 minutes).
- Analysis using chromatography interference. The choice of extraction solvent, and its purity, needs to be considered such that it does not co-elute in the GC run time near the compounds of interest; this may be an issue when using GC detectors other than MS.

3.4 Purge and Trap

Purge and trap is a widely applicable technique for the extraction of VOCs from aqueous samples followed by direct transfer and introduction into the injection port of a GC. An aqueous sample (e.g. 5 ml) is placed into a glass sparging vessel (Figure 3.5a). The sample is then 'purged' with high-purity nitrogen at a flow rate of 40–50 ml min^{-1} for 10–12 minutes. The recovered VOCs are then transferred to a trap, e.g. Tenax™, at ambient temperature. Desorption of the VOCs from the trap takes place by rapidly heating the trap (180–250 °C) and back-flushing the VOCs, in a stream of nitrogen gas, to the GC (Figure 3.5b). The rapid desorption from the trap occurs within 2–4 minutes and with a nitrogen flow rate of 1–2 ml min^{-1} and allows the VOCs to be desorbed in a sharp 'plug'. The VOCs are maintained in the gaseous form by ensuring that the transfer line from trap to GC is independently heated (e.g. 225 °C). The heated transfer line is introduced directly into the injection port of the GC. At the end of each extraction, the trap can be 'baked out' by heating to 230 °C for 8 minutes to remove any residual contaminants.

3.5 Headspace Extraction

Headspace extraction is a generic term applied to a range of techniques that sample the volume of air above an aqueous sample; it could also be done above a solid sample. This technique has been applied for the recovery of VOCs and directly linked to GC detection (see also Section 16.5).

The concentration of compound(s) in the headspace can be calculated using the partition coefficient (K):

$$K = C_s / C_g \tag{3.8}$$

where C_s is the concentration of the compound in the liquid sample phase (or solid) and C_g is the concentration of the compound in the gaseous phase (Figure 3.6a). On that basis, compounds that have a low K value will partition more readily into the gaseous phase and as a result have a high signal response on the GC (and vice versa). Typical K values for some

(a)

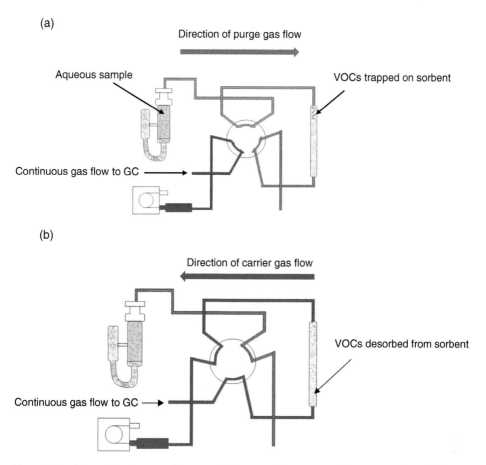

Figure 3.5   Schematic diagram of purge and trap extraction system: (a) purge mode and (b) desorption mode.

common solvents are shown in Table 3.3. The value of K is, however, gratefully influenced by both temperature and the addition of salt (the so-called salting out effect). In the case of an increase in temperature, e.g. raising the temperature of the liquid sample from 40 to 80 °C, will lower the value of K and hence increase the GC signal response. Similarly, the addition of a salt, such as ammonium chloride, ammonium sulphate, sodium chloride, sodium citrate, sodium sulphate or potassium carbonate, will lower the value of K and hence increase the GC signal response. This is because the addition of salt to the aqueous sample solution decreases the solubility of the compounds and as a result the concentration of organic compounds in the headspace increases. The 'salting out' effect is most significant with compounds that have high K values.

As well as K values, the volume of the sample and gaseous phases is also important. These are determined using the phase ratio (β).

$$\beta = V_g / V_s \tag{3.9}$$

where V_g is the volume of the gaseous phase and V_s is the volume of the sample phase. As a result, a large sample volume will result in a low β value producing a higher GC signal

Table 3.3 Typical K values for some common solvents.

Solvent	K value
Cyclohexane	0.077
n-Hexane	0.14
Toluene	2.82
Dichloromethane	5.65
Ethyl acetate	62.4
Isopropanol	825

response (and vice versa). However, by decreasing the β value (by increasing the sample volume), compounds with high K values partition less into the headspace compared with compounds with low K values; in this situation, it is important to optimize the temperature and 'salting out' effect first before adjusting the volumes. Ultimately, by seeking to minimize both K and β values will lead to the highest GC signal response.

As well as seeking to optimize K and β values, another way to increase the GC signal response for specific compounds (e.g. alcohols and acids) is by the addition of a derivatization stage (see Section 16.4.6). By derivatizing, the specific compounds will increase their volatility and hence increased presence in the headspace above the sample. The derivatization stage can be performed in the same sample vial as that used for the headspace analysis. Chromatographic issues can arise due to the presence of increased gaseous compound concentration, specifically poor peak shape, sample carry-over and peak fronting. These issues can be addressed using cryogenic cooling and sample re-focusing at the (head) top of the column. Headspace sampling can be done in a variety of ways including via a gas tight syringe (also referred to as static headspace) or dynamic headspace.

3.5.1 Procedure for Static Headspace Sampling

Static headspace sampling (SHS) (Figure 3.6b) operates by placing the sample vial in a thermostatic oven for a period of time (e.g. an incubation temperature of 60 °C for 5 minutes). To allow the opportunity for equilibrium to be achieved, in the shortest possible time, between the liquid and gaseous phases, shaking of the sample vial is preferred. Then, a known volume (e.g. 250 ml) of the gaseous headspace is removed using a gas-tight syringe and injected into the injection port of the GC. It is important to check the GC injection port septum regularly; the gas-tight syringe has a wider needle than a typical syringe (for GC injection). The potential for a gas leak to occur is therefore enhanced if the septum is not replaced more regularly.

The main issues in the use of SHS are as follows:

- Can the autosampler of the GC system, fitted with an incubator, be used for improved repeatability?
- How can the β value be optimized (i.e. sample to headspace to vial volumes)?
- Is salting out required to improve GC sensitivity?
- What is the optimum incubation temperature?

- What is the optimum volume to be used of the gas-tight syringe?
- What is the potential risk of sample 'carry over' if a heated gas tight syringe is not used?
- How can you minimize sample condensation between the gas tight syringe and the incubator (for the sample vial)?
- What is the influence of the sample injection volume on the resultant chromatography?

3.5.2 Procedure for Dynamic Headspace Sampling

Dynamic headspace sampling (DHS) (Figure 3.6c) operates by placing the sample vial in a thermostatic oven for a period of time (e.g. an incubation temperature of 80 °C) and passing an inert (purge) gas over the surface; the headspace volatile compounds are removed, at a typical purge flow rate of 15 ml min^{-1}, and passed through a trap (e.g. Tenax TA™, held at e.g. 25 °C). The compounds are then rapidly desorbed from the trap (by heating, e.g. 25 to 280 °C

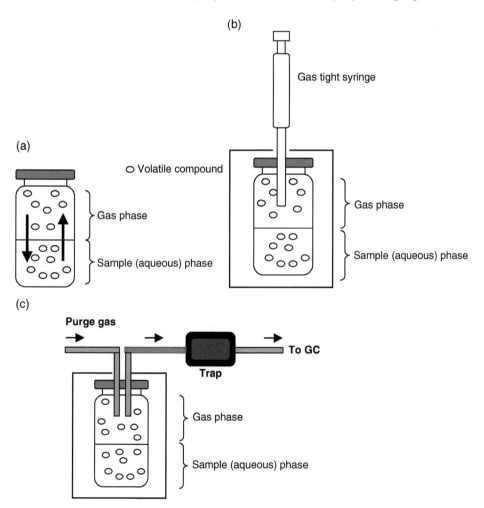

Figure 3.6 Headspace analysis (a) principle of headspace analysis, (b) static headspace and (c) dynamic headspace.

in 1 minute, followed by a hold of 5 minutes) directly into the injection port of the GC. Typically, DHS is done using a dedicated system with an integral heated transfer line to the GC; this can eliminate the potential for condensation of compounds by maintaining a sufficiently high temperature.

The main issues in the use of DHS are as follows:

- How can the β value be optimized (i.e. sample to headspace to vial volumes)?
- Is salting out required to improve GC sensitivity?
- What is the optimum incubation temperature?
- What is the selectivity of the trap (for retaining the compounds)?
- How does sample injection volume affect the resultant chromatography?

3.6 Application

Case Study A   LLE of Organic Pollutants Aqueous-based Media

Background: A procedure to assess the recovery of organic pollutants from aqueous-based media has been done using LLE followed by analysis using gas chromatography-mass spectrometry (GC-MS).

Experimental
Chemical/reagents
The solvents (acetone and dichloromethane) used were certified analytical grade, purchased from Fisher Scientific (Loughborough, Leicestershire). The pesticide standard (TCL Pesticide mix) in toluene:hexane (50:50 v/v), base-neutral-acids (BNAs) (EPA TCLP acids mix) and phenol (aromatic volatiles mix) standards and in methanol were obtained from Sigma-Aldrich Company Ltd (Supelco UK, Dorset, UK). An internal standard, pentachloronitrobenzene, 0.05 g in 10 ml ($5000 \mu g \, ml^{-1}$), was also obtained from Sigma-Aldrich Company Ltd. Derivatizing agent N, O-bis(trimethylsilyl) acetamide (BSA) was obtained from Sigma-Aldrich Company Ltd.

GC-MSD Analysis
The GC-MS (HP G1800A GCD system, Hewlett Packard, Palo Alto, USA) was operated in single-ion monitoring mode with a splitless injection volume of $1 \mu l$. The column was a DB 5ms purchased from Sigma-Aldrich Company Ltd. with dimensions: length $30 \, m \times 0.25 \, mm$ internal diameter $\times 0.25 \, \mu m$ film thickness. For all compounds, the injection port temperature was set at 250 °C and the detector interface temperature was set at 280 °C. Selected standards for the analytes were run daily to assess analytical performance.

Pesticides were analysed using the following temperature program: initial temperature 120 °C for 2 minutes, then to 250 °C at 4 °C min^{-1}. The separation of all the target pesticides was achieved in approximately 34 minutes. Phenols required derivatization prior to analysis. This was done by taking an aliquot of extract/standard (1.0 ml) and transferring it to a tapered tube where internal standard (50 μl) and derivatizing agent BSA (100 μl) were added. The aliquot was vortex spun (15 seconds) prior to analysis by GC-MS. The temperature program used for the separation of phenols was as follows: initial temperature 90 °C for 2 minutes, then to 230 °C at 7 °C min^{-1}. Separation of the phenols was achieved in approximately 22 minutes. Finally, BNAs were separated as follows: initial temperature 70 °C for 2 minutes to 230 °C at 7 °C min^{-1}. Separation of all target BNAs was achieved in approximately 34 minutes.

Extraction Procedures: LLE

Samples were extracted in a 250 ml separating funnel using 10 ml of DCM. The process was repeated a further two times (i.e. 3 × 10 ml DCM). The combined extracts were evaporated, under a flow of nitrogen, prior to transferring the resultant extract to a 25.0 ml volumetric flask. Internal standard (pentachloronitrobenzene, 50 μl) was added and made up to the mark with dichloromethane. Each extract was then analysed by GC-MS.

Results and Discussion

Calibration curves were established with five standards for each group of organic pollutants (pesticides, phenols and base-neutral-acid compounds), with concentrations ranging from 2 to 20 μg ml^{-1} using an internal standard of pentachloronitrobenzene (10 μg ml^{-1}). The detector response was linear over the range of concentration studied for each group with correlation coefficients >0.99 (Table 3.4).

Table 3.4   Analytical data for quantitation of organic compounds.

Compound	Qualifying ions m/z	Quantifying ions m/z	Calibration equation y = mx+c	Correlation coefficient r^2
Pesticides[a, b]				
Lindane	183	181	y = 0.2389x + 0.0803	0.9979
Endosulphan I	237	195	y = 0.0517x + 0.0237	0.9940
Endrin	82	79	y = 0.1373x + 0.0215	0.9963
p,p'-DDE	246	176	y = 0.1770x + 0.0457	0.9972
p,p'-DDD	235	195	y = 0.5291x + 0.0323	0.9970
Endosulphan II	237	195	y = 0.0657x + 0.0210	0.9930
Phenols[a, b]				
Cresol	180	165	y = 1.6497x + 0.6521	0.9941
2,4,6- trichlorophenol	253	93	y = 0.0983x + 0.0326	0.9922
Pentachlorophenol	323	93	y = 0.0528x + 0.0235	0.9930
BNA's[a, b]				
Hexachloroethane	119	117	y = 0.1770x − 0.1081	0.9965
Acenaphthene	154	153	y = 0.7442x − 0.0766	0.9940
Dibenzofuran	168	139	y = 0.4518x − 0.0246	0.9971
Fluorene	166	165	y = 0.3439x − 0.0081	0.9982
Hexachlorobenzene	286	284	y = 0.0930x − 0.0053	0.9925

[a] Internal standard was pentachloronitrobenzene [qualifying ion 249; quantifying ion 237].
[b] Based on a typical concentration range of 0–20 μg ml^{-1} with five calibration data points.

Experiments were undertaken to determine the extraction recovery of pesticides, phenols and BNAs compounds from water at pH 2.5 and 7.0. The organic compound spike level was $10\,\mu g\;ml^{-1}$. The results are shown in Table 3.5. It can be observed that average LLE recoveries, irrespective of the solution media pH, for pesticides ranged from 82% for lindane through to 91% for DDD; phenols from 86% for PCP through to 96% for cresol; and BNA compounds from 87% for hexachloroethane and acenaphthene through to 93% for dibenzofuran (all based on five determinations). The overall mean, based on 150 individual determinations, was $8.86\,\mu g\;ml^{-1} \pm 0.602\,\mu g\;ml^{-1}$ (88.6%).

Table 3.5   Liquid–liquid extraction recoveries of organic pollutants from aqueous matrices.

Compounds	Water at pH 2.5 ($\mu g\;ml^{-1}$) (n = 5)	% Recovery	Water at pH 7.0 ($\mu g\;ml^{-1}$) (n = 5)	% Recovery
Pesticides				
Lindane	8.27±0.52	82.7	8.16±0.32	81.6
Endosulphan I	8.61±0.63	86.1	8.25±0.31	82.5
Endrin	8.74±0.66	87.4	8.42±0.68	84.2
DDE	9.01±0.34	90.1	8.73±0.27	87.3
DDD	9.13±0.87	91.3	8.61±0.47	86.1
Endosulphan II	8.62±0.46	86.2	8.37±0.79	83.7
Phenols				
Cresol	9.56±0.51	95.6	8.94±0.51	89.4
2,4,6-Trichlorophenol	9.43±0.72	94.3	9.13±0.76	91.3
Pentachlorophenol	9.02±0.33	90.2	8.62±0.24	86.2
BNAs				
Hexachloroethane	8.73±0.46	87.3	8.97±0.23	89.7
Acenaphthene	8.72±0.57	87.2	8.85±0.92	88.5
Dibenzofuran	9.31±0.37	93.1	8.92±0.18	89.2
Fluorene	9.24±0.42	92.4	8.98±0.52	89.8
Hexachlorobenzene	8.92±0.55	89.2	9.13±0.59	91.3

4

Solid-Phase Extraction

LEARNING OBJECTIVES
After completing this chapter, students should be able to:
• Understand the principles of solid-phase extraction. • Select an appropriate sorbent for an application. • Be aware of the range of solid-phase extraction formats and their capabilities. • Be able to select an appropriate solvent system for reversed-phase solid-phase extraction.

4.1   Introduction

Solid-phase extraction (SPE) is a common sample preparation method that involves the interaction between an aqueous sample and a solid phase (or sorbent). It was originally developed and applied in the 1970s. The principle is that the compounds of interest, within the aqueous sample, are selectively adsorbed on to the surface of the solid phase. SPE can act as a method for isolation, enrichment, and/or clean-up of compounds of interest from aqueous samples. Subsequently, the retained compounds are selectively eluted for analysis. The solid-phase sorbent is normally packed into small plastic tubes or cartridges. Extensive developments have taken place in terms of SPE technology such that a wide variety of formats, in addition to cartridges, are possible, including disks, pipette tips and 96-well plates. In addition, a wide range of sorbents are commercially available (e.g. silica or polymer-based media and mixed-mode media), as well as extensive developments in terms of its automation using dedicated robotic systems and online systems. However, the essential principles are the same. Whichever SPE experimental set-up is used, the aqueous sample and associated solvents, are forced by pressure or a vacuum through the sorbent. By careful selection of the sorbent, the organic compound(s) are retained by the sorbent in preference to other extraneous material present in the sample. Ideally, and often after extensive method development, the extraneous material can be washed from the sorbent by passing an appropriate solvent or solvent system. Then, the compound(s) of interest are eluted from the sorbent using an alternate solvent system. This final solvent, containing the eluted compounds, is then collected for analysis. Further sample clean-up or pre-concentration can be carried out, if desired (see Chapter 15).

The important variables in SPE are therefore the choice of sorbent and the solvent system used for effective pre-concentration and/or clean-up of the compound in the sample. An example of the effectiveness of SPE clean-up is demonstrated in Figure 4.1. Initially, a sample is separated by high-performance liquid chromatography (HPLC) without sample clean-up using a SPE C18 (Figure 4.1a). Then after spiking the aqueous sample with a

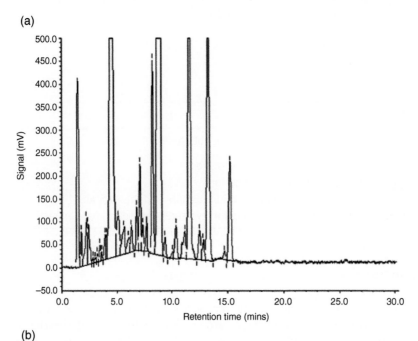

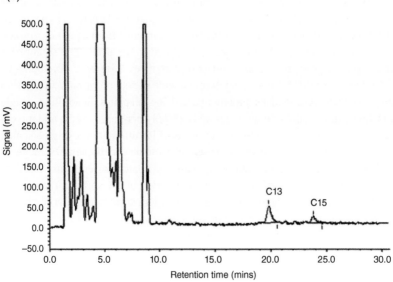

Figure 4.1 Use of solid-phase extraction for sample clean-up. Analysis of a commercial surfactant using reversed-phase high-performance liquid chromatography with fluorescence detection. (a) River water sample pre-SPE, and (b) spiked river water sample post-SPE clean-up.

commercial surfactant, lutensol with an alkyl chain length of C13 and C15, and SPE C18, the resultant chromatogram is obtained (Figure 4.1b). While no potential interferences from the river water are evident at the retention times of the compounds (i.e. 19.5 and 23.8 minutes for the C13 and C15 alkyl chains, respectively), the SPE clean-up procedure has removed a significant amount of extraneous material within the first 16 minutes of the chromatographic run. The process of SPE can therefore allow more affective detection and identification of the compounds in aqueous samples.

4.2 Types of SPE Sorbent

An important variable in any SPE system is the choice of sorbent. Generally, SPE sorbents can be divided into two generic classes: silica-based sorbents and unmodified substrates.

- Silica-based sorbents
  - Reversed phase, e.g. C18, C8, phenyl, cyanopropyl, aminopropyl.
  - Normal phase, e.g. cyanopropyl, aminopropyl, diol.
  - Ion-exchange, e.g. SAX, SCX.

- Unmodified substrates
  - Silica
  - Alumina
  - Florisil
  - Graphitized carbon
  - Resins
  - Macroreticular polymers

The most common sorbents are based on silica particles (irregular-shaped particles with a particle diameter between 30 and 60 μm), with a mass of sorbent between 50 and 3000 mg, and a cartridge reservoir that varies between 1 and 60 ml (Figure 4.2). In a reversed-phase

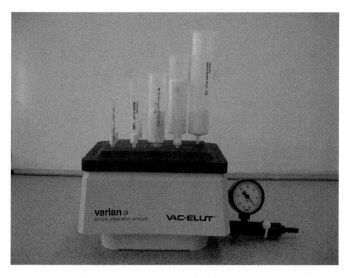

Figure 4.2 Solid-phase extraction cartridges mounted in a vacuum manifold.

system, for example, the silica particles have functional groups bonded to the surface silanol groups to alter their retentive properties. The bonding of the functional groups is not always complete, so unreacted silanol groups can remain. These unreacted sites are polar, acidic sites and can make the interaction with compounds more complex. To reduce the occurrence of these polar sites, some SPE media are 'end-capped'. In end-capping, a further reaction is carried out on the residual silanols using a short-chain alkyl group to remove the hydroxyl groups. It is typical that the addition of a C1 moiety is indicative of end-capping. It is the nature of the functional groups that determine the classification of the sorbent.

Normal-phase sorbents have polar functional groups, e.g. cyano, amino and diol (also included in this category is unmodified silica). The polar nature of these sorbents means that it is more likely that polar compounds, e.g. phenol, will be retained. This is based on the heuristic principle, or 'rule of thumb', that 'like attracts like'. In contrast, reversed-phase sorbents have non-polar functional groups, e.g. octadecyl, octyl and methyl, and conversely are more likely to retain non-polar compounds, e.g. polycyclic aromatic hydrocarbons. Ion exchange sorbents have either cationic or anionic functional groups and when in the ionized form attract compounds of the opposite charge. A cation-exchange phase, such as benzenesulphonic acid, will extract a compound with a positive charge (e.g. phenoxyacid herbicides) and vice versa. A summary of commercially available silica-bonded sorbents is given in Table 4.1.

4.2.1 Multimodal and Mixed-Phase Extractions

SPE normally takes place using one device (e.g. cartridge) with a single sorbent (e.g. C18). However, if more than one type or class of compound is present in the aqueous sample, or if additional selectivity is needed to isolate a specific compound, then multimodal SPE can be used. Multimodal SPE can be done in one of two ways: either by connecting two alternate-phase SPE cartridges in series using an adapter (Figure 4.3) or by having two different functional group sorbents present within one cartridge. In each case, it would be possible, for example, to separate a hydrophobic organic compound and inorganic cations using multimodal SPE. For example, for the multimodal retention of a hydrophobic organic compound and an inorganic cation, the two sorbents selected could be a reversed-phase sorbent (e.g. C18) and a strong cation-exchange sorbent (i.e. SCX) (Table 4.1).

4.2.2 Molecularly Imprinted Polymers (MIPs)

More recently, molecularly imprinted polymers (MIPs) have been developed for use as sorbents in SPE. The use of MIPs has shown to be more selective for the extraction of target compounds from complex matrices such as aqueous samples or organic extracts, as they are engineered cross-linked polymers synthesized with artificial generated recognition sites able to specifically retain a target molecule in preference to other closely related compounds. In addition, MIPs offer more flexibility in analytical methods as they are stable to extreme of pH, organic solvents and temperature. The extraction procedures using MIPs are identical to other SPE media, i.e. the stages of wetting and conditioning of sorbent, sample loading, washing and compound elution have to be carried out. Hence, a careful selection of the most appropriate solvent to be applied in each step is important to separate the compound selectively.

Table 4.1 Some common solid-phase extraction media.

Primary interaction	Phase	Description	Structure
Non-polar	Silica based	C_{18}, octadecyl	$-Si-C_{18}H_{37}$
	Silica based	C_8, octyl	$-Si-C_8H_{17}$
	Silica based	C_6, hexyl	$-Si-C_6H_{13}$
	Silica based	C_4, butyl	$-Si-C_4H_9$
	Silica based	C_2, ethyl	$-Si-C_2H_5$
	Silica based	CH, cyclohexyl	-Si-
	Silica based	PH, phenyl	-Si-
	Silica based	CN, cyanopropyl	$-Si-(CH_2)_3CN$
	Resin based	ENV+	hydroxylated polystyrene divinylbenzene
Polar	Silica based	CN, cyanopropyl	$-Si-(CH_2)_3CN$
	Silica based	Si, silica	$-Si-OH$
	Silica based	DIOL, 2,3-dihydroxypropoxypropyl	$-Si-(CH_2)_3-OCH_2CHOHCH_2OH$
	Silica based	NH_2, aminopropyl	$-Si-(CH_2)_3NH_2$
	Silica based	FL, florisil	$MgO_{3.6}(SiO_2)_{0.1}OH$
	Silica based	Al, Alumina	
Ionic	Silica based – anion	NH_2, aminopropyl	$-Si-(CH_2)_3NH_2$
	Silica based – anion	SAX, quaternary amine	$-Si-(CH_2)_3N^+(CH_3)_3Cl^-$
	Silica based – cation	CBA, propylcarboxylic acid	$-Si-(CH_2)_3COOH$
	Silica based – cation	SCX, benzenesulphonic acid	- Si- SO_3^- H^+
	Silica based – cation	SCX-2 (PRS), propylsulphonic acid	$-Si-(CH_2)_3SO_3^-$ H^+
	Silica based – cation	SCX-3, ethylbenzenesulphonic acid	- Si- $(CH_2)_2$ SO_3^- H^+

To aid with the selection of the appropriate SPE sorbent for a specific analyte, a useful flow chart is provided (Figure 4.4). Figure 4.4a shows the sorbent selection guide for water-soluble compounds, while Figure 4.4b for organic solvent soluble compounds.

Figure 4.3 An adapter to connect solid-phase extraction cartridges in series.

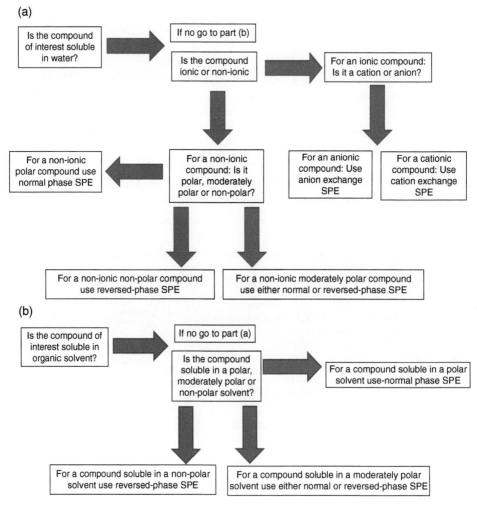

Figure 4.4 Generic solid-phase extraction sorbent selection guide (a) for water-soluble compounds and (b) for organic solvent soluble compounds.

4.3 SPE Formats and Apparatus

The design of the SPE device can vary, with each design having its own advantages related to the number of samples to be processed and the nature of the sample and its volume. The most common arrangement is the syringe barrel or cartridge. The cartridge (Figure 4.5) itself is usually made of polypropylene (although glass and polytetrafluoroethylene, PTFE, are also available) with a wide entrance, through which the sample is introduced, and a narrow exit (male luer tip). The appropriate sorbent material, ranging in mass from 50 mg to 10 g, is positioned between two frits, at the base (exit) of the cartridge, which act to both retain the sorbent material and to filter out particulate matter. Typically, the frit is made from polyethylene with a 20 μm pore size.

Solvent flow through a single cartridge is typically done using a side-arm flask apparatus (Figure 4.6), whereas multiple cartridges can be simultaneously processed (from 8 to 30 cartridges) using a commercially available vacuum manifold (Figure 4.7). In both cases, a vacuum pump is required to affect the movement of solvent/aqueous sample through the sorbent.

If no vacuum pump is available, then a manual SPE approach can be done using a plunger inserted into the cartridge barrel (Figure 4.8). In this situation, the solvent is added to the syringe barrel and forced through the SPE using the plunger. This system is effective if only few samples are to be processed.

An alternate approach to SPE with a cartridge is the use of a disk, not unlike a filter paper. This SPE disk format is referred to by its trade name of an Empore™ disk. The 5–10 μm sorbent particles are intertwined with fine threads of PTFE, which results in a disk approximately 0.5 mm thick and a diameter in the range 47–70 mm. Empore™ disks are placed in a solvent filtration system and a vacuum applied to force the solvent/sample through (Figure 4.9). To minimize dilution effects, it is necessary to introduce a test-tube

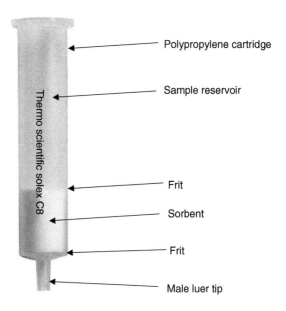

Figure 4.5 The anatomy of a solid-phase extraction cartridge.

Polypropylene cartridge

Sample reservoir

Thermo scientific solex C8

Frit

Sorbent

Frit

Male luer tip

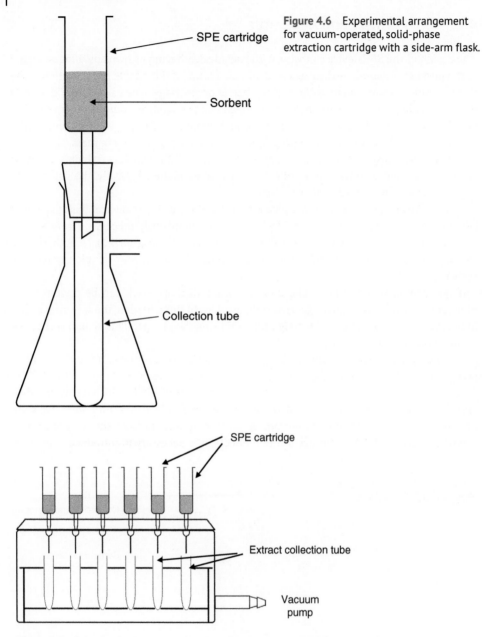

Figure 4.6 Experimental arrangement for vacuum-operated, solid-phase extraction cartridge with a side-arm flask.

SPE cartridge

Sorbent

Collection tube

SPE cartridge

Extract collection tube

Vacuum pump

Figure 4.7 Experimental arrangement for vacuum-operated, solid-phase extraction cartridges on a simultaneous system.

into the filter flask to collect the final extract. Manifolds are also commercially available for multiple sample extraction using Empore™ disks. For automated sample preparation in high-throughput laboratories, a 96-well plate SPE format is also available (Figure 4.10). At the other extreme, microsample extraction/clean-up can be done using a sorbent with a pipette tip (Figure 4.11).

Figure 4.8   A plunger-based system for solid-phase extraction.

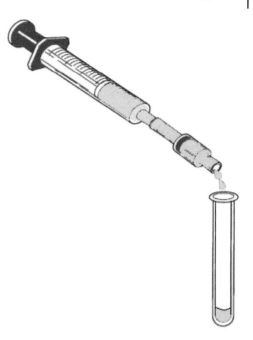

Figure 4.9   Experimental arrangement for vacuum-operated solid-phase extraction cartridge in Empore™ disk format.

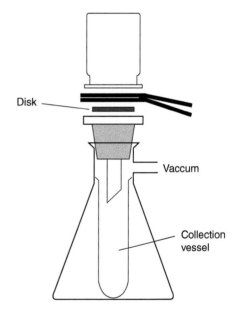

Disk

Vaccum

Collection vessel

Both the cartridge and disk formats have their inherent advantages and limitations. The SPE disk, with its thin sorbent bed and large surface area, allows rapid flow rates of solvent/aqueous sample. Typically, one litre of aqueous sample can be passed through an Empore™ disk in approximately 10 minutes, whereas with a cartridge system the same volume of aqueous sample may take approximately 100 minutes! However, large flow rates can result in poor recovery of the compound of interest due to there being a shorter time for compound–sorbent interaction.

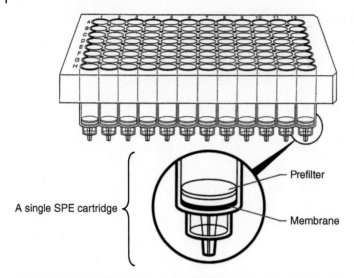

Figure 4.10 Experimental arrangement for vacuum-operated solid-phase extraction cartridge in 96-well plate SPE format.

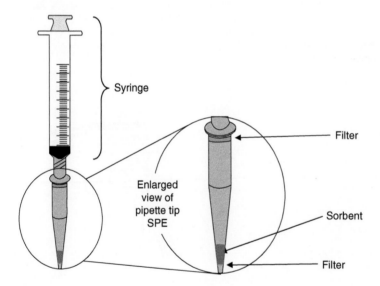

Figure 4.11 Experimental arrangement for vacuum-operated solid-phase extraction cartridge in pipette tip format.

4.4 Method of SPE Operation

Irrespective of SPE format, the method of operation is the same and can be divided into several steps (Figure 4.12a). Each step is characterized by the nature and type of solvent used, which in turn, is dependent upon the characteristics of the sorbent and the sample.

(a)

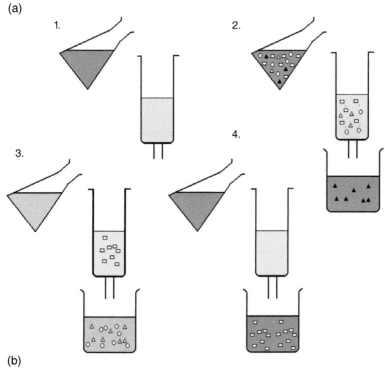

(b)

1. Select appropriate SPE system

> • Manual
> • Single manifold
> • Multiple manifold
> • Automated

Select appropriate SPE sorbent

> • Non-polar sorbent
>   C18 / C8 / C2 / cyclohexyl / phenyl / carbograph / polymeric
> • Polar sorbent
>   Cyanopropyl / silica /diol / aminopropyl
> • Ion exchange
>   SAX / SCX

Pre-condition SPE sorbent (e.g. methanol if using C18).
Rinse sorbent with distilled water (or buffer) to equilibrate sorbent for aqueous sample.

2. Add the aqueous sample (pH adjusted, if necessary) to the sorbent.

3. Add solvent(s) to remove extraneous matrix but not target compounds.

4. Insert collection tube (for retaining eluted target compounds).
   Elute target compounds using appropriate solvent(s).
   Analyse eluted target compounds

Figure 4.12 Solid-phase extraction: (a) schematic diagram of the four stages of the process, (b) key stage descriptors.

The steps are explained in the context of a reversed-phase SPE system using a C18 sorbent, and with a vacuum manifold applied, as follows (Figure 4.12b):

- Wetting the sorbent.
  - By wetting the sorbent with, for example, methanol allows the bonded alkyl chains, which are twisted and collapsed on the surface of the silica, to be solvated so that they spread to form a large surface area. This ensures good contact between the compound(s) and the sorbent in the adsorption step.

It is also important that the sorbent remains wet in the following two steps or poor recoveries can result.

- Conditioning of the sorbent.
  - Water or buffer, similar to the sample solution that is to be extracted, is applied to the sorbent.
- Loading of the sample.
  - The aqueous sample is passed through the sorbent. The compounds of interest are retained by the sorbent in preference to extraneous material and other related compounds that may be present in the sample. Obviously, this ideal situation does not always occur and compounds with similar structures will undoubtedly be retained.
- Rinsing or washing the sorbent to elute extraneous material.
  - The sorbent is washed with a suitable solvent that allows unwanted extraneous material to be removed without influencing the retention of the compounds of interest.
- Elution of the compound(s) of interest
  - The compounds of interest are eluted from the sorbent using the minimum amount of solvent to affect quantitative desorption (e.g. 2 ml methanol) and collected in a sample tube/vial.

By careful control of the amount of solvent used in the elution step, and the sample volume initially introduced on to the sorbent, a pre-concentration of the compounds of interest can be affected.

Successful SPE requires careful consideration of the nature of the SPE sorbent, the solvent systems to be used and their influence on the compound of interest. An important aspect of the compound is its form with respect to pH and the influence it has on the SPE process. In addition, it may be that it is not a single compound that you are seeking to pre-concentrate but a range of compounds. If they have similar chemical structures, then a method can be successfully developed to extract multiple compounds. In this respect, it is necessary to consider the pKa of a compound. The pKa is defined as the pH at which 50% of the compound molecule in solution is charged (ionized) and 50% is unionized. A heuristic, 'rule of thumb' approach to identify the form of the compound is the ± 2 rule. If you adjust the pH by ± 2 pH units from the compound's pKa, it will then be 100% in either ionized or non-ionized form. For example, for the extraction of phenol and a range of substituted phenols (Figure 4.13), using reversed-phase SPE, a consideration of the pKa of each compound is required. So, while the ± 2 rule can be applied (Figure 4.13) to individual compounds, it would not allow application of RP-SPE for all four compounds; in this situation, a compromise pH would be required to separate all four compounds. Once developed, the SPE method can then be used to process large quantities of sample with good precision.

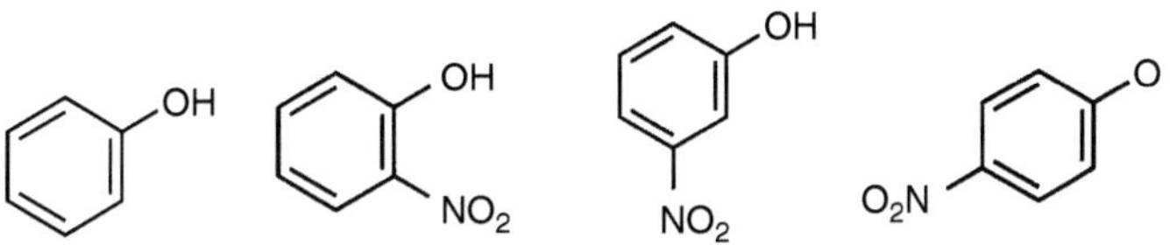

Compound	Phenol	2-Nitrophenol	3-Nitrophenol	4-Nitrophenol
pKa	10.0	7.23	8.35	7.15
RP-SPE pH	8.0	5.23	6.35	5.15
Compromise pH for RP-SPE is ≤ 5.0				

Figure 4.13   Example of the influence of pKa on reversed-phase solid-phase extraction of phenol and a range of substituted phenols.

4.5   Solvent Selection

The choice of solvent directly influences the retention of the compound on the sorbent and its subsequent elution, whereas the solvent polarity determines the solvent strength (or ability to elute the compound from the sorbent in a smaller volume than a weaker solvent). The solvent strength for normal- and reversed-phase sorbents is shown in Table 4.2. Obviously, this is the ideal. In some situations, it may be that no individual solvent will perform its function adequately, so it is necessary to resort to mixed-solvent systems. It

Table 4.2   Solvent strengths for normal- and reversed-phase sorbents.

Solvent strength for normal-phase sorbents	Solvent strength for reversed-phase sorbents	
Weakest	Hexane	Strongest
	Iso-octane	
	Toluene	
	Chloroform	
	Dichloromethane	
	Tetrahydrofuran	
	Ethyl ether	
	Ethyl acetate	
	Acetone	
	Acetonitrile	
	Isopropyl alcohol	
Strongest	Methanol	
	Water	Weakest

should also be noted that for a normal-phase solvent, both solvent polarity and solvent strength are coincident, whereas this is not the case for a reversed-phase sorbent. In practice, however, the solvents normally used for reversed-phase sorbents are restricted to water, methanol, isopropyl alcohol and acetonitrile. For ion exchange sorbents, solvent strength is not the main effect. The key influencing parameter for retention of ions, with ion exchange SPE, is the sorbent; for the subsequent elution of the ions, the pH and ionic strength are critical. As with the choice of sorbent, some preliminary work is required to affect the best solvents to be used.

4.6 Factors Affecting SPE

While the choice of SPE sorbent is highly dependent upon the compound of interest and the sorbent system to be used, certain other parameters can influence the effectiveness of the SPE methodology. Obviously, the number of active sites available on the sorbent cannot be exceeded by the number of molecules of the compound, otherwise breakthrough will occur. Breakthrough refers to the capacity of the sorbent for the compounds under investigation. If the breakthrough is exceeded, then <100% of the compounds will be retained by the sorbent. Therefore, it is important to assess the capacity of the SPE sorbent for its intended application. In addition, the flow rate of sample through the sorbent is important. If the flow rate is too fast, this will allow minimal time for compound–sorbent interaction with resultant poor recoveries. These considerations need to be carefully balanced against the need to pass the entire sample through the cartridge or disk. It is normal therefore for a SPE cartridge to operate with a flow rate of 3–10 ml min^{-1}, whereas 10–100 ml min^{-1} are typical for the disk format.

Once the compound of interest has been adsorbed by the sorbent, it may be necessary to wash the sorbent of extraneous matrix components prior to elution of the compound. The choice of solvent is critical in this stage, as has been discussed previously. For the elution stage, it is important to consider the volume of solvent to be used, as well as its type and nature (e.g. polarity, use of mixed solvents, pH control). For quantitative analysis, by, for example, HPLC or GC, two factors are important: (i) pre-concentration of the compound of interest from a relatively large volume of sample to a small extract volume, and (ii) clean-up of the sample matrix to produce a particle-free and chromatographically clean extract. All these factors require some method development either using a trial-and-error approach or by consultation with existing literature. It is probable that both are required in practice.

4.7 Selected Methods of Analysis for SPE

The general methodology to be followed for off-line SPE will be described using generic examples with emphasis on reversed phase, normal phase, ion exchange, and mixed mode systems.

4.7.1 Application of Reversed-Phase SPE

Reversed-phase (RP) SPE refers to the sorption of organic solutes from a polar mobile phase, such as water or aqueous solvent into a non-polar stationary phase, such as C8 and

(a)

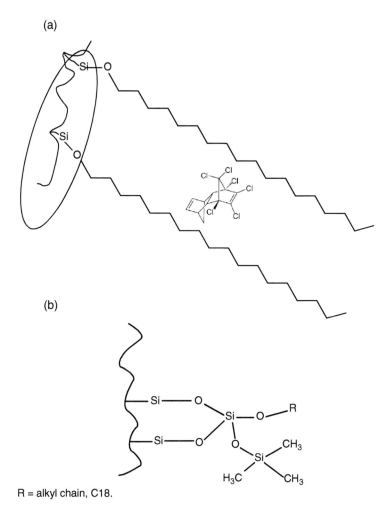

(b)

R = alkyl chain, C18.

Figure 4.14 (a) Reversed-phase mechanism of sorption (partitioning) of the organochlorine pesticide, aldrin in SPE using a C18 sorbent and (b) expanded region at silica particle surface.

C18 sorbent. The sorption mechanism involves the hydrophobic (or partitioning) interaction of the solute within the chains of the stationary phase, i.e. van der Waals or dispersion forces. An example is shown in Figure 4.14a for aldrin with C18 sorbent; Figure 4.14b shows the surface of the silica particle.

The generic procedure for RP SPE is:

1) Wet (or solvate) the silica-bonded phase or polymer packing with six to ten hold-up volumes of methanol or acetonitrile. Then, condition (or flush) the cartridge with six to ten hold-up volumes of water or buffer. Do not allow the cartridge to dry out.
2) Load the sample dissolved in a strongly polar [weak] solvent [typically water].
   Elute unwanted components with a strongly polar solvent.
   Elute weakly retained components of interest with a less polar solvent.

Elute more tightly bound components with progressively more non-polar [stronger] solvents.

3) When you recover all the components of interest, discard the used cartridge in a safe and appropriate manner.

See Table 4.2 for suggestions of organic solvent for RP SPE.

4.7.2 Application of Normal-Phase SPE

Normal phase (NP) SPE refers to the sorption of the functional groups of the compound (solute) from a non-polar solvent to the polar surface of the stationary phase such as silica gel, cyanopropyl, florisil ($MgSiO_3$) and alumina (Al_2O_3). The mechanism of sorption involves polar interactions such as hydrogen bonding, dipole–dipole interaction, π–π interactions, and induced dipole–dipole interactions. An example is shown in Figure 4.15 for 4-chlorophenol with a cyanopropyl sorbent. To achieve retention, the interaction between the solute and the stationary phase must dominate.

The generic procedure for NP SPE is:

1) Wet (or condition) the cartridge with six to ten hold-up volumes of non-polar solvent, usually the same solvent in which the sample is dissolved.
2) Load the sample solution onto the cartridge bed.
   Elute unwanted components with a non-polar solvent.
   Elute the first component of interest with a more polar solvent.
   Elute remaining components of interest with progressively more polar [stronger] solvents.
3) When you recover all your components, discard the used cartridge in a safe and appropriate manner.

See Table 4.2 for suggestions of organic solvent for NP SPE.

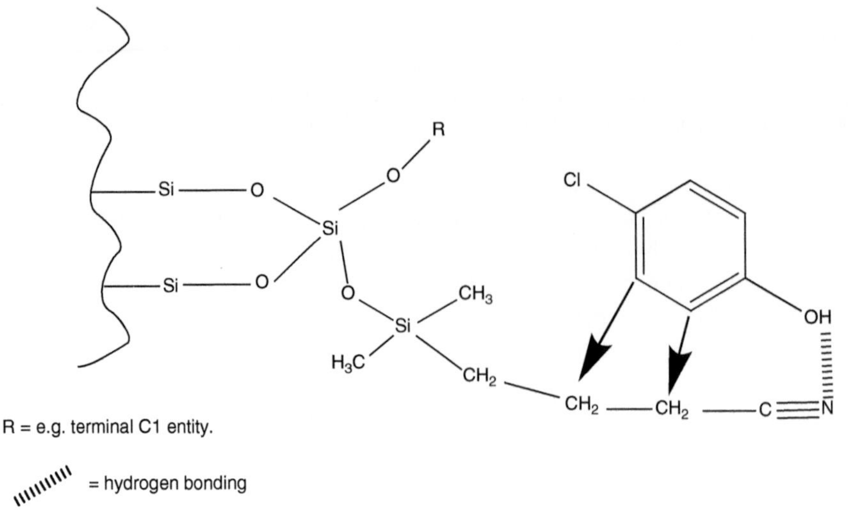

R = e.g. terminal C1 entity.

= hydrogen bonding

Figure 4.15 Normal-phase mechanism of hydrogen bonding for 4-chlorophenol in SPE using a cyanopropyl sorbent.

4.7.3 Application of Ion Exchange SPE

Ion-exchange SPE has been used in the separation of ionic compounds from either a polar or non-polar solvent to the oppositely charged ion exchange sorbent, such as benzenesulphonic acid, propanesulphonic acid and quaternary amine. The separation mechanism involves ionic interaction; hence, polar compound may be effectively separated from polar solvents, including water and less polar organic solvents. An example of the analyte–sorbent interaction, i.e. hydrogen bonding, is shown in Figure 4.16 for an anionic surfactant (sodium 1-decanesulphonate) using an aminopropyl anion exchange SPE.

The generic procedure for ion exchange SPE is:

1) Wet (or condition) the cartridge with six to ten hold-up volumes of deionized water or weak buffer.
2) Load the sample dissolved in a solution of deionized water or buffer.
3) Elute unwanted, weakly bound components with a weak buffer.
   Elute the first component of interest with a stronger buffer (change the pH or ionic strength).
   Elute other components with progressively stronger buffers.
4) When you recover all your components, discard the used cartridge in an appropriate manner.

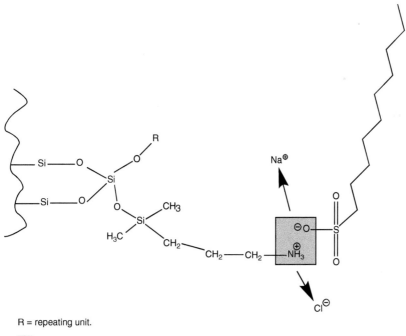

R = repeating unit.

☐ = electrostatic interaction

[sodium 1-decanesulphonate or sodium decylsulphonate is an anionic surfactant that is also used as an ion-associating reagent for HPLC analysis.]

Figure 4.16 An example of separation of an anionic surfactant (sodium 1-decanesulphonate) using an aminopropyl anion exchange SPE.

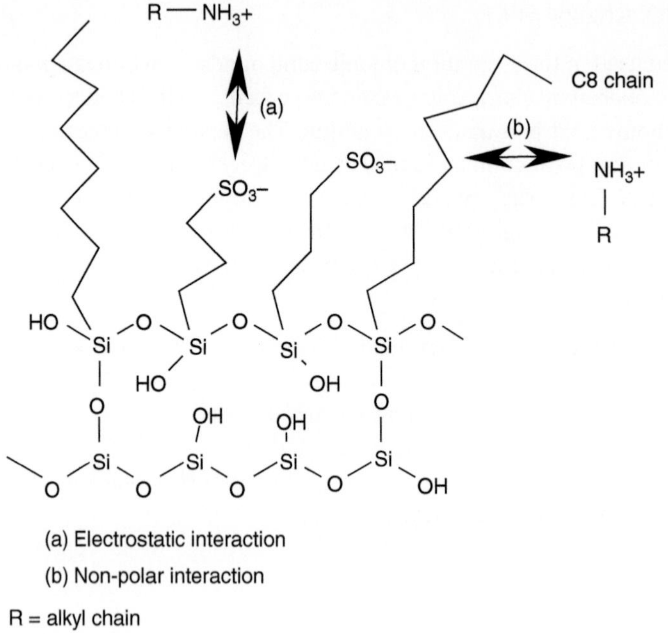

Figure 4.17 An example of the retention mechanisms in a mixed-mode SPE cartridge.

(a) Electrostatic interaction

(b) Non-polar interaction

R = alkyl chain

4.7.4 Application of Mixed-Mode SPE

An example of the analyte–sorbent interactions for an analyte are shown in Figure 4.17; electrostatic interaction between the charged group of the ion exchange sorbent and analyte and non-polar (or hydrophobic/partitioning) between the reversed phase (C8 alkyl chain) and the R-group of the analyte.

4.8 Automation and Online SPE

The use of automated SPE allows large numbers of samples to be extracted routinely with unattended operation. The use of automated SPE should therefore allow more samples to be extracted (higher sample throughput) with better precision. Two categories of automated SPE can be distinguished: the use of instrumentation that imitates the manual offline procedure and an online SPE procedure that utilizes column switching.

The manual offline approach imitates the offline manipulations required for SPE via a robotic arm or autosampler. Thus, it is possible to programme the stages of SPE, i.e. wetting, conditioning, sample loading, washing and elution, followed by collection of compound in an appropriate solvent. The volumes to be used for each stage are programmed into the system as a method. This assumes that the SPE method has been previously well characterized. After completion of this process, the extracted compounds are ready for chromatographic analysis.

Online SPE is the situation where the eluent of the SPE column is automatically directed into the chromatograph (e.g. HPLC) for separation and quantitation of the compounds of interest. This situation is often described as 'column switching' or 'coupled column' techniques. The SPE column or 'pre-column' frequently contains a low efficiency sorbent, which performs a pre-separation of the sample, after which the compound containing fraction is directed on to a second high-efficiency column for separation and quantitation of the compounds of interest. A simplified diagram for column switching is shown in Figure 4.18. The solvent to wet and precondition the sorbent is pumped through the pre-column and then directed to waste. Then, the sample is loaded on to the pre-column (Figure 4.18a) and rinsed with an appropriate solvent. In the elution stage (Figure 4.18b), the high-pressure switching valve is rotated so that the mobile phase passes through the pre-column and flushes the compounds on to the analytical separation column. While the analytical separation takes place, the switching valve returns to the 'load' position for re-conditioning of the pre-column ready to start the next sample. Commercial systems are available that utilize this automated online procedure. The main advantages to a laboratory of online SPE are: the number of manual manipulations decreases, which improves the precision of the data; there is a lower risk of contamination as the system is closed from the

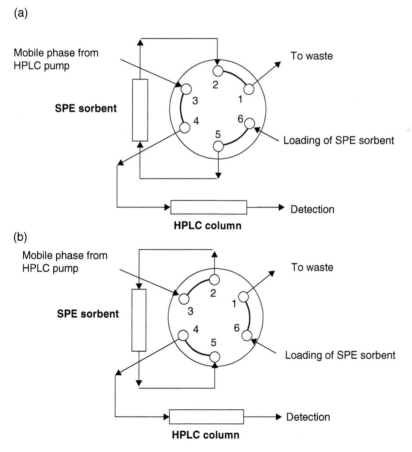

Figure 4.18 Online solid-phase extraction using column switching (a) sample loading, and (b) elution mode.

point of sample injection through to the chromatographic output to waste; and all of the compound loaded on to the pre-column is transferred to the analytical column.

These advantages are off-set by some disadvantages: the initial time taken to develop a method that is both robust and reliable in terms of both the column technology (pre-column and analytical column) and the equipment used, and the additional capital cost involved. It is envisaged that offline SPE is the preferred method of choice for non-routine samples, whereas an automated online SPE system would be used for large numbers of routine samples, process monitoring and the monitoring of dynamic systems.

4.9 Applications

Case Study A   An Evaluation of Compound Retention and Breakthrough Using Reversed-Phase Solid-Phase Extraction

Background: Developing an effective method for solid-phase extraction can be time-consuming, but ultimately rewarding as it allows both the breakthrough solvent composition and wash solvent system to be developed. This case study presents a general format to aid method development of an RP SPE system.

Experimental
Analysis of a compound by HPLC-UV detection is required. A range of solvent mixtures should be prepared varying in composition from 100 water to 100% methanol in 10% solvent changes; therefore, 11 solvent mixtures should be prepared. SPE (C18) cartridges with the same volume and capacity along with a vacuum manifold are required. An aqueous sample, spiked at a detectable concentration, is prepared.

An aliquot of the spiked aqueous sample is applied to 11 × C18 SPE cartridges. Depending on the pKa of the compound, the aqueous sample may need pH adjustment to maintain in a unionized form for this RP SPE system. Elution of each individual SPE cartridge takes place using the solvent system in the order from 100 water to 100% methanol and retaining each fraction for analysis by HPLC.

Results and Discussion
For a reversed-phase SPE system, using a C18 sorbent, a plot is generated as outlined in Figure 4.19, by eluting each cartridge with a defined solvent (solvent mixture). Once the plot has been generated, it can be interpreted for application to sample extracts.

Stage 1: Interpretation of this data infers that the compound can be eluted from the RP-SPE cartridge using a solvent mixture of between 80/20 and 95/10, % methanol to % water. A reasonable assumption would therefore be to use a solvent composition of 90/10, % methanol to % water for elution of the compound.

Stage 2: Data from this stage provides guidance on a solvent mixture for washing off extraneous material from the RP-SPE cartridge without eluting the compound of interest. The generated data indicates that a solvent mixture of between 5/95 and 25/75, % methanol to % water could be used to remove extraneous material without compromising the elution of the compound. A reasonable assumption would therefore be to use a solvent composition of 20/80, % methanol to % water for removal of the extraneous material.

Further method development could apply this approach for multiple compounds, as well as different solvent systems and sample extracts.

Figure 4.19 Method development: An investigation of compound retention and breakthrough using reversed-phase solid-phase extraction.

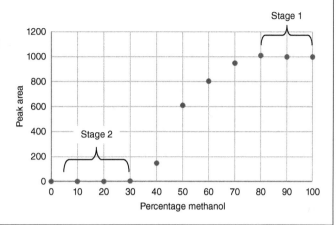

Case Study B Use of a Reversed-Phase Polymeric Solid-Phase Extraction Approach for Determination of Quaternary Ammonium Compounds in Natural Waters Modified from [1]

Background: The quaternary ammonium compounds, i.e. didecyldimethylammonium chloride (didecyldimethyl quat) and dodecylbenzyldimethylammonium chloride (benzyl quat) are a broad range of antimicrobial chemicals, which are used in applications ranging from disinfectants and preservatives to pest control. A method has been developed for their determination in seawater using solid-phase extraction (SPE) with polymeric (Strata-X) SPE cartridges followed by liquid chromatography-mass spectrometry (LC-MS) analysis.

Experimental
Sampling
Environmental seawater samples were taken in duplicate from three locations on the North East coast of England near the mouth of the River Tyne, near Newcastle upon Tyne. A blank sample was also freshly prepared using artificial seawater, which was prepared according to ASTM D1141-98 [2] but omitting the heavy metal stock solution (stock No.3). The samples were collected by fully submersing the 100 ml pre-treated A-grade volumetric borosilicate flasks. The collected samples were then extracted within 4 hours without any pre-filtration.

To minimize compound losses, pre-treatment of the glassware was done by soaking the volumetric flasks overnight in non-cationic surfactant cleaning agent (Decon 90). This was followed by rinsing three times with copious quantities of water, the flasks were then filled to overflowing with a surface treatment solution (i.e. dodecyltrimethylammonium bromide, 50 000 ng ml^{-1}, prepared in artificial seawater) and left overnight. After a final thorough rinse with water, the flasks were ready for sampling.

Chemicals

Analytical standard chemicals were purchased: Dodecyltrimethylammonium bromide 99%, didecyldimethylammonium bromide >98% and dodecylbenzyldimethylammonium chloride >99% from Sigma-Aldrich (Gillingham, UK), tetra-n-butylammonium nitrate and ammonium acetate >97% from Alfa Aesar (Heysham, UK), acetonitrile HPLC Grade and Decon 90 from Fisher Scientific (Loughborough, UK) and acetic acid from VWR/BDH (Lutterworth, UK). Distilled water was purified using a Milli-Q system (18 MΩ cm^{-1}) (Merck, Millipore, Darmstadt, Germany).

Chromatography

High-performance liquid chromatography (HPLC) was performed on a Thermo Surveyor HPLC equipped with a degasser, quaternary pump, autosampler and a C_{18} Xterra MS column 2.5 µm × 4.6 × 50.0 mm with 50 µl injection volume. The chromatography was performed using an isocratic method; 90(A) : 10(B) at 400 µl min^{-1}. Eluent (A) was acetonitrile acidified with 1% v/v acetic acid, while eluent (B) was aqueous 50 mM ammonium acetate solution acidified to pH 3.6 with acetic acid. The HPLC was coupled to an electrospray ion-trap mass spectrometer (MS) (Thermo LCQ Advantage) and operated in positive ion configuration. Optimization of the ion source was performed by the automatic optimization function of the MS software (LCQ Tune) and assisted by direct injection of a benzyl quat solution. The electrospray ion source was operated at 280 °C. The other settings were as follows: sheath gas flow rate: 25 (arbitrary units); sweep gas: 12 (arbitrary units); ion spray voltage: 4.00 kV; capillary voltage: 4.0 V and tube lens offset: 5.0 V.

Sample Preparation: Solid-Phase Extraction

Offline solid-phase extraction (SPE) was performed by passing 100 ml of seawater through a reversed-phase polymeric Strata-X (200 mg / 6 ml) cartridge (Phenomenex, Macclesfield, UK). The SPE procedure was (i) conditioning step: 5 ml acetonitrile followed by 10 ml distilled water; (ii) loading of sample step: 10 ml of sample was added at constant flow rate of 20 ml min^{-1}; (iii) wash step: 20 ml of distilled water with 10% v/v acetic acid and finally (iv) elution step: analytes were eluted using 8.5 ml 90(A) : 10(B) where A is acetonitrile acidified with 10% v/v acetic acid and B is distilled water with 10% v/v acetic acid. The extract was then transferred to a 10 ml perfluoralkoxy (PFA) volumetric flask and 100 µl of internal standard (i.e. tetra-n-butylammonium nitrate, 5000 ng ml^{-1} in the SPE elution solvent was added), and made up to the mark. Out of this 10 ml flask, an aliquot was then transferred to a 2 ml silanized autosampler vial (Chromacol, UK) and analysed using LC-MS.

Results and Discussion

The analytical figures of merit for the method are shown in Table 4.3. A reaction scheme for the formation of the major quantifying ions for the benzyl quat (Figure 4.20a) and didecyldimethyl quat (Figure 4.20b) are shown in Figure 4.20. The applicability of the developed SPE-LC-MS method for monitoring very low levels of quats was assessed by the extraction and analysis of seawater samples. The results (Table 4.4) indicated no discernible (i.e. below LOD) concentration for the benzyl quat, whereas didecyldimethyl quat was measured in samples taken closest to the mouth of the River Tyne.

Table 4.3 Chromatographic and calibration details (50–100 000 ng l^{-1}) ($n = 10$).

Compound	Mass ion	Quantification ion	Retention time (min)	Calibration linearity	LOD[b] (ng l^{-1})	LOQ[c] (ng l^{-1})
Dodecylbenzyldimethyl ammonium cation	304	212	1.93	0.9945	190	640
Didecyldimethylammonium cation	326	186	2.93	0.9989	90	300
Tetra-n-butylammonium nitrate[a]	242	242	1.38	-		

[a] *Internal Standard.*
[b] *Limits of detection (LOD) are based on x3 the standard deviation of the blank.*
[c] Limits of quantification (LOQ) are based on x10 the standard deviation of the blank.

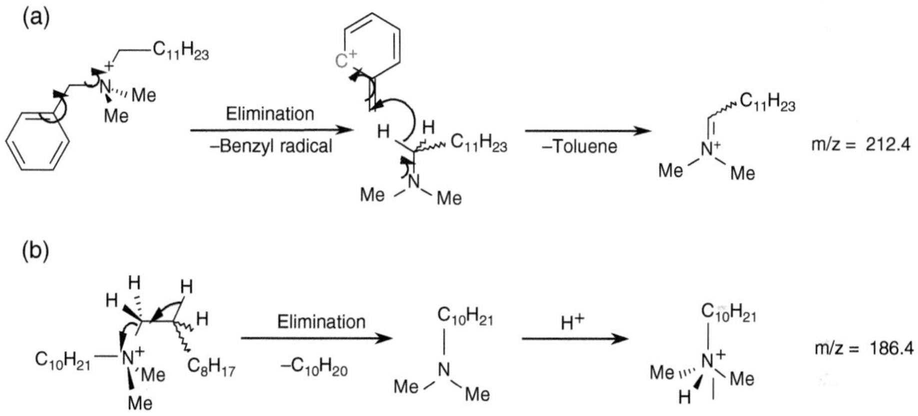

Figure 4.20 Fragmentation pathway for the formation of the major quantification ions as detected by the mass spectrometer of the quaternary ammonium compounds (a) dodecylbenzyldimethylammonium quat cation, and (b) dedecyldimethylammonium cation.

Table 4.4 Determination of quaternary ammonium compounds in seawater samples collected in North East England.

Sites	Dodecylbenzyldimethyl quat (ng l^{-1})[a]	Didecyldimethyl quat (ng l^{-1})[a]
A	ND	197 (121, 272)
B	ND	(133, ND)
C	ND	ND

ND = Not detected.
[a] Mean (individual values).
Note: no benzyl quats were determined in the reagent blanks.

Case Study C   Determination of the Biocide Econea® in Artificial Seawater by Reversed-Phase Solid-Phase Extraction and High-Performance Liquid Chromatography-Mass Spectrometry [3]

Background: Econea®, or 4-bromo-2-(4-chlorophenyl)-5-(trifluoromethyl)-1H-pyrrole-3-carbonitrile, is an environmentally friendly anti-fouling compound used in the immersed coatings of commercial sea going vessels. A method has been developed, using reversed-phase solid-phase extraction, for the determination of the active biocide Econea® in an artificial seawater matrix using HPLC-MS.

Experimental

Chemicals and Reagents

Analytical standard chemicals were purchased as follows: Econea® (PESTANAL®, Sigma-Aldrich, Gillingham, UK); acetonitrile and ammonium acetate (HPLC grade, Fisher Scientific, Loughbrough, UK); acetic acid (VWR, Lutterworth, UK); Tetra Marine SeaSalt (Amazon, Berkshire, UK). Deionized water was prepared from distilled water purified using a Milli-Q-system ($18\,M\Omega\,cm^{-1}$) (Merck Millipore, Darmstadt, Germany).

Instrumentation

High-performance liquid chromatography was performed on a Thermo Surveyor HPLC system (Thermo Fisher Scientific, Hemel Hempstead, UK) equipped with a degasser, quaternary pump and autosampler. Separation was achieved on a Phenomenex Luna C18 $150 \times 2.00\,mm$ i.d.$\times 3\,\mu m$ column with a $10\,\mu l$ injection volume. The chromatography was performed using an isocratic separation, 75% (A) : 25% (B) at $200\,\mu l\,min^{-1}$. Eluant (A) was acetonitrile acidified with 1% (v/v) acetic acid and Eluant B consisted of filtered water with acetonitrile 5% (v/v) and 1% (v/v) acetic acid. The LC was coupled with an electrospray ion-trap mass spectrometer (MS) (Thermo LCQ Advantage) and was operated in negative ion mode. The ion source conditions were optimized using the automatic optimization function of the MS software (LCQ Tune). This was done by introducing a $10\,\mu g\,ml^{-1}$ solution of Econea®, in acetonitrile, by direct infusion into the mass spectrometer at a flow rate of $20\,\mu l\,min^{-1}$ and tuning on the most abundant ion (M+2) at m/z 349. The electrospray ion source was operated at 280 °C. Other source conditions were sheath flow rate 30 (arbitrary units); ionization spray voltage 4.5 kV; capillary voltage −4.00 V; tube lens offset 10.00 V; second octapole offset 11.00 V, first octapole offset 1.75 V and inter-octapole lens 16.00 V.

^{1}H-NMR spectra were obtained using a JEOL ECS 400 NMR spectrometer operating at a frequency of 400 MHz, using 8–32 scans, a relaxation delay of 5 seconds and a flip angle of 45° (5 μs pulse). Spectra were Fourier transformed typically into 32 000 data points using standard exponential window with a line broadening factor of 0.2 Hz. ^{13}C spectra were obtained at a frequency of 100.53 MHz, from 128 to 1048 scans, a relaxation delay of 2 seconds and a flip angle of 30° (2.7 μs pulse). Spectra were Fourier transformed typically into 64 000 data points using standard exponential window with a line-broadening factor of 0.5 Hz. All high-resolution mass spectra (HRMS) were obtained from the EPSRC UK National Mass Spectrometry Service Centre, Swansea, UK.

Solid-Phase Extraction Procedure

Offline SPE was performed by passing 10 ml of an artificial seawater sample through a TELOS C8 (EC, end capped) (200 mg/3 ml) cartridge from Kinesis (St Neots, UK). The SPE procedure was (i) conditioning step: add 6 ml acetonitrile, followed by 6 ml of filtered water containing 5% v/v acetonitrile. Then, a further 6 ml acetonitrile followed by 12 ml of filtered water ensuring that the stationary phase did not run dry; (ii) loading of sample step: 10 ml of sample was added; (iii) wash step: approximately 3 × 3 ml of filtered water, which was also passed through the SPE cartridge ensuring that the stationary phase did not run dry until the last aliquot. Then, vacuum was applied for at least 30 seconds to remove any residual water from the cartridge, and finally (d) elution step: analytes were eluted using 8 ml of acetonitrile and quantitatively transferred to a 10 ml volumetric flask. Finally, 50 μl of internal standard of concentration 50 μg l^{-1}, which was dissolved in acetonitrile, was also added to the volumetric flask prior to making up to volume with acetonitrile. An aliquot of this solution was then transferred to a 2 ml autosampler vial prior to analysis.

Artificial Seawater Preparation

Artificial seawater was prepared from the seawater salt mix, 40 g of seawater salts were added to 1 l of distilled water, and the contents of the beaker were stirred until the seawater salts were dissolved. Any particulates were removed by vacuum through a Qualitative Filter Paper (Fisherbrand Filter 1 QL100 240 mm).

Synthesis of BCCPCA

To methanol (200 ml), Econea® (3.70 g, 1.1 mmol) and sodium hydroxide (5 M, 200 ml, 1000 mmol) were added. The mixture was refluxed for 24 hours. Methanol was removed from the mixture using a rotary evaporator and the solution diluted with water (200 ml). Hydrochloric acid (37%) was added until a solid precipitated as the pH of the solution reached pH 1. The solid was washed with water, then dried to give BCCPCA (compound B, Scheme 4.1) (2.9533 g, 86% yield). m.p. 184–186 °C; ^{1}H-NMR (400 MHz; DMSO) δ 7.83 (d, J = 8.7 Hz, 2H, ArH), 7.63 (d, J = 8.7 Hz, 2H, ArH). ^{13}C-NMR: (100 MHz, DMSO): δ 160.5 (C = O). The HRMS for [M+H]$^+$ was calculated as 324.9374, 325.9408, 326.9351, 327.9385, 328.9324; found 324.9383, 325.9413, 326.9356, 327.9387, 328.9325.

Synthesis of Internal Standard: Methyl Ester of BCCPCA

To 20 ml methanol, Econea® (3.57 g, 1.02 mmol) and powdered sodium hydroxide (0.83 g, 2.08 mmol) were added. The mixture was refluxed for 44 hours. Sodium hydroxide (0.81 g, 2.03 mmol) and 5.51 g of methanol were then added to the reaction flask. After stirring for three days, hydrochloric acid (2 M; 20 ml, 40 mmol) was added until the pH was 1. The resulting solid was collected and washed with water to give an off-white solid. Analysis by ^{1}H-NMR showed the solid to be a mixture of the trimethoxymethyl (compound C, Scheme 4.2) and the methyl ester (see Scheme 4.2, compound D, Scheme 4.2).

The crude white solid was added to a round-bottomed flask to which hydrochloric acid (2 M; 80 ml, 160 mmol) was added, and then stirred overnight; the resulting solid was collected and washed with water. Analysis by ^{1}H-NMR showed the solid to have converted completely to the methyl ester (see Scheme 4.2, compound D). m.p. 165–167 °C; ^{1}H-NMR (400 MHz; DMSO) δ 7.83 (d, J = 8.7 Hz, 2H, ArH), 7.64 (d, J = 8.7 Hz, 2H, ArH), 3.87 (s, 1H, OCH$_3$). ^{13}C-NMR: (100 MHz, DMSO): δ 159.5 (C = O), 52.6 (OCH$_3$); HRMS for [M+H]$^+$ calculated 338.9530, 339.9564, 340.9508, 341.9542, 342.9481; found 338.9533, 339.9563, 340.9506, 341.9537, 342.9476.

Results and Discussion

The biocide Econea®, or 4-bromo-2-(4-chlorophenyl)-5-(trifluoromethyl)-1H-pyrrole-3-carbonitrile (compound A in Scheme 4.1), is available commercially to protect growing plants from infestation and attack by insects and plant mites, as well as a component within paint as an anti-fouling agent. Econea® does not accumulate in the marine environment due to its rapid degradation in seawater (3 and 15 hours at 25 and 10 °C, respectively) to 3-bromo-5-(4-chlorophenyl)-4-cyano-1H-pyrrole-3-carboxylic acid (BCCPCA) (compound B in Scheme 4.1).

The current standard method (ISO method [4]) for the indirect detection and quantification of Econea® in seawater samples uses high-performance liquid chromatography with ultraviolet detection (HPLC-UV) at 280 nm to detect a hydrolysis breakdown product of Econea®, 3-bromo-5-(4-chlorophenyl)-4-cyano-1H-pyrrole-3-carboxylic acid (BCCPCA) (compound B in Scheme 4.1). As seawater may contain a mixture of Econea®, as well as the breakdown product, in the ISO method, any Econea® in the sample is converted to BCCPCA, and the method quantitates this compound. The ISO method requires the samples to be degraded in a thermostatically controlled cabinet at 50±5 °C for between 4 hours and a maximum of 24 hours.

The analytical performance of the new direct method for the determination of Econea® in artificial seawater is shown in Table 4.5. The experimentally determined limit of quantitation (i.e. $0.17 \mu g \, l^{-1}$) is of the order of 10× more sensitive than the ISO method (LOQ $1.87 \mu g \, l^{-1}$) for the breakdown product of Econea®; this is not unexpected as the standard method uses HPLC coupled with a UV detector. An example chromatogram of a sample containing $50 \mu g \, l^{-1}$ of Econea® and $50 \mu g \, l^{-1}$ internal standard is shown in Figure 4.21.

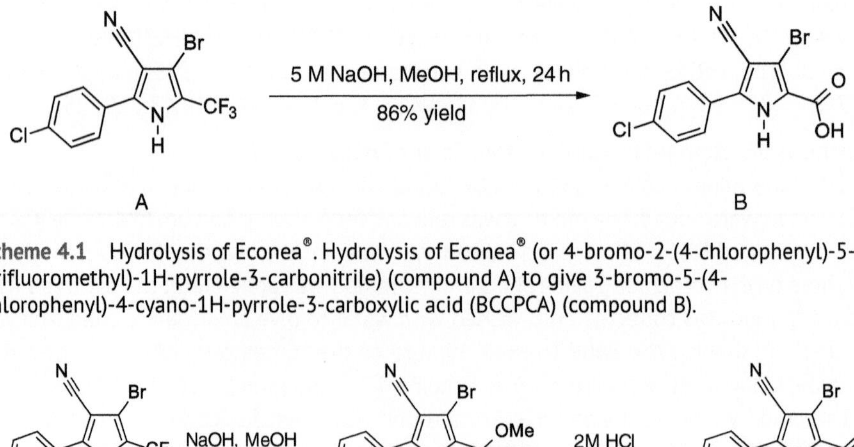

Scheme 4.1 Hydrolysis of Econea®. Hydrolysis of Econea® (or 4-bromo-2-(4-chlorophenyl)-5-(trifluoromethyl)-1H-pyrrole-3-carbonitrile) (compound A) to give 3-bromo-5-(4-chlorophenyl)-4-cyano-1H-pyrrole-3-carboxylic acid (BCCPCA) (compound B).

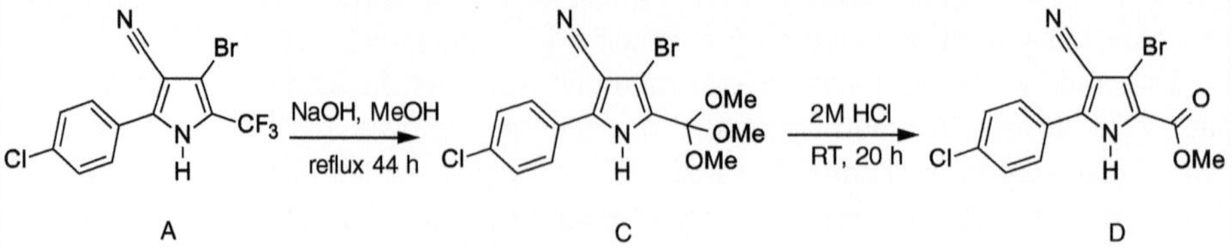

Scheme 4.2 Synthesis of internal standard, the methyl ester of BCCPCA. Compound A, Econea® (or 4-bromo-2-(4-chlorophenyl)-5-(trifluoromethyl)-1H-pyrrole-3-carbonitrile); Compound C, tri(methoxy)methyl of BCCPCA (i.e. 3-bromo-5-(4-chlorophenyl)-4-cyano-1H-pyrrole-3-carboxylic acid); and Compound D, methyl ester of BCCPCA (i.e. 3-bromo-5-(4-chlorophenyl)-4-cyano-1H-pyrrole-3-carboxylic acid).

Table 4.5 Analytical data for the determination of Econea® (4-bromo-2-(4-chlorophenyl)-5-(trifluoromethyl)-1H-pyrrole-3-carbonitrile) or its degradation product BCCPCA (3-bromo-5-(4-chlorophenyl)-4-cyano-1H-pyrrole-3-carboxylic acid).

Compound	Analytical technique	Calibration range ($\mu g\ l^{-1}$)	Number of data points	Equation (y = mx+c)	Correlation coefficient, R^2	LOD ($\mu g\ l^{-1}$)	LOQ ($\mu g\ l^{-1}$)
Econea®	HPLC-MS	0.5–100	6	y = 0.0193x +0.0269	0.9994	0.05	0.17
BCCPCA	HPLC-UV^a						1.87

^{a} data from ISO method [4].

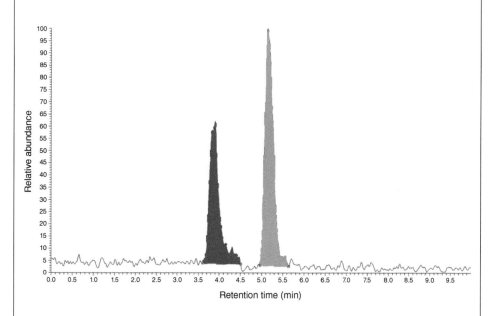

Retention time of internal standard (methyl ester of BCCPCA) is 3.95 minutes

Retention time of Econea® is 5.25 minutes

Figure 4.21 Example HPLC-MS chromatogram of a seawater sample containing 50 $\mu g\ l^{-1}$ Econea® and internal standard.

4.10 Summary

A crossword of the key terms outlined in this chapter, and Chapters 5–8, can be found in Appendix A1, with the solution in Appendix B1.

References

1 Bassarab, P., Williams, D., Dean, J.R. et al. (2011). *J. Chromatogr. A.* 1218: 673–677.

2 American Standard Method of Test (2008). *D1141-98: Standard Practice for the Preparation of Substitute Ocean Water*. London: British Standard Institution SMT.

3 Downs, R.A., Dean, J.R., Downer, A., and Perry, J.J. (2017). *Separations* 4: 34. https://doi.org/10.3390/separations4040034.

4 International Standards Organisation (ISO) 15181-6 (2008). *Paints and Varnishes—Determination of Release Rate of Biocides from Antifouling Paints—Determination of Econea Release Rate by Quantitation of Its Degradation Product in the Extract*. Geneva, Switzerland: International Standards Organisation.

5

Solid-Phase MicroExtraction

LEARNING OBJECTIVES
After completing this chapter, students should be able to:
• Understand the principle of operation of solid-phase microextraction.   • Define the terms adsorption and absorption for retention of organic compounds.   • Appreciate the merits of the different operating modes of solid-phase microextraction.   • Be able to develop a procedure to investigate the optimum solid-phase microextraction fibre type for sampling.

5.1 Introduction

Solid-phase microextraction (SPME) is the process where organic compounds, from an aqueous sample, are retained by a coated-silica fibre as a method of pre-concentration (Figure 5.1). This action is then followed by desorption of the organic compounds for separation and quantitation. The most important stage of this two-stage process is the retention of the compounds onto a suitably coated-silica fibre or phase. The mechanism for retention of the organic compounds could be based on either adsorption (i.e. compounds are physical or chemically trapped on the surface of the fibre coating) or absorption (i.e. compounds are partitioned into a liquid phase, a thin film) (Table 5.1). The choice of phase is critical, in that it must have a strong affinity for the target organic compounds, so that pre-concentration can occur from either dilute aqueous samples (direct insertion of the fibre within a solution) or the gas phase (headspace sampling of the atmosphere above the sample). The range and choice of media available is diverse and varied (Table 5.1).

5.2 Theoretical Considerations for SPME

The partitioning of compounds between an aqueous sample and the SPME fibre coating is the main principle of operation. A mathematical relationship for the dynamics of the absorption process was developed [1]. Normally, SPME extraction is considered to be complete when the

Extraction Techniques for Environmental Analysis, First Edition. John R. Dean.
© 2022 John Wiley & Sons Ltd. Published 2022 by John Wiley & Sons Ltd.

(a)

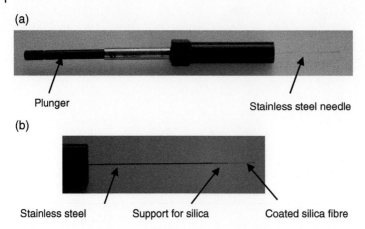

Plunger   Stainless steel needle

(b)

Stainless steel   Support for silica   Coated silica fibre

Figure 5.1 Solid-phase microextraction (a) photograph of a manual holder, and (b) photograph of the silica fibre and its stainless steel protective sheath.

organic compound concentration has reached a distribution equilibrium between the aqueous sample matrix and the SPME fibre coating. The equilibrium conditions are defined as:

$$n = \left(K V_2 V_1 C_o \right) / \left(K V_2 + V_1 \right)$$

(5.1)

where n = amount of organic compound extracted by the coating; K = the SPME fibre coating to aqueous sample matrix distribution constant; C_0 = initial concentration of organic compound in the aqueous phase; V_1 = volume of the aqueous sample; and V_2 = volume of the SPME fibre coating.

As the polymeric stationary phases used for SPME have a high affinity for organic molecules, the values of K are large. These large values of K lead to good pre-concentration of the target compounds in the aqueous sample and a corresponding high sensitivity in terms of the analysis. However, it is unlikely that the values of K are large enough for exhaustive extraction of compounds from the sample. Therefore, SPME is an equilibrium method, but provided proper calibration strategies are followed does provide quantitative analysis. It has been shown [1] that in the case where V_1 is very large (i.e. $V_1 >> K V_2$), the amount of compound extracted by the SPME fibre coating can be simplified to:

$$n = K V_2 C_o$$

(5.2)

and hence independent of the sample volume. This feature can be most effectively exploited in field sampling. In this situation, compounds present in natural waters, e.g. lakes and rivers, can be effectively sampled, pre-concentrated and then transported back to the laboratory for subsequent analysis.

The dynamics of extraction are controlled by the mass transport of the organic compounds from the aqueous sample to the SPME fibre coating. The dynamics of the absorption process have been mathematically modelled [1]. In this work, it was assumed that the extraction process is diffusion limited. Therefore, the amount of sample absorbed plotted as a function of time can be derived by solving Fick's Second Law of Diffusion. A plot of the amount of sample absorbed versus time is termed the extraction profile. The dynamics of extraction can be increased by stirring the aqueous sample (Figure 5.2).

Table 5.1 Typical solid-phase microextraction fibre coatings.

Polarity of phase	Type of phase	Adsorbent versus absorbent fibre type	Phase	Phase thickness (µm)
Non-polar	Bonded	Absorption	Polydimethylsiloxane	7
	non-bonded			30
	non-bonded			100
			Polydimethylsilsiloxane/ divinylbenzene	60
			Polydimethylsilsiloxane/divinylbenzene	65
Medium polarity	Partially cross-linked	Adsorption	Polydimethylsilsiloxane/carboxen	75
			Carbowax/divinylbenzene	65
			Carbowax/template resin	50
Polar	Bonded	Absorption	Polyacrylate	85
			Polyethyleneglycol	60

Adsorbent (particle) fibres: physically traps or chemically reacts with compounds; a porous material with a high surface area. SPME fibres have limited capacity. Absorbent (film) fibres: compounds are extracted by partitioning into a thin-film liquid phase. SPME fibres can have a high capacity.

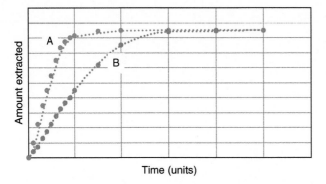

Figure 5.2 A typical extraction profile for solid-phase microextraction: influence of agitation on amount extracted. (A) Good agitation and (B) poor agitation.

5.3 Practical Considerations for SPME

The coated SPME fibre, with its small physical diameter and cylindrical geometry, is incorporated into a syringe-like holder (or barrel) (Figure 5.3). The SPME holder (Figure 5.3) provides two functions, one is to provide protection for the fibre during transport, while the second function is to allow piercing of the rubber septum, of the gas chromatograph (GC) injection port, via its stainless needle. The selective nature of the phase of the SPME fibre precludes the introduction of solvent into the GC. In addition, no instrument modification is required for the GC in terms of, for example, a thermal desorption unit. The heat (e.g. 220 °C) for desorption from the fibre is provided by the injection port of the GC.

In the inoperable mode, the fused-silica coated fibre is retracted within the stainless steel needle of the SPME holder (barrel) (Figure 5.3) for protection. In operation, however, the coated-silica fibre is exposed to the sample in the form of its matrix, liquid, solid or vapour. The sampling, by the 1 cm long fibre, can take place in the headspace of the sample matrix (i.e. above a solid or liquid sample) or by insertion within the matrix, i.e. within a liquid or gaseous sample. In either case, the SPME fibre is exposed to the organic compounds derived from the matrix (solid, liquid or gaseous) for a pre-selected time period. This exposure duration needs to be optimized, to achieve maximum recovery of the organic compounds, and will vary depending on the matrix type (solid, liquid or gas) and the chemical and physical properties of the organic compounds. In addition, the type of SPME-coated fibre needs to be evaluated and selected (Table 5.1). After sampling, the fibre is retracted within its holder for protection until inserted in the hot injection port of the gas chromatograph (typically held between 150 and 270 °C) of the GC. The temperature of the injection port will need to be investigated based on the chemical and physical properties of the target organic compounds, e.g. thermally stability. Once located in the hot injector, the fibre is exposed for a particular time to allow for effective desorption of the compounds; again, the desorption time needs to be investigated (typically in the time zone of 1–3 minutes).

As the coating on the fibre is selective towards the organic compounds (Table 5.1), it is common to find that no solvent peaks are present in the subsequent gas chromatogram. Unless precautions are made, it is important that the delay between the sorption step and the subsequent desorption and analysis step is small. This is because the silica-coated fibre can equally concentrate compounds from the workplace atmosphere as it can from the sample or that

Figure 5.3 Schematic diagram of solid-phase microextraction assembly.

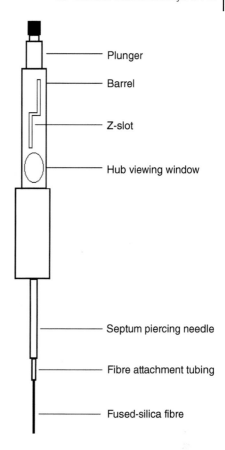

- Plunger
- Barrel
- Z-slot
- Hub viewing window
- Septum piercing needle
- Fibre attachment tubing
- Fused-silica fibre

losses can occur from the fibre. In the first case, the risk of contamination from the workplace environment is potentially high. One way to minimize the risk of contamination for aqueous samples is to operate the SPME using an autosampler on the gas chromatograph. In this case, sealed vials in the autosampler contain the aqueous samples. In operation, the SPME needle can then pierce an individual vial and carry out the sorption stage. This can be immediately followed by analysis. If an automated system is not available, contamination from the atmosphere can only be eradicated by minimizing the time between extraction and analysis and/or working in a clean room environment. Losses of compound from the SPME fibre can be achieved by, for example, storing the SPME fibres in a fridge (4 °C, and in the dark).

5.3.1 SPME Agitation Methods

The available agitation methods for SPME need to be considered including their impact on the extraction process. The possible agitation methods are:

- Static, i.e. no agitation. Useful for headspace SPME and volatile organic compounds.
- Magnetic stirring of the sample matrix; this approach requires a magnetic stirring bar in the aqueous sample vial. A risk of cross-contamination is possible unless thorough cleaning of the stir bar is made. However, the possibility of consistent stirring can aid reproducibility of extraction.

- Vortex/vial movement of the sample matrix; this form of aggressive agitation can put stress on the SPME needle and fibre. For direct insertion of SPME, the agitation will need controlling to avoid cavitation during the mixing. As the vial is externally agitated, the possibility of cross-contamination (from a stir bar) is avoided.
- Movement of the SPME fibre; this form of agitation can put stress on the needle and fibre. The longevity of the fibre will be reduced.
- Sonication using an ultrasonic bath; an aggressive form of agitation that can lead to the aqueous sample solution being heated (involuntarily).

5.3.2 Other SPME Operating Considerations

In addition, the main issues to address in use of SPME are:

- Whether using a manual or automated SPME system.
- Sampling mode, i.e. direct immersion or headspace sampling.
- Type of coating and thickness of the fibre, i.e. the active surface for adsorption or absorption.
- Extraction time, i.e. how long to expose the fibre to the sample.
- Extraction enhancement protocols, e.g. whether to alter the pH and/or ionic strength of the sample solution; and control of sample temperature.
- Desorption mode and time, i.e. use of a GC injection port and the duration; also important to not exceed the operating temperature of the fibre coating when using the injection port of the GC.

5.4 Application of SPME

Case Study A Investigation of (a) Extraction Time and (b) Fibre Type with Headspace Solid-Phase Microextraction Followed by Gas Chromatography-Mass Spectrometry

Background: An investigation of factors that influence the data obtained when performing headspace solid-phase microextraction (HS-SPME). In this particular case study, the samples used are microbiological in type, e.g. *Escherichia coli* in broth (a food source for bacteria growth), but the principles are similar for other forms of environmental analysis using HS-SPME.

Experimental
Instrumentation and Reagents
Standards of ethyl 2-methylbutanoate (99%), butyl acetate (99.5%), 3-methyl butyl acetate (98%), butyl 2-methyl butanoate (>98%), 1-octanol (98%), decyl acetate (≥ 95%), 1-decanol (99%), 9-decen-1-ol (97%), 1-dodecanol (98%), 1-tetradecanol (97%) and indole (≥ 99%) were purchased from Sigma-Aldrich (Dorset, UK). Brain heart infusion (BHI) broth (CM1135) was obtained from Oxoid (Basingstoke, UK). The broth is composed of brain heart infusion solids (12.5 g), beef heart infusion solids (5.0 g) and proteose peptone (10.0 g) as the sources of nitrogen, vitamins and carbon. In addition, glucose (2.0 g) is added as the carbohydrate (energy) source. The Gram-negative bacteria *Escherichia coli* NCTC 10418 was obtained from the National Collection of Type Cultures (NCTC), Colindale, UK. SPME fibres evaluated for extracting bacterial VOCs

were 100 μm polydimethylsiloxane (PDMS), 85 μm polyacrylate (PA) and 50/30 μm divinylbenzene/carboxen/polydimethylsiloxane (DVB/CAR/PDMS) (Supelco Corp., Bellefonte, PA, USA). All fibres were conditioned in the GC injection port prior to use as directed by manufacturers guidelines. All fibres were used with a manual holder.

Gas chromatography-mass spectrometry (GC-MS) analysis, using electron impact ionization, was performed on either a Thermo Finnigan Trace GC ultra with a Polaris Q ion trap mass spectrometer fitted with a polar GC column or a TraceGC with DSQ quadrupole mass spectrometer fitted with a non-polar GC column; both GC-MS systems were operated with Xcalibur software. A VF-WAXms 30 m × 0.25 mm × 0.25 μm (Varian) was used as the polar column. Separation was achieved using the following temperature programme: initial 50 °C with a 2-minute hold, ramped to 220 °C at 10 °C min^{-1} and then held for 2 minutes. The split-splitless injection port was held at 230 °C for desorption of volatiles in split mode at a split ratio of 1:10. Helium was used as the carrier gas at a constant flow rate of 1.0 ml min^{-1}. MS parameters were as follows: full-scan mode with scan range 50–650 amu at a rate of 0.58 scan s^{-1}. The ion source temperature was 250 °C with an ionizing energy of 70 eV and a mass transfer line of 250 °C.

Procedure for Growth of Bacteria and Sample Preparation
All bacteria were sub-cultured overnight at 37 °C on Columbia blood agar one day prior to preparation for volatile organic compound (VOC) analysis. After overnight incubation on blood agar at 37 °C, bacteria were inoculated in sterile brain heart infusion (BHI) broth and incubated at 37 °C. The BHI broth was made up according to manufacturer's guidelines. Sterilization of media was achieved by autoclaving at 126 °C for 11 minutes. Samples were prepared by measuring the absorbance of the incubated bacterial suspension at $OD_{600\,nm}$. At an absorbance reading of 0.132 (equivalent to 0.5 McFarland units), i.e. $1-1.5 \times 10^8$ organisms per ml, an aliquot of 100 μl of bacterial suspension (1.5×10^7 organisms) was added to a 20 ml clear vial with PTFE septum and screw cap containing 10 ml sterile BHI broth. Inoculated broths were incubated for 18 hours at 37 °C and then subjected to volatile profiling via HS-SPME-GC-MS.

HS-SPME-GC-MS Procedure for the Analysis of Bacterial VOCs
Bacterial VOCs were extracted from the headspace of static inoculated and uninoculated broths and concentrated via SPME before desorption in the hot GC injection port. All samples and blanks were held at 37 °C in a water bath for 30 minutes prior to VOC extraction and kept at this temperature throughout sampling. The SPME fibre pierced the PTFE septum (of the sample vial) and was exposed in the headspace for 10 minutes. After VOC extraction, the SPME fibre was exposed in the hot GC injection port for 2 minutes for desorption of bacterial VOCs. All experiments were carried out in triplicate.

Results and Discussion
Calibration graphs of all VOCs were prepared by spiking standards of known concentration into blank culture medium, followed by incubation at 37 °C, and subsequent extraction of VOCs. The VOCs were quantified using external calibration, and the limit of detection (LOD) and limit of quantification (LOQ) were determined as the peak areas of three times the signal-to-noise ratio and 10 times the signal to noise ratio, respectively. Analytical data for the quantification of VOCs is shown in Table 5.2 with an example chromatogram in Figure 5.4.

Table 5.2 Quantitative data analysis of VOCs using the polyacrylate SPME fibre.

VOC	t_R (min)	Quantitative m/z	Linear range (μg ml^{-1})	y	r^2	n	LOD (ng ml^{-1})	LOQ (ng ml^{-1})
Ethyl 2-methyl butanoate	4.0	74, 102	0.25–50	6590x + 2878	0.9985	6	13	43
Butyl acetate	4.2	56, 61	0.25–50	571x−123	0.9993	6	210	690
3-Methyl butyl acetate	4.9	55, 70	0.25–50	10486x − 355	0.9986	6	11	37
Butyl 2-methyl butanoate	6.4	103, 85	0.025–10	16487x − 630	0.9997	6	8	25
1-Octanol	10.8	55, 69	0.125–10	17140x + 6327	0.9942	6	10	33
Decyl acetate	12.4	69, 83	0.005–0.5	372756x + 2742	0.9970	6	1.0	3
1-Decanol	13.2	55, 69	0.0125–5	175302x + 3385	0.9999	6	2	6
9-Decen-1-ol	13.8	67, 81	0.0125–2	65758x + 8184	0.9839	6	2	8
1-Dodecanol	15.4	55, 69	0.0125–1	313351x + 4223	0.9945	6	1	4
1-Tetradecanol	17.4	55, 69	0.0125–1	106434x + 6401	0.9693	6	4	12
Indole	20.1	117, 89	5–200	33841x + 3890	0.9942	6	8	26

n = number of points on calibration curve.

The extraction profile, with respect to sampling time, was investigated for indole from the headspace of a standard culture (i.e. *E. coli* NCTC 10418 in BHI broth). As is typical with HS-SPME analyses, the amount of volatile organic compound (i.e. indole) extracted increased as the extraction time increased (Figure 5.5). However, this increasing linearity can become an issue if the capacity of the SPME fibre is exceeded, i.e. overloading of the fibre can occur if the sample is too concentrated or extraction times are excessively long.

The influence of solution agitation was investigated; agitation versus static (i.e. no shaking) of the sample solution (culture medium) on the volatile organic compound (indole) concentration generated from the sample (i.e. *E. coli* NCTC 10418). Agitation was achieved using a magnetic stir bar, placed in the sample solution and onto a stirrer hot-plate (no heat applied) for an extraction time of 10 minutes. The results (Table 5.3) indicate that agitation had no consequence to the concentration of the VOC determined. A static matrix, that is, no agitation, is effective for extraction when analytes are volatile and thus is recommended for headspace analysis. It is also observed (Table 5.3) that desorption time within the GC injection port had no impact on recovery of the compound.

A major issue in using SPME for analysis is the selection of the fibre type: the heuristic ('rule of thumb') viewpoint would be that a polar SPME fibre should extract polar com-pounds etc. Data shown in Table 5.4 shows the recovery of different volatile organic com-pounds, based on SPME fibre type, from *E. coli* cultured in BHI broth. While some similarities can be seen (the same VOC recovered by all three fibre types), differences are also noted. Additionally, and using a different example, the recovery of indole from *E. coli* with HS-SPME-GC-MS is shown in the form of a chromatogram (Figure 5.6). It is obvious that the recovery of indole is greater, in this example, using a DVB/CAR/PDMS SPME fibre, compared

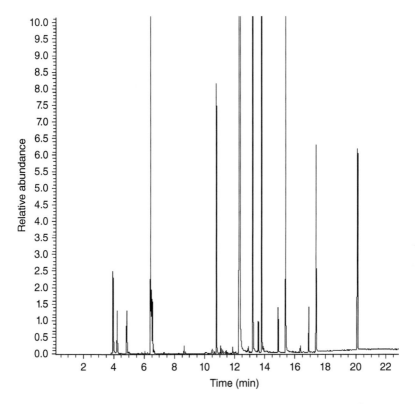

Compound identification (retention time): ethyl-2-methylbutanoate (4.0 minutes); butyl acetate (4.2 minutes); 3-methyl butyl acetate (4.9 minutes); butyl 2-methyl butanoate (6.4 minutes); 1-octanol (10.8 minutes); decyl acetate (12.4 minutes); 1-decanol (13.2 minutes); 9-decen-1-ol (13.8 minutes); 1-dodecanol (15.4 minutes); 1-tetradecanol (17.4 minutes); and indole (20.1 minutes).

Figure 5.4 Headspace solid-phase microextraction of volatile organic compounds (all 10 µg ml^{-1}).

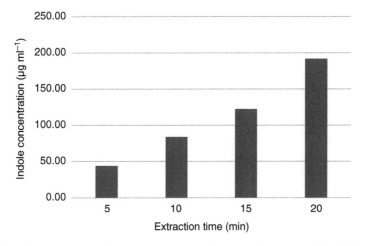

Figure 5.5 Influence of extraction time on recovery of an organic compound (indole) using headspace solid-phase microextraction gas chromatography mass spectrometry.

Table 5.3 Influence of solution (BHI broth) agitation, and fibre desorption time in the gas chromatography injection port, on the recovery of a volatile organic compound (indole) generated by *E. coli* NCTC 10418.

Stirring[a]	Mean indole concentration (μg ml^{-1})[c]
Yes	132.3 ± 44.8
No	132.8 ± 48.0
Desorption time (min)[b]	
2	92.3 ± 8.4
3	85.1 ± 24.7

[a] N = 5.
[b] N = 3.
[c] mean ± standard deviation.

Table 5.4 Investigation of fibre type for the determination of volatile organic compounds evolved from *E. coli* cultured in brain heart infusion broth using HS-SPME-GC-MS.

Volatile organic compound	Aqueous sample of *E. coli*[a]		
Fibre type	PDMS	PA	DVB/CAR/PDMS
3-Methyl butanal	ND	ND	ND
Ethyl 2-methyl butanoate	ND	ND	ND
Butyl acetate	ND	ND	ND
3-Methyl butyl acetate	ND	ND	ND
3-Methyl-1-butanol	ND	7.3 ± 2.0	71 ± 4
Butyl 2-methyl butanoate	ND	ND	ND
1-Octanol	ND	ND	0.67 ± 0.06
3-Methyl-butanoic acid	ND	ND	ND
Butanoic acid	ND	ND	ND
Decyl acetate	ND	ND	ND
1-Decanol	1.43 ± 0.12	1.33 ± 0.16	1.90 ± 0.11
9-Decen-1-ol	ND	ND	ND
1-Dodecanol	0.14 ± 0.02	0.18 ± 0.04	0.16 ± 0.04
1-Tetradecanol	0.14 ± 0.01	0.15 ± 0.01	0.15 ± 0.01
Indole	80 ± 18	72 ± 45	83 ± 58

[a] Mean VOC concentration (μg ml^{-1}) (n = 3) ± 1 S.D.
ND = not detected.

with the PDMS fibre; this comment is based on the observation that the noise on the baseline of the two chromatographs (on the same scale) is greater in Figure 5.6a than Figure 5.6b. The key learning objective is to investigate the optimum fibre type for its specific application.

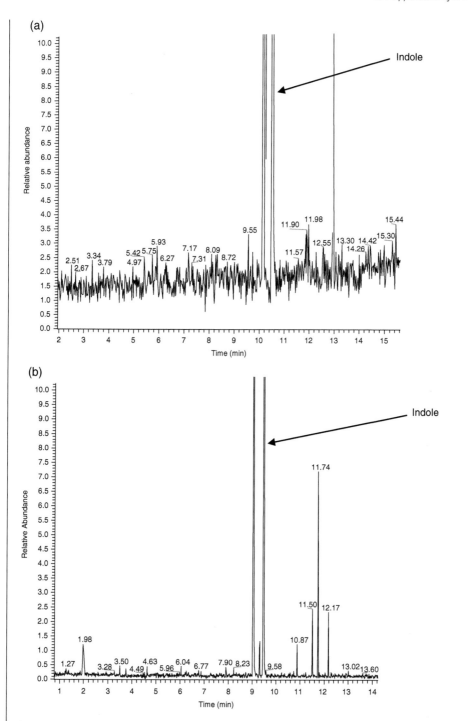

Figure 5.6 Influence of fibre type on extraction profile for indole using headspace solid-phase microextraction gas chromatography mass spectrometry, (a) polydimethylsiloxane fibre (PDMS, 100 μm) and (b) divinylbenzene/carboxen/polydimethylsoloxane fibre (DVB/CAR/PDMS, 50/30 μm).

5.5 Summary

A crossword of the key terms outlined in this chapter and Chapters 4, 6–8 can be found in Appendix A1, with the solution in Appendix B1.

Reference

1 Louch, D., Motlagh, S., and Pawliszyn, J. (1992). *Anal. Chem.* 64: 1187–1199.

6

In-Tube Extraction

LEARNING OBJECTIVES
After completing this chapter, students should be able to: • Understand the operation and function of in-tube extraction techniques. • Be aware of the different approaches for in-tube extraction. • Be able to develop a procedure for analysis of VOCs from an aqueous sample.

6.1 Introduction

Within the context of microextraction, a wide range of alternate formats, to solid-phase microextraction (SPME) (Figure 6.1a), have been developed and applied based on in-needle extraction. These alternate approaches are based on the inclusion within the internal surface of a GC syringe needle of either a short length of GC capillary column or by inclusion of a sorbent coating. Examples include:

- Solid-phase dynamic extraction (SPDE); based on the insertion of a short length of GC capillary column (e.g. 2–5 cm of DB-1 or DB-5), to act as an absorption trap, within the needle (Figure 6.1b). The use of the protective syringe needle provides protection for the stationary phase, as well as faster extraction times due to the mechanical action of multiple pumping of the sample using the gas-tight syringe. An issue can arise due to the thermal desorption mechanism used to extract the volatile organic compounds from the length of capillary stationary phase if an excessive length (e.g. 5 cm) is used. The use of an excessive length of capillary tubing can lead to the formation of a thermal gradient and as a result not all the organic compounds do not desorb at the same time.
- Needle trap (NT) extraction is based on the entrapment of a sorbent within the hollow needle (Figure 6.1c). In this case, it is possible to have more sorbent located within the needle. A range of options have been developed commercially and include (i) microextraction by packed sorbent (MEPS) that can use sorbents material amenable to non-volatile compounds (e.g. C18), solvent desorption and analysis by HPLC (or GC) (Figure 6.2), and (ii) in-tube extraction (ITEX) in which the needle is packed with sorbent (e.g. Tenax™), desorption is by heat and analysis of volatile organic compounds by GC (Figure 6.3). In both cases, multiple pumping of the gas-tight syringe can enhance pre-concentration of the compounds by exposing the sorbent to more of the sample.

Extraction Techniques for Environmental Analysis, First Edition. John R. Dean.
© 2022 John Wiley & Sons Ltd. Published 2022 by John Wiley & Sons Ltd.

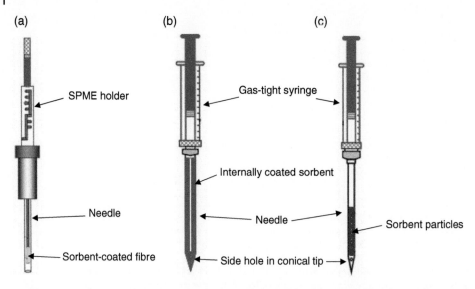

(a)

SPME holder

Needle

Sorbent-coated fibre

(b)

Gas-tight syringe

Internally coated sorbent

Needle

Side hole in conical tip →

(c)

Sorbent particles

Figure 6.1 Comparison of (a) solid-phase microextraction, (b) solid-phase dynamic extraction and (c) needle trap extraction.

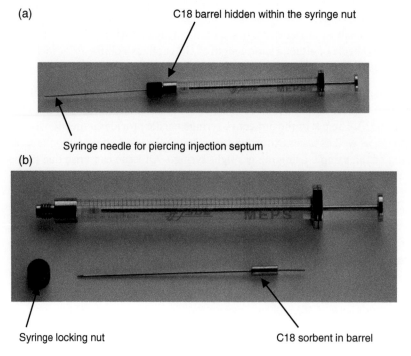

(a)

C18 barrel hidden within the syringe nut

Syringe needle for piercing injection septum

(b)

Syringe locking nut

C18 sorbent in barrel

Figure 6.2 Microextraction by packed sorbent (MEPS). (a) Complete MEPS syringe and (b) unassembled syringe.

(a)

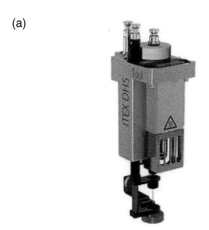

(b)

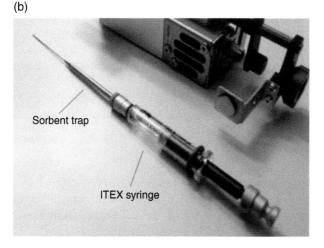

Sorbent trap

ITEX syringe

Figure 6.3 Automated in-tube extraction (ITEX). (a) Full assembly for mounting on autosampler and (b) disassembled.

6.2 Microextraction in a Packed Syringe (MEPS)

Microextraction in a packed syringe (MEPS) is a type of in situ solid-phase extraction (SPE) in which the sorbent is in a chamber at the top of a syringe needle (Figure 6.2b). The typical sorbents include C18, C8, C2 and polystyrene-divinylbenzene copolymer (PS-DVB). The MEPS device can be directly used instead of a conventional syringe for introduction of samples into an HPLC (or GC).

6.2.1 Procedure for MEPS

In operation, an aqueous sample is drawn up (and down) the gas-tight syringe needle (of the MEPS device) to fill (and empty) the sorbent chamber. This process is repeated multiple times to affect pre-concentration of the compounds from the aqueous sample. As well as the compounds of interest of other extraneous material will be retained on the sorbent, i.e. pre-concentrated. To remove extraneous material, a wash stage can be incorporated in the process, e.g. 50 μl of water. Finally, the compounds are eluted with an organic solvent (e.g. 20–50 μl methanol) directly into the injection port of the GC or Rheodyne valve of the HPLC. The process can be fully automated using the autosampler of the GC or HPLC. In the case of GC, a large volume of sample (up to 50 μl of extract) can be introduced using a programmed temperature vaporizer (PTV) injector (Figure 16.11), instead of the more conventional spilt/splitlesss injector (see Figure 16.9).

6.2.2 Main Issues in MEPS

The main issues to address in use of MEPS are:
Sampling

- Dilution of samples may be required depending on their viscosity. A dilution of 1 in 5 in water, or 0.1% formic acid, is recommended for moderately viscous samples and 1 in 25 for highly viscous samples.
- Centrifugation or filtration of samples maybe necessary to remove macro particles, e.g. centrifugation can be done at 3000 rpm for 2 minutes or filtration via a 0.25 μm filter.
- Use of extraction enhancement protocols to aid recovery of organic compounds, e.g. adjustment of aqueous sample pH and/or its ionic strength.
- Syringe plunger speed for sampling. A typical sampling rate is between 10 and 20 μl s^{-1}.
- Sample loading. The extraction efficiency can be improved by using multiple MEPS loading cycles, e.g. >10.

Washing

- Removal of extraneous material from MEPS sorbent. Use small volumes of water (50–100 μl) with 5–10% organic solvent (e.g. methanol, isopropanol or acetonitrile) to elute unwanted matrix components whilst preventing compound loss. This will need to be investigated to ensure the compounds of interest are not lost (see Chapter 4, Case Study A).

Drying

- It may not be necessary to dry the MEPS sorbent, due to the small chamber volume, and small scale of sorbent bed.

Elution solvent

- A careful selection of an appropriate solvent or mixture is required. Typical organic solvents used include methanol, isopropanol or acetonitrile, individually or as a mixture with an acid (e.g. acetic or formic acids) or a base (e.g. triethylamine) in the range 0.1–3% to elute the organic compounds in a small volume, e.g. 20–50 μl. This will need to be investigated to ensure the compound of interest are eluted (see Chapter 4, Case Study A).

Cleaning post-elution to reduce/eliminate sample carry over

- This can be done using a combined strong and weak wash procedure, i.e.:

Strong wash: Use either methanol or acetonitrile with 10–20% isopropanol. The use of 0.2%, v/v formic acid or 0.2%, v/v ammonium hydroxide may be required.
Weak wash: Use either water or 5% methanol in water.

The development of the MEPS device has largely been focused on applications in pharmaceutical analysis (drugs and their metabolites) in blood, plasma and urine and analysis by HPLC. Applications in the environmental context have also been reported (e.g. endocrine disrupting chemicals (EDCs), organophosphate flame retardants (OPFRs), polycyclic aromatic hydrocarbons (PAHs), polychlorinated biphenyls (PCBs) and pesticides) in natural waters, urine and soil with analysis by GC.

6.3 In-Tube Extraction (ITEX)

Automated in-tube extraction (ITEX) uses a gas-tight syringe, fitted with a microtrap, to sample the headspace above a sample located in an autosampler vial. An image of the automated ITEX system is shown in Figure 6.3a, with the disassembled syringe and in situ trap in Figure 6.3b. A range of sorbent trap are available including Carbopack C, Carboxen, Carbosieve S-III, molecular sieve, Tenax™ GR, Tenax™ TA and a mixture of Tenax™ GR/Carbosieve S-III. As a result, optimization of the most appropriate trap for recovery of volatile organic compounds is required. Note: Tenax™ is the trade name for the polymer resin, poly(2,6-diphenyl-p-phenylene oxide) or PPPO.

6.3.1 Procedure for ITEX-DHS

In operation (Figure 6.4), the aqueous sample (e.g. 1 ml), in a suitable sealed vial (e.g. a volume of 20 ml), is both heated (e.g. 50–100 °C) and agitated (250–750 rpm) for a

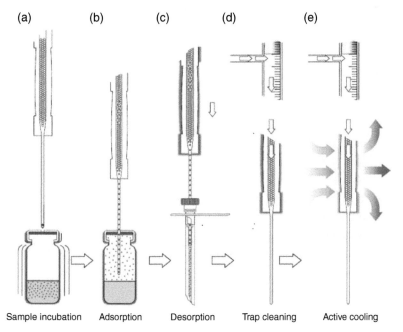

Figure 6.4 Method of operation of ITEX-DHS. (a) Sample incubation, (b) adsorption, (c) desorption, (d) trap cleaning and (e) active cooling.

predetermined time (e.g. 5–20 minutes) (Figure 6.4a). Then, the gas-tight syringe needle pierces the septum of the sealed vial; adsorption of the volatile organic compounds in the aqueous sample headspace are recovered by the mechanical action of the syringe (e.g. using 1–50 extraction strokes) (Figure 6.4b). Desorption of the retained organic compounds from the trap (e.g. Tenax) is done rapidly and with heat (i.e. 150–350 °C) and transferred into the GC injection port (Figure 6.4c). After injection, the trap is cleaned, by heat (Figure 6.4d) and then cooling (Figure 6.4e) with an inert gas, prior to loading of the next sample.

6.4 Application of ITEX-DHS

Case Study A   Analysis of Odorants in Tap Water Using Automated ITEX-Dynamic Headspace-Gas Chromatography-Mass Spectrometry

Background: The presence of volatile organic compounds, geosmin, isoborneol, 2-methyl-isoborneol and 2,4,6-trichloroanisole (Figure 6.5) can impart malodorous components to drinking water to the consumer. The sensitivity of the human nose to this type of compound means extremely sensitive methods of detection are required for their identification and analysis. The use of ITEX-DHS-GC-MS can facilitate their analysis in drinking water.

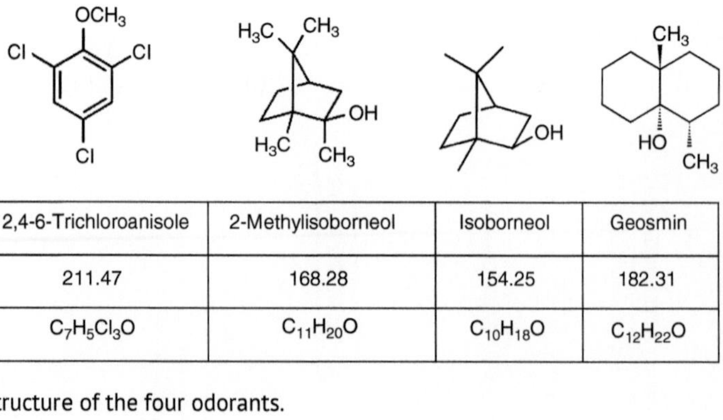

Compound name	2,4-6-Trichloroanisole	2-Methylisoborneol	Isoborneol	Geosmin
Molar mass (g mol^{-1})	211.47	168.28	154.25	182.31
Chemical formula	$C_7H_5Cl_3O$	$C_{11}H_{20}O$	$C_{10}H_{18}O$	$C_{12}H_{22}O$

Figure 6.5 Chemical structure of the four odorants.

Experimental

Chemicals

Standards of 2,4,6-trichloroanisole, isoborneol, geosmin and 2-methylisoborneol were purchased from Sigma-Aldrich, Dorset, UK. Ultrapure water was obtained from an ELGA water purification system. Tap water samples were collected from different locations.

Instrumentation

Gas chromatography (GC) analysis was performed using a Thermo Scientific Trace 1300 GC with a split/splitless injector operating in splitless mode. The GC was coupled to a

Thermo Scientific™ ISQ™ 7000 Single Quadrupole equipped with an electron impact ionization source. Separation was obtained using a Thermo Scientific TraceGOLD TG-5MS column 30 m × 0.25 mm × 0.25 μm. The temperature programme was: start at 60° C and then 25 °C min^{-1} until 120 °C (hold 1 minutes), then 10 °C min^{-1} until 160 °C, then hold for 1 minute. Carrier gas flow: 1.0 ml min^{-1}; split flow: 10 ml min^{-1}; splitless mode for 0.5 minutes; injector temperature: 250 °C; ion source temperature: 300 °C; scan mode: selected-ion monitoring. The transfer line temperature was fixed at 300 °C.

Dynamic headspace extraction and enrichment were performed using the TriPlus RSH autosampler equipped with a vial incubation oven and agitator, and the ITEX-DHS sampling tool. The ITEX-DHS was operated as follows: using a vial size of 20 ml to which 1 ml of sample was placed. Incubation was done at 800 °C for 10 minutes using an agitator speed of 250 rpm. The trap, Tenax GR : Carbosieve S-III (1:1) was pre-cleaned at 250 °C for 20 seconds. Headspace sampling took place with the trap and syringe at 50 °C. The headspace was sampled (1 ml) at an aspirate flow rate of 50 μl s^{-1} for 10 extraction strokes. Desorption of the retained VOCs took place at a dispensing rate of 500 μl s^{-1} at 150 °C; injection onto the column was done at 100 μl s^{-1}. After each sample episode, the trap was cleaned for 60 seconds, to prevent carrier over.

Results and Discussion
The number of extraction strokes will influence the recovery of the four malodorous compounds from water. An investigation of the influence of the extraction strokes from 5 to 40 was done. It can be seen (Figure 6.6) that the overall profile of recovery of the compounds increases uniformly over the extraction strokes 5–30, and then reaches a steady plateau (at 30–40 extraction strokes). Using a compromise of 10 extraction strokes, assessed in terms of the total analysis time, chromatographic separation was

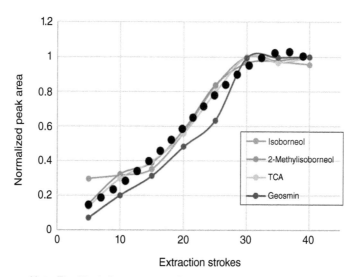

Note: The black dots represent the average extractionstroke profile

Figure 6.6 Influence of ITEX extraction strokes on recovery of organic compounds.

Case Study A   (Continued)

achieved in 8.25 minutes. An ITEX-DHS-GC-MS chromatogram, in selected-ion monitoring mode, for a standard solution is shown in Figure 6.7a. Calibration of the four malodourous compounds, in the range 0–100 ng l^{-1}, illustrated the sensitivity of the approach (Figure 6.8). The developed method was used to analyse the compounds in tap water. An example of ITEX-DHS-GC-MS chromatogram, in single-ion monitoring mode, is shown in Figure 6.7b. Method detection limits ranged from 0.008 ng l^{-1} (2-methylisoborneol) to 1.1 ng l^{-1} (2,4,6-trichloroanisole).

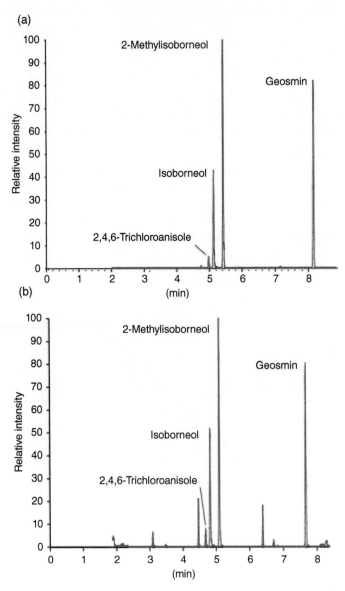

Figure 6.7 ITEX-DHS-GC-MS chromatogram for odorants. (a) A standard (5 ng l^{-1}) in water and (b) a sample of tap water.

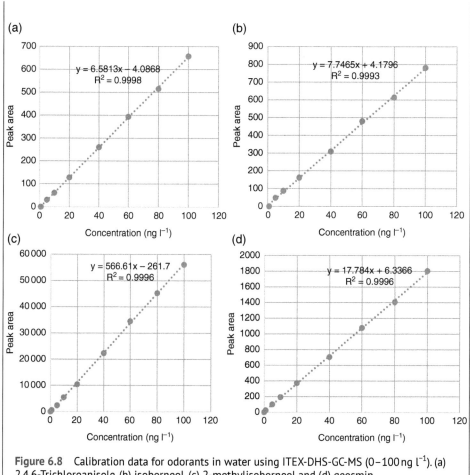

Figure 6.8 Calibration data for odorants in water using ITEX-DHS-GC-MS (0–100 ng l^{-1}). (a) 2,4,6-Trichloroanisole, (b) isoborneol, (c) 2-methylisoborneol and (d) geosmin.

6.5 Summary

A crossword of the key terms outlined in this chapter and Chapters 4, 5, 7 and 8 can be found in Appendix A1, with the solution in Appendix B1.

7

Stir-Bar Sorptive Extraction

LEARNING OBJECTIVES

After completing this chapter, students should be able to:

- Understand the operating principle and function of stir-bar sorptive extraction.
- Be aware of the underlying theoretical considerations of stir-bar sorptive extraction.
- Be aware of the key practical considerations when using stir-bar sorptive extraction.
- Be able to calculate the theoretical percentage recovery of a compound using stir-bar sorptive extraction.

7.1 Introduction

For stir-bar sorptive extraction (SBSE), the sorbent (or coating) is on the outside of a glass-jacketed magnetic stir bar (or magnetic 'flea') of 10 mm length (Figure 7.1). Sampling is done by placing the coated stir-bar either in the aqueous sample (SBSE) or in the headspace (HS-SBSE) above the sample (Figure 7.2); agitation is achieved by use of a stirrer hotplate. After SBSE sampling, the stir-bar is removed from the aqueous sample, with tweezers, rinsed in distilled water to remove salts and other sample components prior to drying on tissue paper. Finally, the extracted compounds are recovered, from SBSE or HS-SBSE, using either thermal desorption coupled to gas chromatography or liquid desorption followed by gas chromatography or high-performance liquid chromatography.

7.2 Theoretical Considerations for SBSE

The basis of SBSE is that organic compounds are partitioned between an aqueous phase (of the sample) and a polymer coating. The partitioning of the organic compounds between the two phases is proportional to the octanol–water partition coefficient (K_{ow}). On that basis, the theoretical percentage recovery (R) for SBSE can be calculated as follows:

$$R(\%) = \left(M_{PDMS}/M_o\right) \times 100 = \left(\alpha/1+\alpha\right) \times 100 \qquad (7.1)$$

Extraction Techniques for Environmental Analysis, First Edition. John R. Dean.
© 2022 John Wiley & Sons Ltd. Published 2022 by John Wiley & Sons Ltd.

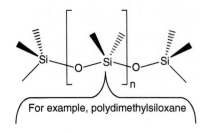

For example, polydimethylsiloxane

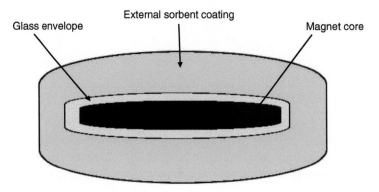

Figure 7.1   Schematic diagram of a stir-bar sorptive extraction device.

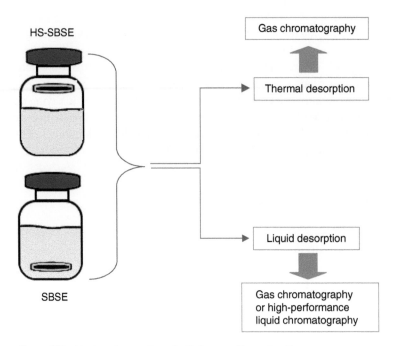

Figure 7.2   Modes of operation of stir-bar sorptive extraction.

where M_{PDMS} is the mass of the compound present on the polydimethylsiloxane (PDMS), M_0 is the total amount of organic compound originally present in the aqueous (water) sample and α is determined using the formula K_{ow}/β where β is the volume of the aqueous (water) divided by the volume of the PDMS coating (i.e. the phase ratio).

On this basis, it can be inferred that the lower the PDMS volume (i.e. a lower β value) will produce a higher recovery of the organic compound. In addition, the higher the hydrophobicity (ability to repel water) of the organic compounds, assessed in terms of a greater value of K_{ow}, will also lead to a higher recovery of the organic compound.

7.3 Practical Issues for SBSE

A range of sorbents have been used for SBSE. While initial developments were done using a non-polar coating, i.e. polydimethylsiloxane, with typical volumes of the coating between 24 and 126 µl. Other coatings with a medium-to-high polarity have been used, i.e. polyethyleneglycol-modified silicone and polyacrylate/polyethyleneglycol.

Thermal desorption is the obvious procedure for desorption of compounds from the coated stir-bar. The compounds are desorbed in the temperature range 150–300 °C for up to 15 minutes at flow rates of up to 100 ml min^{-1}. Longer desorption times are required for SBSE, compared to SPME, due to the much higher coating volumes used, i.e. a minimum of 24 µl compared with 0.5 µl. In contrast, liquid desorption requires the use of organic solvents to effectively remove the compounds from the stir-bar. This not only raises issues of solvent desorption compatibility, the potential contamination risk, but ultimately the significant effect of dilution of the compounds into the organic solvent.

7.3.1 Main Issues in SBSE

The main issues to address in use of SBSE are:

- Type and thickness of the coated stir-bar, i.e. the active surface for adsorption.
- Sample volume: typically, between 10 and 50 ml.
- Equilibrium time between sample (liquid or vapour) and the stir-bar.
- Agitation speed of sample solution (e.g. 750–1500 rpm).
- Extraction enhancement protocols:
  - Aqueous sample pH: SBSE is possible in the range pH 2–11.
  - Aqueous sample ionic strength: up to 20% possible, e.g. by using NaCl.
  - Aqueous sample temperature: in the range 20–60 °C.
  - Addition of an organic modifier (e.g. up to 20% methanol) to minimize the 'wall-effect' phenomena where organic compounds are retained by the glass walls of the sample vial; this is not uncommon for hydrophobic (water repelling compounds) or non-polar compounds, i.e. $\log K_{ow} \geq 3$.
- Desorption mode:
  - Use of a thermal desorption unit (TDU) connected to the GC injection port. In this situation, it is also important to not exceed the operating temperature of the stir-bar coating when using the TDU.

  o Consideration needed of the desorption temperature, purge flow rate and programmed temperature vaporizer, PTV (injector) parameters.
- For liquid desorption, the key aspects are the choice of solvent (e.g. methanol, acetonitrile or combination of solvents), the amount of solvent (ideally <1.5 ml), the 'soak' time
under mechanical or sonication conditions (<30 minutes) and the number of desorption
steps required to attain maximum extraction efficiency.

7.4 Application of SBSE

**Case Study A Calculate the Theoretical Percentage Recovery of an Organic
Compound (e.g. Pentachlorophenol) Using SBSE**

Background: Pentachlorophenol (Figure 7.3) is a non-naturally occurring biocide. It is
currently classified as a probable human carcinogen. However, previously it was widely
used as a wood preservative. It is extremely toxic to humans from acute (short-term)
ingestion and inhalation exposure.

Calculation:
Using a re-configured version of Eq. (7.1) will produce:

$$R(\%) = (K_{ow}/\beta)/(1+(K_{ow}/\beta)) \times 100 \tag{7.2}$$

Now using Eq. (7.2), with a sample volume of 50 ml and an organic compound (e.g. pentachlorophenol) with a log K_{ow} of 5.1, and a stir bar with a PDMS coating of 50 μl, calculate the % recovery.
 For a log K_{ow} of 5.1, the K_{ow} is 125 893, while the phase ratio, β, the volume of water/
volume of PDMS is 1000.
 Therefore,

$$R(\%) = (125893/1000)/(1+(125893/1000)) \times 100$$
$$R = 99.2\%$$

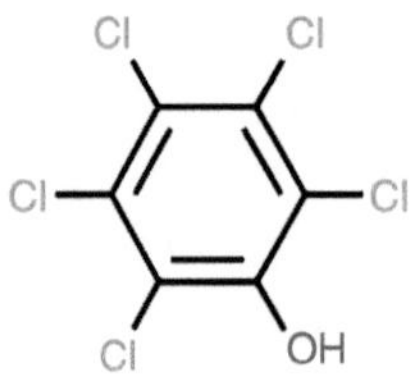

Figure 7.3 Chemical structure of
pentachlorophenol.

7.5 Summary

A crossword of the key terms outlined in this chapter and Chapters 4–6 and 8 can be found
in Appendix A1, with the solution in Appendix B1.

8

Membrane Extraction

LEARNING OBJECTIVES

After completing this chapter, students should be able to:

- Understand the principles of membrane extraction.
- Be aware of the diverse range of devices used for passive sampling in membrane extraction.
- Appreciate and comprehend the models used to explain membrane extraction.
- Be able to calculate the time-weighted average concentration for a compound.

8.1 Introduction

This chapter considers the use of membrane extraction devices for passive sampling of organic compounds in natural waters. Ensuring good water quality is essential to ensure the fit-for-purpose of potable water, river and lake water, as well as estuarine and coastal waters. Often these natural waters are monitored for their quality, in terms of the concentration of pollutants, by local and regional government agencies and private suppliers. Often however, the organic pollutants are present at low concentration in natural waters, and their concentration can vary according to temporal (i.e. time–hours, days) and spatial (i.e. influx from a discharge) variation. Sampling can be done by taking a sample at a point in time (See Section 2.4) using a grab or spot sampler. However, this approach may provide misleading information due to the sample being taken at a specific point in time, and the natural water sample not being homogeneous due to thermal variation or fluctuation due to influx from a discharge. A range of water monitoring devices are currently available for measurement of pH, dissolved organic carbon, temperature, turbidity etc. The focus here is on passive sampling devices. A large range of membrane extraction devices have been developed and explored for use in natural waters as environmental monitors. Passive sampling infers that a time duration will go by between exposing the in situ device, to the natural water environment, and its removal ready for analysis.

In general terms, all passive sampling devices consist of a receiving phase, selected on the basis that it is capable of retaining the pollutants to be extracted (e.g. a liquid or solid absorbent), and which is separated from the natural water by a diffusion-limiting layer (e.g. a porous membrane) (Figure 8.1).

Extraction Techniques for Environmental Analysis, First Edition. John R. Dean.
© 2022 John Wiley & Sons Ltd. Published 2022 by John Wiley & Sons Ltd.

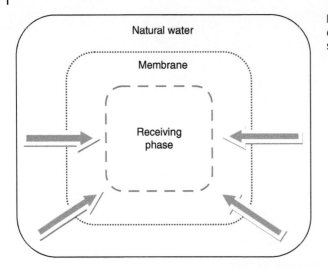

Figure 8.1 Generic characteristics of a passive sampling device.

8.2 Theoretical Considerations for Membrane Extraction

The device can be exposed to the natural environment for varied amounts of time from minutes, hours, days, weeks, months and years. It is therefore possible to determine a time-weighted average (TWA) concentration (Figure 8.2) of the organic compound pollutants in the selected environment. It is noted (Figure 8.2) that the TWA provides the more realistic assessment of the concentration of the pollutant over time, as compared with the grab (spot) sampling, which provides a concentration at a fixed time only. While the focus here is on water samples, passive sampling can also take place to monitor the air or soil samples. The TWA concentration (C_w) can be determined (in ng l^{-1}) using the equation:

$$C_w = \left(M_s - M_o\right)/\left(R_s \times t\right) \tag{8.1}$$

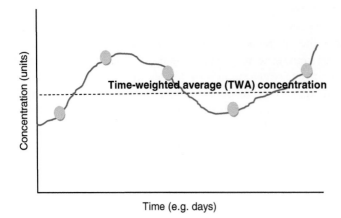

Concentration in individual spot (grab) water samples.

Figure 8.2 Variation in concentration of organic pollutant, in natural waters, over time.

where M_s is the mass of organic pollutant (ng), M_o the mass of organic pollutant on in situ blank (ng), R_s is the sampling rate of organic pollutant (e.g. l day^{-1}) and t = amount of time sampling device deployed in situ (e.g. days).

To determine the concentration of the organic pollutants in natural waters, it is necessary to consider the path of the compounds before they are collected by the receiving phase (with the passive sampling device). Two theoretical models have been applied: the mass transfer coefficient (MTC) model and the chemical reaction kinetic (CRK) model.

8.2.1 Mass Transfer Coefficient Model

The MTC model is based on the mass transfer of the pollutants across the sampler compartments, i.e. from the natural water (bulk water) to the receiving phase of the passive sampler (Figure 8.3). This diffusion-based model considers the mass transfer across the multiple compartments, within the sampler. Consideration of Figure 8.3 allows three barriers (or boundary layers) to be identified. The first boundary layer is the requirement for the pollutant to be transported across the water boundary layer, which is present on the surface of the passive sampler. The second layer is the transport of the pollutant across the biofilm layer; biofilms accumulate on the surface of any entity that is left in natural waters for an extended period. Finally, the transport of the pollutants across a membrane and/or receiving phase layer of the passive sampler. In some cases, the

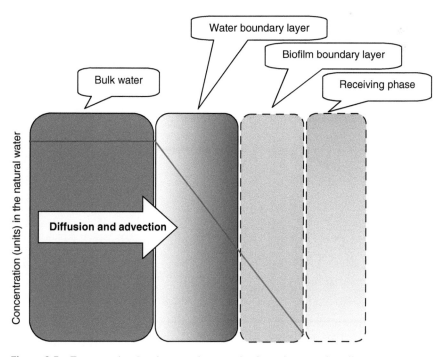

Figure 8.3 Transport barriers in a passive sampler for polar organic pollutants.

passive sampler may have a membrane boundary layer and a receiving phase boundary layer. The MTC can be expressed as:

$$MTC = \left(N_s / A \times t\right) = \lambda_0\left(C_w - \left(C_s / K_{sw}\right)\right) \tag{8.2}$$

where N_s is the amount of pollutants in the receiving phase (ng), A is the surface area (m^2), t is the exposure time of the sampler (days), λ_0 is the overall mass transfer coefficient (m s^{-1}), C_w is the average concentration of pollutant in the water or the TWA (ng l^{-1}), C_s is the concentration of the pollutant in the sampler as a function of time (ng g^{-1}), K_{sw} is the partition coefficient between the water and receiving phase (l g^{-1}).

It is possible to simplify Eq. (8.2), if the concentration of the pollutant when exposed to the passive sampling device has not reached equilibrium, i.e. $C_s < C_{s(eq)}$, where $C_{s(eq)}$ is the concentration of the pollutant in equilibrium with the passive sampler to:

$$N_s = \lambda_0 \times A \times C_w \times t \tag{8.3}$$

Subsequently, it is also possible to determine the overall mass transport coefficient:

$$\lambda_0 = R_s / A \tag{8.4}$$

where R_s is the water volume analysed by the passive sampling device per unit of time (e.g. l day^{-1}).

In general terms, pollutants with an octanol–water partition coefficient (log K_{ow}) > 3 are more limited by the water boundary layer. A possible solution to this rate limiting step is to have the passive sampler in faster flowing natural waters.

8.2.2 Chemical Reaction Kinetic Model

In the CRK model, the uptake by the passive sampler is modelled on first-order kinetics by considering only two compartments: the bulk water and the passive sampler receiving phase. Using this model, it is possible to identify (Figure 8.4) three possible regions: a linear

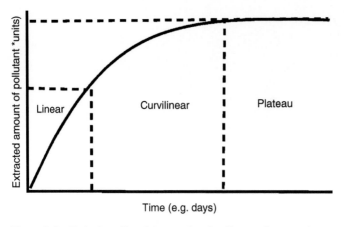

Figure 8.4 Typical profile of the uptake of pollutants by a passive sampling device.

region, a curvilinear region, and a plateau region. The uptake rate, by a pollutant, can be defined as:

$$C_s = C_w \times K_{sw}\left(1 - e^{(-K_e \times t)}\right) \tag{8.5}$$

where C_s is the concentration of the pollutant in the sampler as a function of time (ng g^{-1}), C_w is the average concentration of pollutant in the water or the TWA (ng l^{-1}), K_{sw} is the partition coefficient between the water and receiving phase (l g^{-1}), K_e is the elimination rate constant of the pollutant from the receiving phase (l day^{-1} g^{-1}) and, t is the exposure time of the sampler (days).

Depending on the exposure time, the passive sampling device can follow either a kinetic uptake (i.e. in the linear region) or an equilibrium uptake (i.e. in the plateau region) (Figure 8.4). In the equilibrium mode, the concentration of the pollutant is independent of the volume of natural water analysed and can be expressed as:

$$C_w = \left(C_s / K_{sw}\right) = N_s / \left(K_{sw} \times V_s\right) \tag{8.6}$$

where C_w is the average concentration of pollutant in the water or the TWA (ng l^{-1}), K_{sw} is the partition coefficient between the water and receiving phase (l g^{-1}), N_s is the amount of pollutants in the receiving phase (ng) and V_s is the volume (ml) of the receiving phase. In this mode, as the concentration of the pollutant in the water and passive sampler is the same, C_w is referred to as the equilibrium sampling concentration, $C_{w(eq)}$. The partition coefficient, K_{sw}, can be determined using:

$$K_{sw} = C_s / C_{w(eq)} = K_e / K_u \tag{8.7}$$

where C_s is the concentration of the pollutant in the sampler as a function of time (ng g^{-1}), $C_{w(eq)}$ is the average concentration of pollutant in the water or the TWA (ng l^{-1}), K_e is the elimination rate constant of the pollutant from the receiving phase (l day^{-1} g^{-1}) and K_u is the uptake rate constant (l day^{-1} g^{-1}).

In the kinetic mode, the average concentration of pollutant in the water (C_w) or the TWA (ng l^{-1}) can be expressed as:

$$C_w = N_s / \left(R_s \times t\right) \tag{8.8}$$

where N_s is the amount of pollutant in the receiving phase (ng), R_s is the water volume analysed by the passive sampling device per unit of time (e.g. l day^{-1}) and t is the exposure time of the sampler (days). The term R_s, the passive sampler – specific sampling rate can be determined experimentally in either the laboratory or in situ at the measuring site. This approach is the most commonly used to assess pollutants in natural waters as it provides a direct measure of the TWA, i.e. the concentration of transient pollutants.

8.3 Passive Sampling Devices

Semi-permeable membrane: A semi-permeable membrane (SPM) device (Figure 8.5) typically consists of low-density polyethylene (LDPE) tubing or a membrane. Inside the tubing (or sandwiched between the membrane) is a high-molecular-weight lipid (e.g.

triolein), which will retain compounds that transfer across the LDPE tubing/membrane. For this process to occur, the compounds must be both highly soluble in water and non-ionized; the use of triolein makes the SPMD highly effective for compounds with a log $K_{ow} > 3$.

Passive in situ concentration/extraction sampler (PISCES): The PISCES (Figure 8.6) has two polyethylene membranes which increases the sampling rate. Pollutants diffuse, from natural waters, through the membranes into the receiving phase (e.g. hexane). Sampling takes place over weeks or days, depending upon the receiving phase.

Diffusive gradients in thin film (DTG): The DTG (Figure 8.7) consists of two layers of acrylamide gel, one containing a binding agent, the receiving phase, e.g. activated charcoal, the other acting as the diffusion layer (e.g. agarose). The layers are protected, within

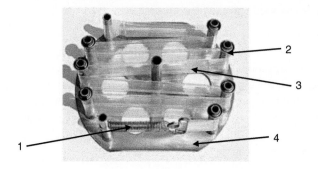

1. Spring, 2. Pegs, 3. Low-density polyethylene lay flat tubing, 4. Deployment rack: wall thickness: 70–95 μm, and Triolein (receiving phase): ultra-high purity and weight: 4.4–4.6 g.

Figure 8.5 A membrane extraction device for passive sampling of aqueous samples: semi-permeable membrane device [1].

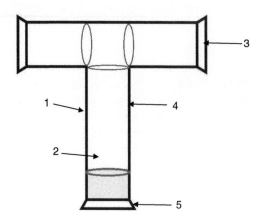

1. Polyethylene membrane, 2. Receiving phase, e.g. hexane or iso-octane, 3. Teflon nut, 4. Metallic T-pipe, and 5. Metallic nut.

Figure 8.6 A membrane extraction device for passive sampling of aqueous samples: passive in situ concentration/extraction sampler (PISCES) Based on [1].

the sampling device, by a filter membrane, i.e. polytetrafluoethylene (PTFE) or polyether-sulphone (PES). Most of its development was done for metals in natural waters but has been applied for organic pollutants using Oasis HLB or Strata-X, as the receiving phase. For organic pollutants, the device is often pre-fixed as o-DTG. Sampling times are in the range hours to weeks, when deployed.

Polar organic chemical integrative sampler (POCIS): The POCIS consists of a solid sorbent positioned between two microporous polyethersulphone (PES) diffusion-limiting membranes (Figure 8.8), sealed with stainless steel rings. The choice of sorbent influences the selectivity of the device. The configurations are often referred to as the pharmaceutical POCIS when using HLB Oasis as sorbent or the pesticide POCIS when

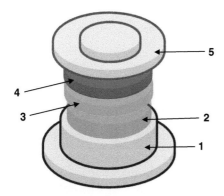

1. Base, 2. Receiving phase, e.g. Oasis HLB sorbent, Strata-X, activated charcoal etc., 3. Diffusive gel (e.g. polyacrylamide or agarose), 4.Filter membrane: PTFE or PES, and 5. Cap.

Figure 8.7 A membrane extraction device for passive sampling of aqueous samples: diffusive gradients in thin film (DTG) Based on [1].

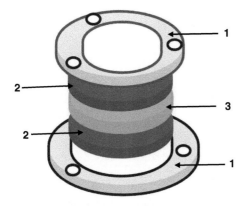

1. Stainless steel ring, 2. Two sheets of microporous polyethersulfone, and 3. Receiving phase: (a) Oasis HLB resin or (b) solute ENV + BioBeads S-X3 with surface dispersed powdered Ambersorb 1500 carbon.

Figure 8.8 A membrane extraction device for passive sampling of aqueous samples: polar organic chemical integrative sampler (POCIS) Based on [1].

using Isolute® ENV+, polystyrene divinylbenzene and Ambersorb® 1500 carbon dispersed on S-X3 Biobeads® as sorbents. Sampling times are in the range weeks to months, when deployed.

Thin-film solid-phase microextraction (TF-SPME): The TF-SPME (Figure 8.9) consists of a modified solid-phase microextraction device (see Chapter 5). The TF-SPME uses a thin film of ethylene vinyl acetate coated on a glass surface as the solid-phase sampler. A variety of extraction phases have been used, as those for conventional SPME. The TF-SPME can be used in two modes: as a retracted thin-film device with, for example, a HLB coating (Figure 8.9i), or an open-bed configuration with, for

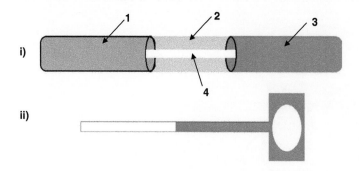

i) Retracted TF-SPME TWA sample or ii) Open-bed TF-SPME. 1. Teflon rod assembly, 2. Teflon spacer, 3. Tube with a drilled hole as diffusion path, and 4. TF-SPME (e.g. HLB or C18).

Figure 8.9 A membrane extraction device for passive sampling of aqueous samples: thin-film solid-phase microextraction (TF-SPME) Based on [1].

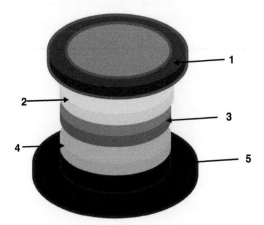

1. Cap, 2. Stainless mesh, 3. Limiting membrane materials, e.g. Polysulfone, polyethylene, polyethersulfone (PES), and low-density polyethylene (LDPE), 4. Receiving phase, e.g. C18 disk, cellulose acetate, SDB-RPS and SDB-XC and 5. PTFE housing.

Figure 8.10 A membrane extraction device for passive sampling of aqueous samples: Chemcatcher® Based on [1].

example, a C18 coating (Figure 8.9ii). Sampling times are in the range days to weeks, when deployed.

Chemcatcher®: The Chemcatcher® consists of a stainless steel mesh, followed by a diffusion-limiting membrane (e.g. low-density polyethylene) and a receiving phase (e.g. C_{18} Empore disk) to retain the pollutants (Figure 8.10). The device is retained within a PTFE housing. Sampling times are in the range days to weeks, when deployed.

Membrane enclosed-sorptive coating (MESCO): The MESCO (Figure 8.11) consists of a bar coated with polydimethylsiloxane (PDMS), i.e. the receiving phase, enclosed in a dialysis membrane composed of regenerated cellulose as the diffusion-limiting barrier.

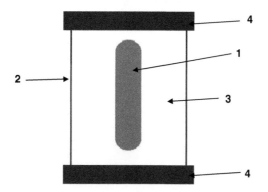

1. A Gerstel-Twister bar used for stir-bar sorptive extraction (SBSE), 2. A dialysis membrane bag made from regenerated cellulose, 3. The dialysis membrane bag is filled with 3 ml of bi-distilled water and 4. Sealed at each end with Spectra Por enclosures.

Figure 8.11   A membrane extraction device for passive sampling of aqueous samples: membrane-enclosed sorptive coating (MESCO) Based on [1].

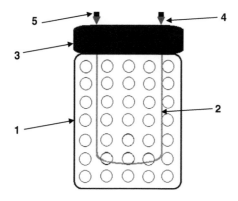

1. Disposable plastic bottle, 2. Polypropylene hollow fibre with the receiving phase (e.g. 1-octanol, ethyl decanoate, toluene), 3. Cap, 4. Syringe needle and 5. Silicone cap.

Figure 8.12   A membrane extraction device for passive sampling of aqueous samples: hollow fibre liquid-phase microextraction (HF-LPME) Based on [1].

The MESCO passive sampling device is an adaptation of the stir-bar sorptive extractor (SBSE) (Chapter 7). Sampling times are in the range days to weeks, when deployed.

Hollow-fibre liquid-phase microextraction (HF-LPME): The HF-LPME consists of a hollow fibre, filled with the receiving phase (i.e. an organic solvent) that is immersed in an aqueous sample (the donor phase) using a 'U' configuration (Figure 8.12). The pollutants are collected in the receiving phase and after sampling, is removed and a portion analysed by GC or HPLC. Sampling times are in the range days to weeks, when deployed.

8.4 Application of Passive Sampling Using Chemcatcher®

Case Study A   Determination of Non-Steroidal Anti-inflammatory Drugs in River Water

Background: Non-steroidal anti-inflammatory drugs (NSAIDs) are a range of medicines used to relieve pain, reduce inflammation and bring down a high temperature. Common NSAIDs include ibuprofen, naproxen and diclofenac (Figure 8.13). Their use and

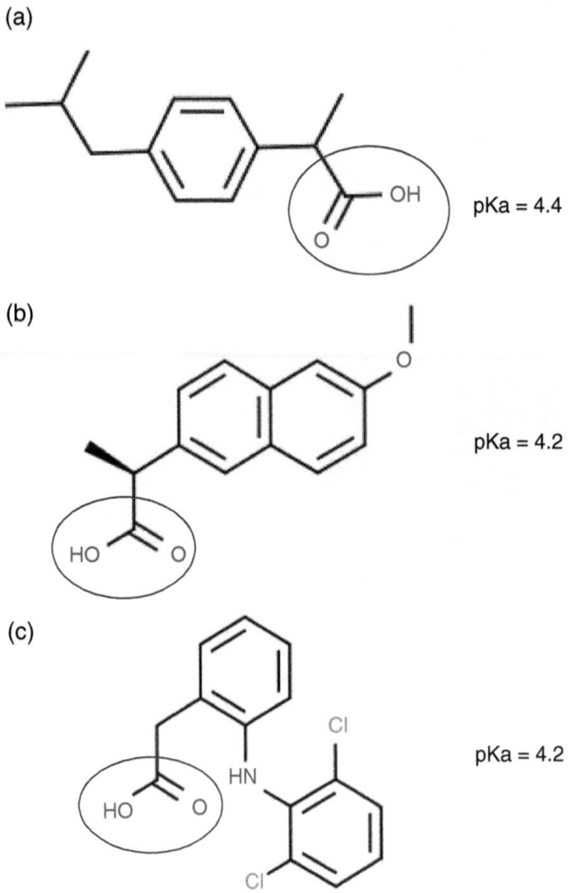

(a)

pKa = 4.4

Figure 8.13   Chemical structures of NSAIDs (a) ibuprofen, (b) naproxen and (c) diclofenac.

(b)

pKa = 4.2

(c)

pKa = 4.2

disposal mean their occurrence in river water systems is not uncommon. In this case study, an approach to sample river water for NSAIDs using the Chemcatcher® is proposed.

Sampling: Passive sampling was done using the Chemcatcher®. The receiving phase was an Empore™ anion exchange disk, for recovery of the NSAIDs in the surface water. Anion exchange media are proposed for the extraction and recovery of negatively charged ions such as compounds with the carboxylic acid functionality (Figure 8.13). The pKa of ibuprofen, naproxen and diclofenac is 4.4, 4.2 and 4.2, respectively; this means at the typical pH of natural water (pH 6.5–9.5), these compounds will be in their carboxylate form, i.e. exist as a negatively charged ion. The passive sampling device was immersed in the river water as shown (Figure 8.14) for a period of between 10 and 14 days. Data obtained after analysis (by LC-MS) indicated typical masses on the Chemcatcher® of ibuprofen (30 ng), naproxen (52 ng) and diclofenac (2 ng).

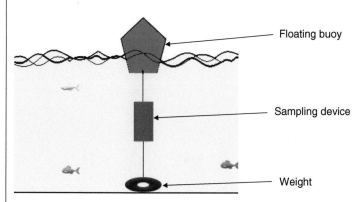

Figure 8.14 Sampling approach using Chemcatcher® for non-steroidal anti-inflammatory drugs in river water.

8.5 Summary

A crossword of the key terms outlined in this chapter and Chapters 4–7 can be found in Appendix A1, with the solution in Appendix B1.

Reference

1 Valenzuela, E.F., Menezes, H.C., and Cardeal, Z.L. (2020). *Environ. Chem. Lett.* 18: 1019–1048.

Section D

Extraction of Solid Samples

9

Classical Approaches for Extraction of Solid Samples

LEARNING OBJECTIVES
After completing this chapter, students should be able to: • Be aware of the main types of liquid–solid extraction. • Understand the principles and theory of liquid–solid extraction. • Be aware of the different solvents, and their properties, used for liquid–solid extraction.

9.1   Introduction

Extraction is done using a range of classical methods. The classical approaches are normally based on large volumes of organic solvent, glassware and some form of heating. They can also be described as labour intensive and relatively slow. The most common is Soxhlet extraction (and modifications thereof), while shake-flask and sonication provide alternates. In addition, new instrumental approaches have led to developments in terms of speed of extraction, reduction in organic solvent consumption and the introduction of automation. However, these instrumental approaches come with a higher price tag, than the conventional approaches. The overall context of extraction from semi-solid or solid samples is illustrated in Figure 9.1. These methods of liquid–solid extraction can be conveniently divided into those for which heat is required (Soxhlet and Soxtec), those methods for which no heat is added, but which utilize some form of agitation, i.e. shaking or ultrasonic extraction, and those which require heat and pressure, i.e. pressurized liquid extraction, microwave-assisted extraction and supercritical fluid extraction.

9.2   Theory of Liquid–Solid Extraction

The extraction of organic compounds from solid matrices is a process in which the solutes desorb from the sample matrix and then dissolve in the extraction solvent. The efficiency of his process is controlled by three important, but inter-related, factors: solubility, mass transfer and matrix effects.

Extraction Techniques for Environmental Analysis, First Edition. John R. Dean.
© 2022 John Wiley & Sons Ltd. Published 2022 by John Wiley & Sons Ltd.

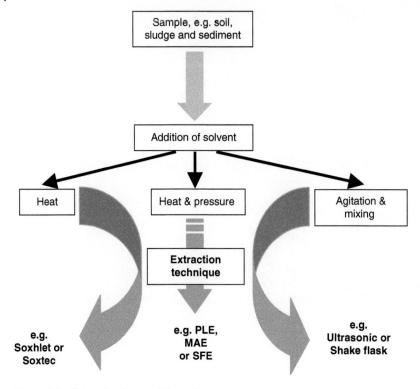

Figure 9.1 Extraction from solid matrices.

Solubility is the maximum amount of a substance that can be dissolved into another at a given temperature. In this situation, the solubility of the organic compound in a solvent is important, the heuristic ('rule of thumb') viewpoint of 'like dissolves like' does work to a point, i.e. a non-polar organic compound will be soluble in a non-polar organic solvent. However, the solubility of an organic compound, in an organic solvent, is also affected by temperature and pressure. A list of common solvents used for extraction of organic compounds from solid matrices is shown in Table 9.1. It is important to note that sometimes combinations of solvents are used to assist with extraction efficiency.

Mass transfer refers to the organic compound transport from the interior of the solid matrix to the extraction solvent. This process requires the solvent to penetrate the matrix and subsequent removal of the organic compounds from the adsorbed sites. Mass transfer is dependent upon the diffusion coefficient, as well as the particle size and structure of the sample matrix.

Matrix effects relate to the potential of the sample matrix to retain the organic compound. The term 'matrix effects' can relate to the surface functionality of the matrix having a greater retentive capability on the organic compound compared to its solubility in the extraction solvent.

Table 9.1 Some common solvents used for extraction, and their properties.

Solvent	Molecular weight	Density (g ml^{-1}) at 25 °C	Boiling point (°C)	Viscosity (cP) at 25 °C	Dielectric constant at 20 °C	Dipole moment
Acetone	58.08	0.791	56	0.306	21.01	2.880
Acetonitrile	41.05	0.786	82	0.369	36.64	3.924
Chloroform	119.38	1.492	61	0.537	4.81	1.040
Dichloromethane	84.93	1.325	40	0.413	8.93	1.600
Ethyl acetate	88.11	0.902	77	0.423	6.08	1.78
Hexane	86.18	0.659	69	0.300	1.89	0
Methanol	32.04	0.791	65	0.544	33.0	1.70
2-Propanol	60.10	0.785	82	2.038	20.18	1.560
Tetrahydrofuran	72.11	0.889	66	0.456	7.52	1.750
Toluene	92.14	0.865	111	0.560	2.38	0.375
Water	18.02	1.000^{a}	100	1.000^{b}	80.10	1.82

^{a} at 3.98 °C.
^{b} at 20 °C.

The rate of extraction in liquid-solid extraction can be considered using Fick's Law:

$$dC_e / dt = \left(A \times D \times V_e / d \right) \times \left(C_s - C_e \right) \tag{9.1}$$

where C_e is the concentration of the organic compound (or solute) in the extraction solvent, t is the extraction time, A is the surface area of the solid sample, D is the diffusion coefficient of the solute in the sample (soaked with the extraction solvent), V_e is the volume of the extraction solvent, C_s is the concentration of the solute in the sample and d is the diffusion layer thickness (i.e. the minimum depth where the concentration of the solute is C_s).

On that basis, it is possible to interpret Eq. (9.1) as follows.

- A larger difference between C_s (concentration of the organic compound (or solute) in the extraction solvent) and C_e (concentration of the organic compound (or solute) in the extraction solvent) leads to a higher extraction efficiency (of the organic compound). [Note: $C_s - C_e$ term in Eq. (9.1).]
- The rate of extraction (dC_e/dt term in Eq. (9.1)) is higher when the surface area of the solid sample is large (A term in Eq. 9.1). Therefore, a solid sample with a large surface area is best. This can be achieved by grinding and sieving the solid sample to create a small particle size.
- The rate of extraction (dC_e/dt term in Eq. 9.1) is higher when the diffusion coefficient of the solvent is large (D term in Eq. 9.1). A higher diffusion coefficient of the solvent is evident when the solvent is less viscous. The more viscous organic solvents tend to have lower diffusion coefficients.

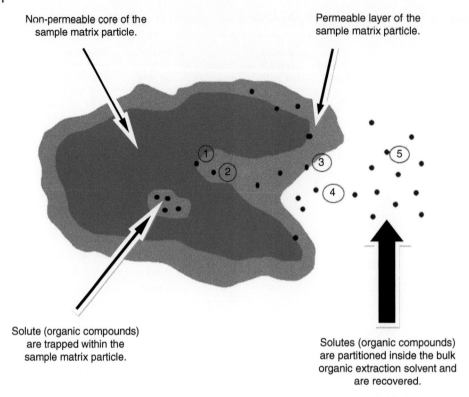

Non-permeable core of the sample matrix particle.

Permeable layer of the sample matrix particle.

Solute (organic compounds) are trapped within the sample matrix particle.

Solutes (organic compounds) are partitioned inside the bulk organic extraction solvent and are recovered.

Figure 9.2 Theory of liquid–solid extraction. Extraction processes: 1. desorption from the solid matrix, 2. diffusion inside of the sample matrix, 3. solubilization by the extraction solvent, 4. diffusion into the extractant solvent and 5. organic compounds (solutes) to be collected.

- The rate of extraction (dC_e/dt term in Eq. (9.1) is higher when the volume of the extraction solvent is large (V_e term in Eq. 9.1).
- The rate of extraction (dC_e/dt term in Eq. 9.1) is higher when the diffusion layer thickness is small (d term in Eq. 9.1). To allow the diffusion layer to be small, a small sample particle size is preferred to increase the extraction efficiency. To maximize the concentration of the organic compound in the extraction solvent, C_s, then the extraction solvent should have the minimal penetration depth. This can be achieved by having a solid sample that has a small particle size and hence a large surface area.

An illustration (Figure 9.2) diagrammatically explains the extraction process of organic compounds from a solid matrix with the extraction solvent.

9.3 Soxhlet Extraction

The utilization of Soxhlet extraction for the removal of compounds from sample matrices is the oldest of the methods discussed. Soxhlet extraction was introduced by Franz Ritter von Soxhlet in 1879 [1]. The basic Soxhlet extraction apparatus consists of a solvent reservoir, extraction body (e.g. Soxhlet extractor), a heat source (e.g. isomantle) and a water-cooled reflux condenser (Figure 9.3). Soxhlet uses a range of organic solvents to remove organic compounds, primarily from (semi-) solid matrices, and the procedure is done inside a fume cupboard. A single Soxhlet extractor can only extract one sample at a time, as only one sample can be extracted per set of apparatus. However, it is possible to operate with multiple Soxhlet extractors, and associated heating devices, if space allows in a fume cupboard.

9.3.1 Experimental

For a Soxhlet extractor of volume 150–200 ml, a solid sample (approx. 10 g, accurately weighted) and a similar mass of anhydrous sodium sulphate (a drying agent) are placed in the cellulose (porous) thimble and placed within the inner tube of the Soxhlet apparatus. The apparatus is then fitted to a round-bottomed flask of appropriate volume (e.g. 250 ml) containing the organic solvent of choice (e.g. for soil samples the solvents used are: DCM only; acetone:hexane (1:1, v/v); DCM:acetone (1:1, v/v); toluene:methanol (10:1, v/v)) and to a reflux condenser (Figure 9.3). The solvent is then boiled gently using an isomantle. The hot solvent vapour passes up through the outer tube marked (#), is condensed by the reflux

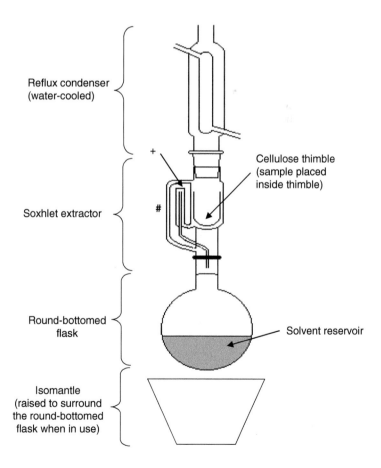

Figure 9.3 Soxhlet extraction apparatus.

Reflux condenser (water-cooled)

Soxhlet extractor

Round-bottomed flask

Isomantle (raised to surround the round-bottomed flask when in use)

Cellulose thimble (sample placed inside thimble)

Solvent reservoir

(a) (b) (c) (d) (e)

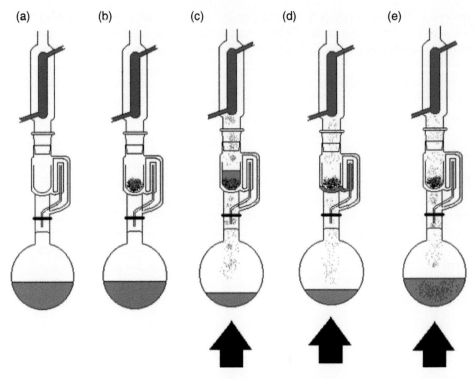

Figure 9.4 Procedure for Soxhlet extraction. (a) Assembled apparatus for Soxhlet extraction, (b) sample placed in thimble inside Soxhlet, (c) heat applied to organic solvent, (d) organic solvent vapour condenses through the sample (in the thimble), and (e) solvent-containing extracted organic compounds returns to round-bottomed flask.

condenser, and the condensed solvent falls into the thimble and slowly fills the body of the Soxhlet apparatus. When the solvent reaches the top of the tube (+), it syphons into the round-bottomed flask that has the organic solvent containing the extracts from the sample. A complete cycle (as just described) can take 15 minutes or 4 cycles hour^{-1}. The whole process is repeated frequently until the pre-set extraction time is reached, typically 6, 12, 18 or 24 hours. As the extracted components will normally have a higher boiling point than the solvent, they are preferentially retained in the round-bottomed flask and fresh solvent recirculates. This ensures that only fresh solvent is used to extract the compound from the sample in the thimble. The procedure for Soxhlet extraction is illustrated in Figure 9.4. Soxhlet extraction is normally regarded as the benchmark technique in liquid–solid extraction against which all over extraction techniques are compared. This is because while the process is slow (up to 24 hours) and uses large volumes of organic solvent, the extraction recoveries are high.

9.4 Soxtec Extraction

Automated Soxhlet extraction or Soxtec utilizes a three-stage process to obtain rapid extractions (Figure 9.5). In stage 1, a thimble containing the sample is immersed in the boiling solvent for approximately 60 minutes. After this (stage 2), the thimble is elevated above the

(a) (b) (c)

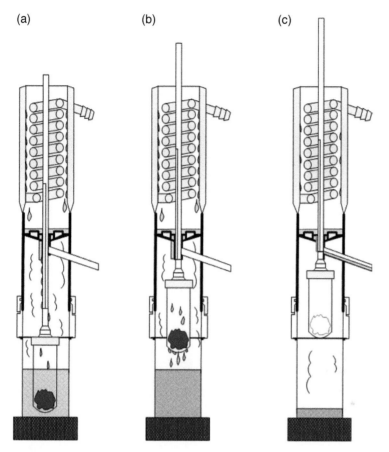

Figure 9.5 Apparatus and procedure for Soxtec extraction. (a) Stage 1 boiling: rapid solubilization of compounds in boiling solvent, (b) stage 2 rinsing: efficient removal of remaining soluble matter and (c) stage 3 recovery: automatic collection of distilled solvent for re-use.

boiling solvent and the sample extracted as in the Soxhlet extraction approach. This is done for up to 60 minutes. The final stage (stage 3) involves the evaporation of the solvent directly in the Soxtec apparatus (10–15 minutes). This approach has several advantages including speed (it is faster than Soxhlet, approximately 2 hours per sample), the fact that it uses only 20% of the solvent volume of Soxhlet extraction and the sample can be concentrated directly in the Soxtec apparatus.

9.5 Ultrasonic Extraction

Ultrasonic extraction or sonication uses sound waves (20 kHz) to agitate a sample, in a container, immersed in organic solvent. The effect of sonication is derived principally from acoustic cavitation, i.e. the formation, growth and implosive collapse of bubbles in a liquid. Cavitation serves as a means of concentrating the diffuse energy of sound. In using a solvent to extract organic compounds from a solid matrix, it is this cavitation, and the shock waves it creates, that accelerates the solid particles to high velocities. The now moving solid particles collide causing

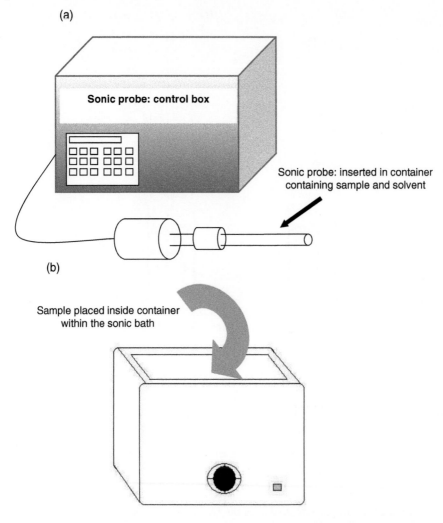

(a)

Sonic probe: control box

Sonic probe: inserted in container
containing sample and solvent

(b)

Sample placed inside container
within the sonic bath

Figure 9.6 Ultrasonic extraction using either (a) sonic probe or (b) an ultrasonic bath.

disruption at their surface in terms of changes in morphology, composition and extractability of organic compounds. Two approaches for sonication are possible: a sonic probe or a sonic bath (Figure 9.6). In the case of the sonic probe, a localized effect is evident from the probe, whereas in the case of the sonic bath a more disperse effect is observed. In addition, the probe comes into contact with the sample and solvent, whereas in the case of the bath, no such contact occurs, as propagation of the sound waves takes place through the sample container.

9.5.1 Experimental

An accurately weighed soil sample (e.g. 1 g) is placed in a suitable glass container. Then the organic solvent is added, e.g. 10 ml of acetone:dichloromethane, 1:1, v/v. The sonic probe (Figure 9.6a) is placed into the solvent and switched on for a short period of time, e.g.

3 minutes. Then, the extract-containing organic solvent is removed and stored. Then, fresh organic solvent is added to the sample and the whole process repeated (three times in total). The combined solvent extracts are evaporated using either a rotary evaporator or a nitrogen blow down (see Chapter 15) to allow concentration of the extract. Finally, the resultant concentrated extract is transferred to a glass volumetric flask and analysed as soon as possible (any storage would take place by locating the volumetric flask in a fridge at 4 °C).

9.6 Shake Flask Extraction

In this extraction method, agitation is either provided by hand or via a mechanical shaker. The action of the mechanical shaker can be as follows:

- An orbital shaker which allows the sample/solvent to fall over itself by the rotating action of the shaker.
- A horizontal shaker which allows the sample/solvent to interact primarily at the point of contact by the forward/back action of the shaker.
- A rocking shaker which allows the sample/solvent to interact at the point of contact by the twisting action of the shaker.

9.6.1 Experimental

An accurately weighed soil sample (e.g. 1 g) is placed in a suitable glass stoppered container. Then the organic solvent is added e.g. 10 ml of acetone:dichloromethane, 1:1, v/v. After securing the stopper the sample is extracted by rotating on an end-over-end shaker for 30 minutes (Figure 9.7). Then, the extract-containing organic solvent is removed and stored. Then, fresh organic solvent is added to the sample and the whole process repeated (three times in total). The combined solvent extracts are evaporated using either a rotary

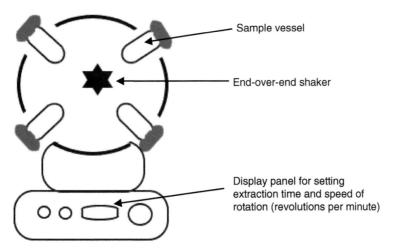

Figure 9.7 Apparatus for shake-flask extraction using a mechanical orbital shaker.

evaporator or a nitrogen blow down (see Chapter 15) to allow concentration of the extract. Multiple extractions can be easily carried out using the shake-flask approach using mechanical laboratory shakers. Finally, the resultant concentrated extract is transferred to a glass volumetric flask and analysed as soon as possible (any storage would take place by locating the volumetric flask in a fridge at 4 °C).

9.7 Application

Case Study A   Soxhlet Extraction of Organic Pollutants from Soil

Background: Organochlorine insecticides are used throughout the world for the control of a wide variety of pests in food and non-food crops. These compounds are classed as environmentally persistent organic pollutants (POP's) and tend to accumulate in wildlife due to their lipophilicity. Soxhlet extraction is often referred to as the benchmark technique against which other extraction techniques are compared. This application evaluates Soxhlet extraction for the recovery of persistent organic pollutants from a certified reference material.

Experimental

Apparatus and Reagents

A Hewlett Packard gas chromatography HP G1800A GCD (Palo Alto, CA, USA) with a Hewlett Packard HP-5ms capillary column (30 m × 0.25 mm i.d., 0.25 μm film thickness), equipped with a quadrupole mass spectrometer detector, was used for analysis.

Standard solutions were prepared in dichloromethane, with α-endosulphan (99.6% w/w) and β-endosulphan (99.9% w/w) provided by Riedel-de Haën (Steinhem, Germany). A TCL Pesticides Mix provided by Supelco (Bellefonte, PA, USA) was used for reference material determination. Pentachloronitrobenzene (PCNB) (99% w/w) was employed as the internal standard (Aldrich, Steinheim, Germany). Acetone and dichloromethane were provided by Fisher Chemicals (Loughborough, UK) and anhydrous Na_2SO_4 by BDH Laboratory Supplies (Poole, UK). A Resource Technology Corporation (Laramie, USA) certified reference material CRM805-050 was used to assess the accuracy and precision of analysis.

Procedure for Soxhlet Extraction of Soil

In a cellulose extraction thimble, approximately 5 g (accurately weighed) of soil and 5 g of anhydrous Na_2SO_4 were added. The sample was extracted with 220 ml of acetone:dichloromethane 1:1 (v/v) for 24 hours. The extract was evaporated under a nitrogen flow to <10 ml, then 20 μl of internal standard of 5 mg ml^{-1} was added. The final extract solution (10.0 ml) was analysed by GC-MS.

GC/MS Analysis

For GC/MS determinations, an injection volume of 1 μl was employed in split mode (1:4 ratio). The injector temperature was 250 °C and helium was used as the carrier gas in constant flow mode of 1 ml min^{-1}. The temperature program of the oven was as follows: 60 °C, held for 1 minute, increased at a rate of 15 °C min^{-1} to 180 °C, then a second rate of 3 °C min^{-1} to 250 °C and finally held 1 minute. The detector temperature was 280 °C and measurements were carried out in selected-ion monitoring (SIM) acquisition mode. Retention times and main ions selected for each compound, with their relative abundances, are summarized in Table 9.2. Detector tune tests were performed daily with perfluorotributylamine.

Results and Discussion

Calibration curves were established with five standards, with concentrations ranging from 2 to 20 µg ml^{-1} using an internal standard of concentration 10 µg ml^{-1}. Limit of detection values were established using the expression $3 \cdot s_{blank}/b$, where s_{blank} is the standard deviation of five measurements of a standard solution of 2 µg ml^{-1} and b the slope of the calibration curve. Table 9.2 shows the limit of detection values for each compound. The detector response was linear over the range of concentration studied, with correlation coefficients ranging from 0.9790 to 0.9990. Limits of detection in soil ranged from 0.4 to 0.6 µg g^{-1}.

A certified reference material was analysed by Soxhlet extraction. In this procedure, a pre-concentration step was needed to increase sensitivity. It involved subjecting the extract to a flow of nitrogen to achieve partial solvent evaporation. Recovery tests were carried out to check that minimal compound losses occurred under the cited conditions. Typical recoveries ranged from 82% for DDT to 94% for methoxychlor after Soxhlet extraction (solvent volume reduced from 220 to <10 ml). No correction was made to subsequent data to adjust for these solvent evaporation losses. The recovery of the organic pollutants from the certified reference material, CRM 805-050, as determined by Soxhlet, followed by GC-MS is shown in Table 9.3. In each case, the results agreed with certificate values.

Table 9.2   GC-MSD parameters and limits of detection for organic pollutant determination.

Compound	RT	Quantifier ion (m/z, %)	Qualifier ion (m/z, %)	LOD (µg ml^{-1}) in solution	LOD (µg g^{-1}) in soil
Lindane	13.84	108.95 (100)	180.90 (98)	0.3	0.4
α-Endosulphan	21.01	194.90 (100)	169.90 (75)	0.5	0.6
DDE	22.24	245.95 (100)	246.95 (60)	0.6	0.7
Endrin	23.29	67.15 (100)	81.05 (34)	0.4	0.5
β-Endosulphan	23.75	194.90 (100)	158.90 (80)	0.3	0.4
DDD	24.31	235.05 (100)	237.50 (64)	0.3	0.4
Endrin aldehyde	24.72	67.05 (100)	249.85 (21)	0.6	0.7
DDT	26.22	235.05 (100)	236.95 (64)	0.5	0.7
Methoxyclor	29.37	227.15 (100)	228.15 (17)	0.4	0.5
PCNB^a	13.95	141.95 (100)	236.80 (90)	—	—

^{a} internal standard.

Table 9.3   Determination of organic pollutants in a certified reference soil sample (CRM 805-050) using Soxhlet extraction followed by GC-MS.

Compound	Persistent organic pollutant concentration ($\mu g\ g^{-1}$)	
	Reference value[a]	Soxhlet extraction (mean ± SD, n = 3)
Lindane	11 ± 5	11.5 ± 0.5
α-endosulphan	7 ± 4	3.2 ± 0.6
DDE	19 ± 9	26.6 ± 0.6
Endrin	13 ± 8	18.3 ± 0.4
β-endosulphan	6 ± 3	4.3 ± 0.9
DDD	20 ± 9	22.0 ± 0.5
Endrin aldehyde	0.1 ± 0.2	<LOD
DDT	0.8 ± 0.3	<LOD
Methoxyclor	16 ± 8	17.7 ± 0.3

[a] Resource Technology Corporation.
LOD = limit of detection.

Reference

1 Soxhlet, F. (1879). Die gewichtsanalytische Bestimmung des Milchfettes. *Dingler's Polytechnisches Journal* 232: 461–465.

10

Pressurized Liquid Extraction

> **LEARNING OBJECTIVES**
>
> After completing this chapter, students should be able to:
>
> - Understand the operating principles and instrumentation for pressurized liquid extraction.
> - Comprehend the theoretical basis of pressurized liquid extraction.
> - Be able to select appropriate conditions for *in situ* clean-up in PLE.

10.1   Introduction

The development of pressurized liquid extraction (PLE) can be traced back to 1995 when Dionex Corporation (now part of Thermo Fisher Scientific Inc.) launched the accelerated solvent extraction (ASE) system. Since then, the use and application of PLE has expanded considerably. The technique is also sometimes referred to as pressurized fluid extraction (PFE) or pressurized solvent extraction (PSE). The term used throughout this chapter is pressurized liquid extraction. The basic principle of PLE is that organic solvents, at high temperature and pressure, are used to extract compounds from sample matrices.

10.2   Theoretical Considerations Relating to the Extraction Process

Pressurized liquid extraction uses organic solvents at elevated pressures and temperatures to enhance the recovery of organic compounds from environmental, as well as food, pharmaceutical and industrial samples. The use of organic solvents at elevated pressures and temperatures is advantageous compared to their use at atmospheric pressure and room (or near room) temperature as it results in enhanced solubility and mass transfer effects, and disruption of surface equilibria.

Extraction Techniques for Environmental Analysis, First Edition. John R. Dean.
© 2022 John Wiley & Sons Ltd. Published 2022 by John Wiley & Sons Ltd.

10.2.1 Solubility and Mass Transfer Effects

- Solubility: A higher temperature (e.g. 100 °C) increases the capacity of solvents to solubilize organic compounds. An example of this is shown in Figure 10.1a in which the effect of temperature on the solubility of glycine in water. As the temperature increases, so does the solubility of glycine. It is seen that as a result of the increased temperature, faster diffusion rates occur.

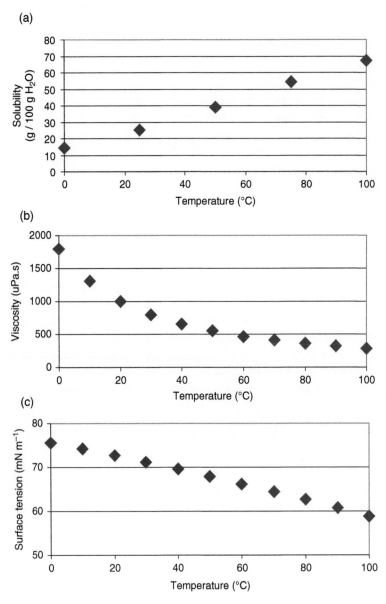

Figure 10.1 Theoretical examples of the extraction process in PLE: (a) influence of temperature on the solubility of glycine, (b) influence of temperature on the viscosity of water and (c) influence of temperature on the surface tension of water [1]/with permission from Taylor & Francis.

- Mass transfer: Improved mass transfer (and hence increased extraction rates) occur when fresh solvent is introduced into the extraction vessel. This is because the concentration gradient is greater between fresh solvent and the surface of the sample matrix.

10.2.2 Disruption of Surface Equilibrium (By Temperature and Pressure)

- Temperature effects: An increase in temperature (e.g. 100 °C) can disrupt the strong organic compound – matrix interactions (e.g. van der Waal's forces, hydrogen bonding and dipole attractions). In addition, a high extraction temperature results in a decrease in viscosity and surface tension of the organic solvent. An example of this is shown in Figure 10.1b and c, for viscosity and surface temperature, respectively. This allows the organic solvent to penetrate the matrix more effectively resulting in improved extraction efficiency.
- Pressure effects: At elevated pressure (e.g. 2000 psi), an organic solvent can remain in liquid form well above its boiling point (at atmospheric pressure). In addition, an elevated pressure allows the organic solvent to penetrate the sample matrix, which will result in improved extraction efficiency.

10.3 Instrumentation for PLE

The essential components of a PLE instrument (Figure 10.2) are:

- An organic solvent mixture (e.g. dichloromethane:acetone 1 : 1, v/v). Other solvent systems are also to be considered (Table 10.1).
- A pump (to deliver the organic solvent) capable of operating up to 70 ml min^{-1}.

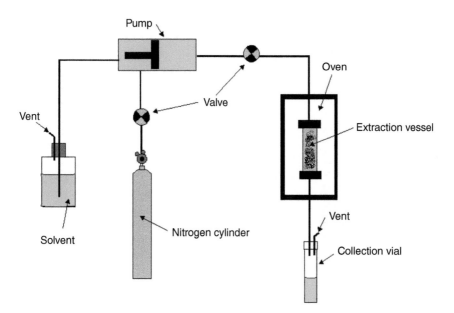

Figure 10.2 Schematic diagram of a pressurized liquid extraction system.

Table 10.1 PLE solvent extraction systems (from EPA Method 3545A). Data from [2].

Class of compound to be extracted	Proposed solvent system[a]
Organochlorine pesticides	acetone:hexane (1:1, v/v) or acetone:dichloromethane (1:1, v/v)
Semi-volatile organics	acetone:dichloromethane (1:1, v/v) or acetone:hexane (1:1, v/v)
Polychlorinated biphenyls	acetone:dichloromethane (1:1, v/v) or acetone:hexane (1:1, v/v) or hexane
Organophosphorus pesticides	dichloromethane or acetone:dichloromethane (1:1, v/v)
Chlorinated herbicides	acetone:dichloromethane:phosphoric acid (250 : 125 : 15, v/v/v) or acetone:dichloromethane:trifluoroacetic acid (250 : 125 : 1, v/v/v)
Dioxins and furans	Toluene or toluene:acetic acid (5% glacial acetic acid in toluene)
Diesel range organics	Acetone:dichloromethane (1:1, v/v) or acetone : hexane (1:1, v/v) or acetone : heptanes (1:1, v/v)

[a] Other solvent systems can be applied provided they are tested for their efficient recovery of the compounds of interest.

- A supply of nitrogen gas (for flushing excess solvent out of the extraction vessel at the end of the process).
- An extraction vessel. These are made of stainless steel and should be capable of withstanding high pressures (up to 4000 psi) safely. Ideally, a range of extraction vessel sizes should be available (e.g. 1, 5, 10, 22, 34, 66 and 100 ml). For ease of use, the sample vessel should have removable end caps that allow for ease of cleaning and sample filling (Figure 10.3) and contain two finger-tight caps with compression seals for high-pressure closure.

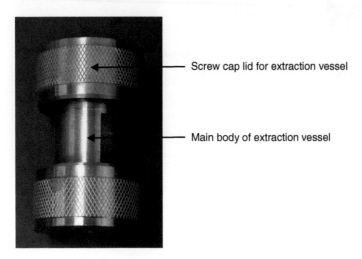

Screw cap lid for extraction vessel

Main body of extraction vessel

Figure 10.3 A photograph of an extraction vessel used for pressurized liquid extraction.

- An oven capable of precise temperature control (up to 200 °C).
- A system for creating pressure within the system capable of operating up to 6.9–20.7 MPa or 1000–3000 psi.
- An extract collection system that allows the solvent (and extractants) to be collected. If necessary, the solvent can be evaporated under a gentle stream of nitrogen gas to dryness (see Chapter 15) and re-constituted with an appropriate solvent prior to analysis or directly analysed without concentration.

10.4 A Typical Procedure for PLE

Extraction Vessel: It is important that the extraction vessel is full to its capacity to allow maximum solvent–matrix–compound interaction. To fill a sample vessel, one end cap is attached and screwed on to finger-tightness. Then, a filter paper (e.g. Whatman, grade D28, 1.98 cm diameter) is introduced into the base of the extraction vessel, followed by the sample. Figure 10.4 shows two sample packing arrangements for the extraction vessel. Option (a) involves the sample (accurately weighed) being mixed with a similar quantity of high-purity diatomaceous earth (hydromatrix), while option (b) is a layer approach. In either option, any residual space in the extraction vessel is then filled with more hydromatrix, and finally another filter paper is placed on top prior to extraction vessel closure. Then, the end cap is screwed on to finger-tightness and the extraction vessel is ready for extraction.

The sealed extraction vessel is placed in the oven and connected in series, and the extraction programme started. In selecting PLE operating conditions, the following can be used:

- An organic solvent mixture of dichloromethane:acetone (1:1, v/v).
- A pressure of 2000 psi.
- A temperature of 100 °C.
- An extraction time of 10 minutes.

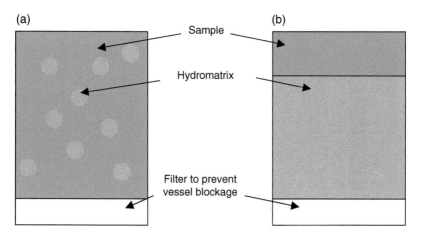

Figure 10.4 Schematic diagram of packing arrangements for the sample in extraction vessel for pressurized liquid extraction. (a) Randomly distributed and (b) ordered, layered structure.

After extraction, the solvent (dichloromethane:acetone, 1:1 v/v) is evaporated under a gentle stream of nitrogen gas to dryness (see Chapter 15) and re-constituted with an appropriate solvent prior to analysis or analysed directly. Figure 10.5 contextualizes the typical procedure for PLE in terms of instrument operation, while Figure 10.6 explains the procedure in a workflow diagram.

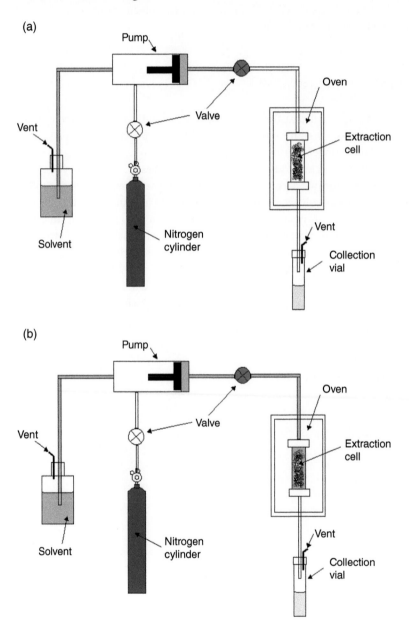

Figure 10.5 Operation of a pressurized liquid extraction system: (a) system set-up established with sample pre-loaded and flow of solvent started, (b) organic solvent fills the extraction cell (vessel), (c) extracted compounds are recovered from the sample matrix and collected in vial, (d) extracted compounds are recovered from sample matrix, but some residual solvent / compounds remains in vial, (e) a nitrogen flush removes all residual solvent/compounds from extraction cell (vessel) and (f) extraction complete.

(c)

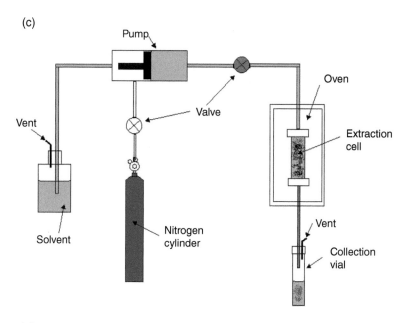

(d)

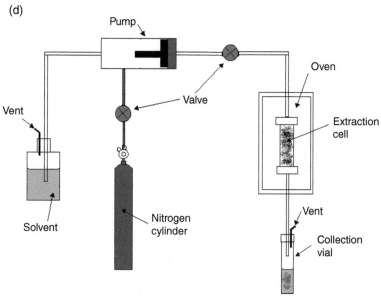

Figure 10.5 (Continued)

(e)

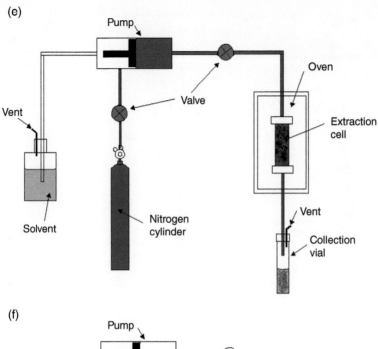

(f)

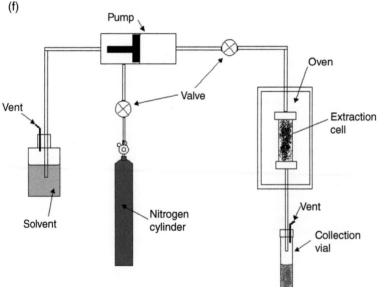

Figure 10.5 (Continued)

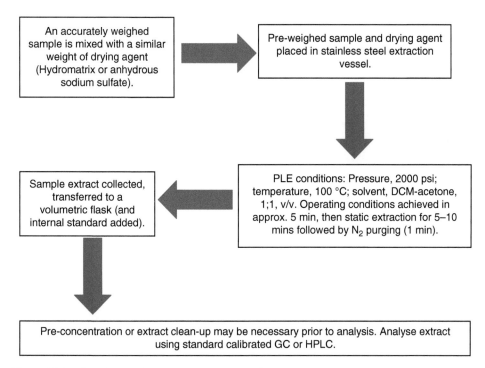

Figure 10.6 Typical procedure for pressurized liquid extraction.

10.5 In Situ Clean-Up or Selective PLE

The extraction of organic compounds using the conditions identified is neither selective nor gentle. The aggressive nature of the process means that unwanted extraneous material will be removed from the sample matrix. This material will often interfere with the subsequent analysis step, e.g. chromatography. Therefore, it has become more common practice to perform in situ clean-up within the extraction vessel. This important development in PLE allows the removal of potential contaminants from the resultant analysis step, e.g. chromatography.

In situ PLE can be done (Figure 10.7) by adding alumina as an adsorbent, though others are also available (Table 10.2) for this purpose. This is followed by a drying agent (e.g. anhydrous sodium sulphate) to remove excess water. In designing an *in situ* clean-up approach, it is important to consider the following:

- What do you hope to remove?
- How is it done currently off-line?

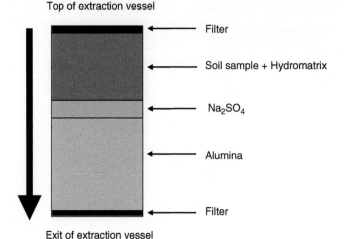

Top of extraction vessel

Exit of extraction vessel

Figure 10.7 In situ packing of extraction vessel in pressurized liquid extraction.

Table 10.2 *In situ* absorbents for sample clean-up.

Absorbent	Potential interference	Compound under investigation
Carbon	Organics	Non-polar compounds, dioxins
Copper	Elemental sulphur	Multi-residue pesticides
Ion exchange resins	Organics, metals and ionic interferences	Anions, cations, arsenic speciation
C18	Organics, lipids	Non-polar compounds
Acid impregnated silica gel	Lipids and oils	PCB and bromine flame retardants
Alumina[b]	Lipids, petroleum waste	Amines, perchlorates and PCBs
Florisil[a]	Oils, lipids and waxes	Pesticides and aromatics

[a] Florisil™ is magnesium silicate with basic properties and allows selective elution of compounds based on elution strength.
[b] Alumina is a highly porous and granular form of aluminium oxide, which is available in 3 pH ranges (basic, neutral and acidic).

The answers to these questions will enable you to design an appropriate *in situ* clean-up procedure. The major potential advantages of in situ PLE are:

- Increased level of automation of the sample preparation stage.
- Elimination of the need for off-line clean-up.
- Reduced solvent use.
- Considerably faster than off-line clean-up.
- Less sample manipulation.

10.6   Method Development for PLE

10.6.1   Pre-Extraction Considerations

The following steps should be considered.

1) What extraction solvent (or solvent mixture) is required? This can often be determined by considering what solvent system has been used previously (e.g. in Soxhlet extraction).
2) What are the organic compounds to be recovered? This is linked to the choice of extraction solvent. Also, it is important to know whether the compounds are soluble in the proposed extraction solvent(s).
3) What is the sample matrix? If the sample is wet or moisture-laden, it may require to be either pre-dried or that a moisture removing adsorbent (e.g. anhydrous sodium sulphate) is added into the extraction vessel along with the sample.
4) What is the sample particle size? The smaller the sample particle size the greater the interaction with the extraction solvent. It may therefore be necessary to grind the sample (with a mortar and pestle) and then sieve (e.g. <250 μm) the sample. Alternatively, the sample may require to be freeze-dried prior to grinding and sieving. The reduced particle size combined with enhanced extraction temperatures and pressure will lead to optimum recoveries.

10.6.2   Packing the Extraction Vessel

The following steps should be considered.

1) How much sample do I have?
2) What size of extraction vessel should be used?
3) How should the extraction vessel be packed with the sample? To maximize sample surface area, it is appropriate to mix the sample with a dispersing agent, e.g. hydromatrix or diatomaceous earth at a suggested ratio of 1 part sample to 1 part hydromatrix. In addition, if the sample is wet or moisture laden, it is appropriate to mix the sample with anhydrous sodium sulphate. Similarly, if the sample contains significant levels of sulphur (often found at high levels in soils from former gas/coal works), it is necessary to add copper or tetrabutylammonium sulphite powder. The addition of copper or tetrabutylammonium sulphite powder will complex out the sulphur preventing it from blocking the stainless steel tubing within the PLE system. Additionally, if the sample is likely to lead to significant co-extractives that could interfere with post-extraction analysis, e.g. chromatography, it may be appropriate to consider in situ sample clean-up using alumina, florisil or silica gel.
4) Ensure that the extraction vessel is full (i.e. remove the dead-volume of the vessel). If necessary, add hydromatrix or similar to remove the void volume.

Note: Florisil is magnesium silicate with basic properties and allows selective elution of compounds based on elution strength. In contrast, alumina is a highly porous and granular form of aluminium oxide, which is available in 3 pH ranges (basic, neutral and acidic). Finally, silica gel, which allows selective elution of compounds based on elution strength.

10.7 Applications of PLE

Case Study A   Extraction of Persistent Organic Pollutants from Environmental Matrices [3–4]

Background: Recovery of organic pollutants from environmental samples is important in terms of assessing the levels of contamination. This information is important in determining environmental risk as it is based on total recovery of organic compounds.

Experimental
Chemicals/Reagents
The solvents used were certified analytical grade, obtained from Fisher Scientific (Loughborough, Leicestershire). The pesticide standard (18 component mixture) in toluene:hexane (50:50 v/v), BTEX, BNA's and phenol standards (in methanol) were obtained from Sigma-Aldrich Company Ltd (Supelco UK, Dorset, UK). Internal standards (pentachloronitrobenzene and n-propyl benzene) were obtained from Sigma-Aldrich Company Ltd. Hydromatrix (Varian Ltd, Surrey, UK) was used to fill the head space of the extraction vessels (Dionex). A range of certified reference materials were purchased from the Laboratory of the Government Chemist (LGC), Teddington, London: Natural Matrix Certified Reference Material, CRM306-030, for BTEX; Natural Matrix Certified Reference Material, CRM401-225, for phenols; Natural Matrix Certified Reference Material, CRM805-050, for pesticides; and Natural Matrix Certified Reference Material, CRM107-100, for BNAs.

GC-MS Analysis
The GC-MS (HP G1800A GCD system, Hewlett Packard, Palo Alto, USA) was operated in single-ion monitoring mode with a splitless injection volume of $1\,\mu l$. The column was a DB-5ms purchased from Sigma-Aldrich Company Ltd with dimensions: length $30\,m \times 0.25\,mm$ internal diameter $\times 0.25\,\mu m$ film thickness. For all analyses, the injection port temperature was set at $250\,°C$ and the detector temperature was set at $280\,°C$. Selected standards for the compounds were run daily to assess analytical performance.

 Pesticides were analysed using the following temperature programme: initial temperature $120\,°C$ for 2 minutes then to $250\,°C$ at $4\,°C/min^{-1}$. The separation of all the target pesticides was achieved in approximately 34 minutes. BTEX's were analysed under isothermal conditions ($70\,°C$). The separation of BTEX was achieved in approximately 8 minutes. Phenols required derivatization prior to analysis. This was done by taking an aliquot of extract/standard ($1\,ml$) and transferring to a tapered tube were internal standard ($50\,\mu l$) and derivatizing agent (N, o-bis (trimethylsilyl) acetamide) ($100\,\mu l$) was added. The extract was vortex spun (15 seconds) prior to analysis by GC-MSD. The temperature program used for separation of phenols was as follows: initial temperature $90\,°C$ for 2 minutes then to $230\,°C$ at $7\,°C\,min^{-1}$. Separation of all the target phenols was achieved in approximately 22 minutes. Finally, base neutral and acid compounds (BNA's) were separated as follows: initial temperature $70\,°C$ for 2 minutes to $230\,°C$ at $7\,°C\,min^{-1}$. Separation of all the target BNAs was achieved in approximately 34 minutes.

Extraction Procedure

All soils were extracted using PLE using an Accelerated Solvent Extractor, ASE 200 (Dionex (UK) Ltd., Camberley, Surrey). Sample vessels (11 ml) were used for all the extractions. The ASE 200 is an automated system capable of 24 sequential extractions. Typical extraction conditions are based on a pressure of 2000 psi (1 psi = 6894.76 Pa), a temperature of 100 °C and a total extraction time of 10 minutes (5 minutes plus 5 minutes static). A single static flush cycle was used. Additional time was required for rinsing with fresh solvent and N_2, hence the total extraction time was approximately 13 minutes per sample. The extract was transferred to a 25 ml volumetric flask, internal standard (pentachloronitrobenzene or n-propyl benzene) was added (50–100 µl) and made up to the mark with solvent. The extract was then analysed by GC-MSD.

Results and Discussion

All organic compounds were run on the GC-MS to allow quantitation of recoveries from the soil reference material matrices. The results are shown in Table 10.3. A variety of solvents were investigated for the extraction of the organic compounds; the final optimum solvents were then used to extract the different classes of organic compounds from their respective reference materials. In the case of BTEX compounds (Table 10.4), the optimum selected solvent was heptane, whereas for the phenols (Table 10.4), the optimum selected solvent was acetonitrile. In the case of pesticide compounds (Table 10.4), acetone was determined to be the optimum solvent, while for the base, neutral and acidic (BNA) compounds, acetone was used (for acenaphthene and fluorine, the optimum solvent used was toluene). The mean concentration (mg kg^{-1}) obtained using PLE with an optimized solvent system was within the prediction internal of the CRMs, with appropriate precision obtained on replicate extractions (Table 10.4).

Table 10.3 Analytical data for quantitation of a range of organic pollutants.

Compound	Qualifying ions m/z	Quantifying ions m/z	Calibration equation Y = mx + C	Correlation coefficient r^2
BTEX^c				
Benzene	78	77	Y = 0.0137x + 0.0108	0.9991
Toluene	92	91	Y = 0.0173x + 0.0521	0.9914
Ethylbenzene	106	91	Y = 0.0253x + 0.0293	0.9982
o, m-Xylene	106	91	Y = 0.0485x + 0.0464	0.9985
Phenols^b				
o, m-Cresol	180	165	Y = 0.0786x + 0.107	0.9976
p-Cresol	180	165	Y = 1.3661x + 1.217	0.9976
2,4,6-Trichlorophenol	253	93	Y = 0.0965x + 0.091	0.9986
Pentachlorophenol	323	93	Y = 0.0645x + 0.025	0.9987

Table 10.3 (Continued)

Compound	Qualifying ions m/z	Quantifying ions m/z	Calibration equation Y = mx + C	Correlation coefficient r^2
Pesticides[a]				
p,p'- DDD	235	195	$Y = 0.5429x - 0.0561$	0.9924
p,p' - DDE	246	176	$Y = 0.038x + 0.0107$	0.9940
Endosulphan I	237	195	$Y = 0.0537x + 0.0012$	0.9968
Endrin	82	79	$Y = 0.1388x + 0.0042$	0.9962
Lindane	183	181	$Y = 0.2437x + 0.0252$	0.9996
BNA[d]				
Hexachloroethane	119	117	$Y = 0.177x - 0.1081$	0.999
2-Nitroaniline	138	65	$Y = 0.1739x - 0.6593$	0.9801
Acenaphthene	154	153	$Y = 0.5844x - 0.6908$	0.998
Fluorene	166	165	$Y = 0.5589x - 0.9042$	0.9963
3-Nitroaniline	138	65	$Y = 0.0891x - 0.632$	0.941
Hexachlorobenzene	286	284	$Y = 0.1904x - 0.1005$	0.9997

[a] Based on a typical concentration range of 0–20 µg ml^{-1} with five calibration data points. Internal standard used was pentachloronitrobenzene (qualifying ion 249; quantifying ion 237).
[b] Based on a typical concentration range of 0–120 µg ml^{-1} with seven calibration data points. Internal standard used was pentachloronitrobenzene (qualifying ion 249; quantifying ion 237).
[c] Based on a typical concentration range of 0–100 µg ml^{-1} with five calibration data points. Internal standard used was N-propylbenzene (qualifying ion 249; quantifying ion 237).
[d] Based on a typical concentration range of 0–70 µg ml^{-1} with six calibration data points. Internal standard used was pentachloronitrobenzene (qualifying ion 249; quantifying ion 237).

Table 10.4 Analysis of BTEX, phenols, pesticides and BNAs in Certified Reference Materials using pressurized liquid extraction.

CRM	Compound	Certificate information			Mean concentration ± SD (mg kg^{-1})
		Certificate value (mg kg^{-1}) ± SD	Prediction interval (mg kg^{-1})	Confidence interval (mg kg^{-1})	
	Benzene	12.2 ± 2.26	7.07–17.3	10.9–13.5	5.8 ± 0.2
CRM 306-030[a]	**Ethylbenzene**	12.8 ± 2.06	8.18–17.5	11.6–14.0	11.6 ± 0.5
	Toluene	68.2 ± 9.94	45.9–90.4	63.1–73.2	66.3 ± 10.8
	Xylene (total)	69.5 ± 9.53	47.9–91.1	63.6–75.5	57.9 ± 0.6

Table 10.4 (Continued)

CRM	Compound	Certificate value (mg kg⁻¹) ± SD	Prediction interval (mg kg⁻¹)	Confidence interval (mg kg⁻¹)	Mean concentration ± SD (mg kg⁻¹)
	Cresol (total)	2657.8 ± 889.3	716–4600	2158–3157	1816 ± 82
CRM 401-225[b]	2,4,6-Trichlorophenol	58.7 ± 19.4	16.6–101	48.8–68.6	40.9 ± 5.1
	Pentachlorophenol	117.1 ± 41.7	26.7–207	96.5–138	151 ± 8
CRM 805-050[c]	Lindane	10.6 ± 4.85	0.0–21.5	7.96–13.3	9.3 ± 0.6
	Endosulphan I	6.9 ± 3.79	0.0–15.48	4.91–8.89	6.00 ± 1
	p,p′- DDE	18.6 ± 9.10	0.0–39.02	13.8–23.4	12.0 ± 1.4
	Endrin	12.97 ± 7.51	0.0–29.85	8.79–17.1	13.3 ± 0.5
	p,p′-DDD	19.45 ± 8.64	0.0–38.95	14.3–24.6	20.4 ± 1.4
CRM 107-100[d]	Hexachloroethane	2.31 ± 0.72	0.89–3.74	2.21–2.42	1.14 ± 0.02
	2-Nitroaniline	15.1 ± 3.74	7.69–22.4	14.5–15.6	11.8 ± 0.8
	Acenaphthene[e]	61.9 ± 15.5	31.3–92.4	59.6–64.1	34.0 ± 2.1
	Fluorene[e]	30.8 ± 7.36	16.3–45.3	29.8–31.9	22.1 ± 0.6
	3-Nitroaniline	4.27 ± 2.53	0.00–9.24	3.89–4.64	4.7 ± 0.3
	Hexachlorobenzene	42.9 ± 10.3	22.6–63.2	41.4–44.4	23.1 ± 0.8

[a] For CRM 306-030 (BTEX), extraction solvent was heptane (n = 3).
[b] For CRM 401-225 (phenols), extraction solvent was acetonitrile (n = 5).
[c] For CRM 805-050 (pesticides), extraction solvent was acetone (n = 5).
[d] For CRM 107-100 (BNAs), extraction solvent was acetone (n = 5).
[e] For CRM 107-100 (BNA's), extraction solvent was toluene (n = 5).

Case Study B *In Situ* **Pressurized Liquid Extraction of Polycyclic Aromatic Hydrocarbons [5 – 7]**

Background: Polycyclic aromatic hydrocarbons (PAHs) are derived from a number of sources including anthropogenic (i.e. industrial processes and combustion of fossil fuels) or natural (i.e. forest fires, volcanic activity and geological sources). The 16 PAH priority pollutants are known for their carcinogenic effect and mutagenic characteristics. The aim of this work is to establish a robust and effective procedure for the recovery of PAHs from contaminated soil prior to analysis by gas chromatography-mass spectrometry (GC-MS). In this study, off-line column chromatography has been evaluated for soil clean-up following PLE. The influence of two different absorbents (florisil and alumina) on extract clean-up have been investigated with respect to PAH recovery. This approach has been compared with an *in situ* PLE procedure.

Experimental
Chemicals
A PAH standard solution was obtained from Thames Restek UK Ltd, Buckinghamshire, UK (2000 μg ml^{-1} in dichloromethane). Alumina and 4,4'-difluorobiphenyl were obtained from Sigma-Aldrich Ltd, Dorset, UK, while florisil was purchased from Fluka (Sigma-Aldrich Ltd, Dorset, UK). All the solvents (dichloromethane, acetone, hexane) were analytical reagent grade and obtained from Fisher Scientific Ltd (Loughborough, UK). High-purity diatomaceaous earth (hydromatrix) was obtained from Varian Inc. (Harbor City, CA, USA). Certified reference materials (LGCQC3008 Sandy soil and CRM 123-100) were obtained from LGC Standards, Teddington, UK. Filter papers (ASE200) made from glass fibre cellulose were obtained from Dionex Corporation (Sunnyvale, USA).

Instrumentation
Extraction was performed with PLE on an ASE200 (Dionex UK Ltd., Camberley, Surrey). The operating conditions were organic solvent: dichloromethane:acetone (50:50, v/v); pressure: 2000 psi; temperature: 100 °C; and extraction time: 10 minutes. The GC-MS instrument included a Trace GC Ultra coupled with a Polaris Q Ion trap MS (Thermo Scientific, UK) and a Triplus autosampler injector. The system was controlled from a PC with Xcalibur software. Separation was performed using a capillary column Rtx®- 5MS (5% diphenyl-95% dimethylpolysiloxane, 30 m × 0.25 mm ID × 0.25 μm) supplied from Thames Restek UK Ltd. The temperature programme was as follows: start at 70 °C for 2 minutes and then 7 °C min^{-1} until 180 °C, then 3 °C min^{-1} until 280 °C, then hold for 3 minutes. The transfer line temperature was fixed at 300 °C.

Procedures
Soil and its slurry spiking
The soil used in the spiking procedure was collected from a former industrial site. The soil was stored in a Kraft bag and was air-dried in a fume cupboard during one week before grinding (using pestle and mortar) below 2 mm and sieving below 250 μm. The soil was sealed in a plastic bag, labelled and stored in the fridge until further analysis. It was assessed to contain a low amount of polycyclic aromatic hydrocarbons. All data was blank subtracted.

A known quantity of soil (1.3 g) was placed inside a beaker. Then, 10 ml of dichloromethane containing 50 μl of the PAH standard solution (2000 μg ml^{-1}) was added to the soil. The sample was then left exposed, in a fume cupboard, for five days prior to PLE. After the PLE (without clean-up), the solution was reconstituted with 25 ml of dichloromethane and 50 μl of internal standard at 1000 μg ml^{-1}.

PLE and Off-Line Clean-Up
The soil sample (1.3 g) was mixed with a similar quantity of hydromatrix and added into the extraction vessel (11 ml) on top of a filter paper. Additional hydromatrix was added to fill the vessel and a final filter paper was placed on top prior to closure. After PLE, the solvent (dichloromethane: acetone 1:1, v:v) was evaporated under a gentle stream of nitrogen gas to dryness and reconstituted with 2 ml of hexane. Then, the extract was treated as per column clean-up, prior to GC-MS.

Column clean-up: A column (200 × 18 mm) was prepared with either 10 g of alumina (Sigma-Aldrich, 150 mesh) or florisil (Fluka, 60–100 mesh) as absorbent with an additional 11 g of anhydrous Na_2SO_4 placed on top. Then, the column was eluted with 50 ml of hexane. The eluate was discarded and just prior complete elution avoiding Na_2SO_4 powder to expose to the air, 2 ml hexane from the PLE procedure was added on top of the column for elution. For the spiking procedure, a PAH standard was added (50 µl of a 2000 µg ml^{-1} standard) in the 2 ml hexane solution. Again, and just prior complete elution and dryness of the sorbent, two times 15 ml of hexane were added and again the eluate was discarded. Finally, the column was eluted with approximately 30 ml of dichloromethane into a flask and then the solvent was retained. Then 60 µl of the internal standard (2 µg ml^{-1}) was added to give a final volume of 30 ml.

PLE with *In Situ* Clean-Up

Either 2 g of florisil or alumina were added into the PLE extraction vessel on top of a filter paper. Then, the soil and hydromatrix were added according to the procedure described above (PLE and off-line clean-up), with a filter paper placed before closure. After *in situ* PFE, the solvent (dichloromethane: acetone 1:1, v:v) was evaporated under a gentle stream of nitrogen gas to dryness and reconstituted with 2 ml of dichloromethane containing the internal standard (20 µl of a 1000 µg ml^{-1} solution), prior to GC-MS. To observe the influence of the absorbent amount, the soils have been spiked (50 µl of a 2000 µg ml^{-1} standard solution) directly in the extraction cell with 0.5, 1 and 2 g of sorbent (alumina and florisil). In the case of no evaporation after extraction, the final solution was reconstituted with 25 ml of dichloromethane with the internal standard (50 µl of a 1000 µg ml^{-1} solution).

Results and Discussion

Separation of the 16 PAHs was achieved (Figure 10.8) and quantitation was then achieved alongside analytical figures of merit, with typical calibration curve correlation coefficients >0.9950 (Table 10.5).

PLE with Off-Line Clean-Up

The results, using PLE followed by off-line clean-up, with both absorbents (i.e. alumina and florisil) gave average recoveries for mid-molecular weight PAHs (fluorene to pyrene) of approximately 80%, whereas for the heavier molecular weight PAHs, i.e. benzo(a) anthracene to benzo(ghi)perylene the average recoveries were typically 50%. For the lightest, i.e. small molecular weight PAHs, recoveries of <5% for naphthalene, <30% for acenaphthylene and <40% for acenaphthene were obtained (Figure 10.9). Typical RSDs for the recovery of PAHs, using alumina and florisil, ranged from 11.1 to 61.4% and 3.3 to 68.9%, respectively.

PLE with *In Situ* Clean-Up

Soil samples were spiked directly into the PLE vessel to assess the impact on PAH recovery using *in situ* clean-up with either alumina or florisil. It can be seen (Figure 10.10a) that good recoveries (~90%) were obtained for all PAHs when no further sample concentration takes place (no solvent evaporation post-extraction). Typical RSDs for the recovery of PAHs, using alumina and florisil, ranged from 4.0 to 10.5% and 1.1 to 22.4%,

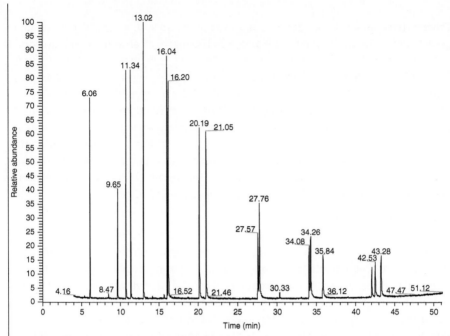

Naphthalene (6.06 minutes); acenaphthylene (10.76 minutes); acenaphthene (11.34 minutes); fluorene (13.02 minutes); phenanthrene (16.04 minutes); anthracene (16.20 minutes); fluoranthene (20.19 minutes); pyrene (21.05 minutes); benzo (a) anthracene (27.57 minutes); chrysene (27.76 minutes); benzo (b)fluoranthene (34.08 minutes); benzo (k)fluoranthene (34.26 minutes); benzo (a)pyrene (35.84 minutes); indeno(1,2,3-cd)pyrene (42.09 minutes); dibenzo(a,h)anthracene (42.53 minutes); benzo(g,h,i)perylene (43.28 minutes); and internal standard (4,4-difluorobiphenyl) (9.65 minutes).

Figure 10.8 Separation of the 16 polycyclic aromatic hydrocarbons by gas chromatography mass spectrometry (GC-MS).

respectively. No specific influence is noted in terms of the use of florisil and alumina on recovery of PAHs. This is not the case in Figure 10.10b in which post-extraction evaporation under a stream of N_2 results in significant losses of naphthalene (>80%), and to a smaller extent for acenaphthylene and acenaphthene. Appropriate recoveries are noted for alumina for the other PAHs, whereas elevated recoveries are noted for the mid-range PAHs when using florisil as the in situ adsorbent. Typical RSDs for the recovery of PAHs, using alumina and florisil, ranged from 2.7 to 25.7% and 3.8 to 22.2%, respectively.

Analysis of a Certified Reference Material
An in situ PLE-GC-MS procedure, using alumina, was applied for the analysis of PAHs from two certified reference material (CRM), a sandy soil (LGC QC 3008) and a soil (CRM 123-100). The data (Table 10.6) showed that the measured results were all within the certified values (± standard deviation) for LGC QC 3008, except the indicative value for dibenzo (a, h) anthracene where the measured value was above the indicative value of <2 mg kg^{-1}) and all within the prediction interval for CRM 123-100.

Table 10.5 Calibration data for analysis of PAHs by GC-MS: based on a five-point graph (0–5 µg ml^{-1}).

PAH structure	PAHs	Retention time (min)	MS Ion for quantitation	LOD in dust (mg kg^{-1}) (S/N > 3)	LOD in dust (mg kg^{-1}) after evaporation (S/N > 3)	Calibration Y = mx + c	Linear regression coefficient R^2
	Naphthalene (NAP)	6.06	128	0.7	0.04	Y = 1.3313x + 0.0865	0.9972
	Acenaphthylene (ACY)	10.76	152	0.2	0.02	Y = 1.3079x + 0.0981	0.9982
	Acenaphthene (ACE)	11.34	154	0.4	0.02	Y = 0.8795x + 0.0880	0.9988
	Fluorene (FLU)	13.02	166	0.1	0.01	Y = 0.9513x + 0.1655	0.9968
	Phenanthrene (PHE)	16.04	178	0.4	0.03	Y = 1.3456 X + 0.1703	0.9998
	Anthracene (ANT)	16.20	178	0.2	0.01	Y = 1.0494 X + 0.1035	0.9980
	Fluoranthene (FLUH)	20.19	202	0.1	0.01	Y = 1.1869x + 0.1665	0.9986
	Pyrene (PYR)	21.05 05	202	0.5	0.03	Y = 1.2741x + 0.1632	0.9975
	Benzo(a)anthracene (BaA)	27.57	228	1.7	0.11	Y = 0.7502x + 0.1146	0.9951

(Continued)

Table 10.5 (Continued)

PAH structure	PAHs	Retention time (min)	MS Ion for quantitation	LOD in dust (mg kg^{-1}) (S/N > 3)	LOD in dust (mg kg^{-1}) after evaporation (S/N > 3)	Calibration Y = mx + c	Linear regression coefficient R^2
	Chrysene (CHY)	27.76	228	2.5	0.17	Y = 0.9428x + 0.1368	0.9990
	Benzo(b) fluoranthene (BbF)	34.08	252	0.5	0.03	Y = 0.7314x + 0.1042	0.9949
	Benzo(k) fluoranthene (BkF)	34.26	252	1.9	0.12	Y = 0.9363x + 0.1443	0.9949
	Benzo(a)pyrene (BaP)	35.84	252	0.2	0.01	Y = 0.6183 X + 0.0613	0.9971
	Indeno(1,2,3-cd) pyrene (IDP)	42.09	276	0.2	0.01	Y = 0.5309x + 0.0790	0.9980
	Dibenzo(a,h) anthracene (DBA)	42.53	278	1.2	0.08	Y = 0.4932x + 0.0782	0.9977
	Benzo(g,h,i)perylene (BgP)	43.28	276	0.2	0.01	Y = 0.6554x + 0.0583	0.9968

Note: internal standard (4,4-difluorobiphenyl) has a retention time of 9.65 minutes.

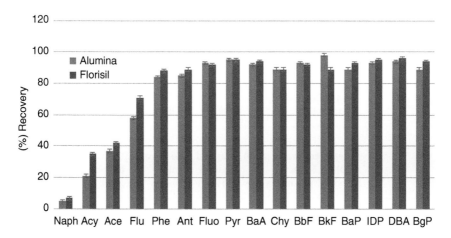

Figure 10.9 Mean recoveries of polycyclic aromatic hydrocarbons, from slurry spiked soil, using PLE with off-line clean-up (mean ± sd, n = 3).

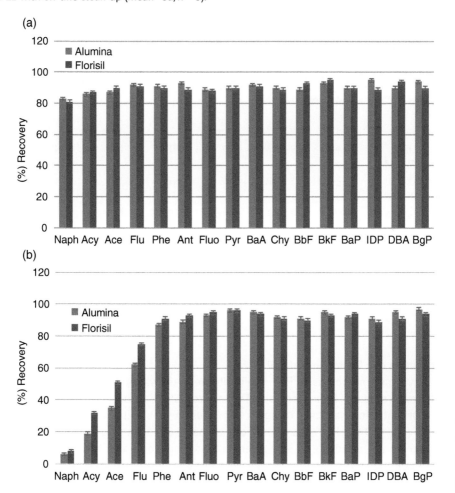

Soil was spiked in-situ within the PLE extraction cell.

Figure 10.10 Mean recoveries of polycyclic aromatic hydrocarbons from soil using PLE with *in situ* clean-up (mean ± sd, n = 3) (a) without further solvent evaporation, and (b) with solvent evaporation.

Table 10.6 Determination of PAHs using in situ PLE-GC-MS from two certified reference materials (CRM LGC QC 3008 and CRM 123-100).

| PAH | CRM LGC QC 3008 (sandy soil 2) | | CRM 123-100 (BNAs in soil) | | | |
	Measured (± SD) n = 3 (mg kg^{-1})	Certificate value (± SD) n = 3 (mg kg^{-1})	Measured (± SD) n = 3 (mg kg^{-1})	Certificate value (mg kg^{-1})	Confidence Interval (mg kg^{-1})	Prediction Interval (mg kg^{-1})
Naphthalene	3.4±0.1	3.1±0.9	6.4±0.8	9.73	8.49–11.0	4.84–14.6
Acenaphthylene	3.9±0.5	3.4±1.6	2.9±0.3	7.24	5.75–8.73	1.37–13.1
Acenaphthene	1.5±0.3	<2	5.0±0.6	7.52	6.20–8.84	2.31–12.7
Fluorene	6.7±0.4	7.7±1.7	4.2±0.3	6.88	5.91–7.85	3.05–10.7
Phenanthrene	28.7±3.8	34±7.1	4.9±0.4	7.94	6.96–8.92	4.07–11.8
Anthracene	8.0±0.8	5.9±2.1	3.9±0.4	6.94	5.90–7.98	2.83–11.1
Fluoranthene	29.2±6.0	32±6.4	6.2±0.6	9.31	8.08–10.5	4.44–14.2
Pyrene	20.6±3.5	24±6.5	4.1±0.3	6.75	5.79–7.71	2.98–10.5
Benzo(a)anthracene	10.2±1.8	11±2.5	5.1±0.2	8.38	7.24–9.52	3.87–12.9
Chrysene	9.1±1.1	9.9±2.1	7.6±0.5	11.3	10.0–12.6	6.23–16.4
Benzo(b)fluoranthene	10.4±1.8	9±3.3	ND	ND	ND	ND
Benzo(k)fluoranthene	6.1±1.3	5.8±2.2	ND	ND	ND	ND
Benzo(a)pyrene	8.3±1.5	8.2±1.8	4.6±0.4	7.77	6.79–8.75	3.92–11.6
Indeno(1,2,3-cd)pyrene	6.6±1.4	5.2±1.8	ND	ND	ND	ND
Dibenzo(a,h)anthracene	3.7±0.2	<2	ND	ND	ND	ND
Benzo(g,h,i)perylene	6.1±1.1	5.2±1.8	ND	ND	ND	ND

ND = not determined.

Application to a Contaminated Soil

The developed method, using PLE with in situ clean-up, was applied to a contaminated soil sample obtained from a local site. The results are shown in Figure 10.11. The major PAH concentration was 254±16 mg kg^{-1} for pyrene, with smaller quantities of fluoranthene (239±10 mg kg^{-1}), benzo(b)fluoranthene (166±17 mg kg^{-1}), benzo(a)pyrene (147±6 mg kg^{-1}), indeno(1,2,3-cd)pyrene (106±4 mg kg^{-1}) and benzo(a)anthracene (103±8 mg kg^{-1}). The reduced concentration/near absence of low molecular weight PAHs is not surprising from a contaminated land site.

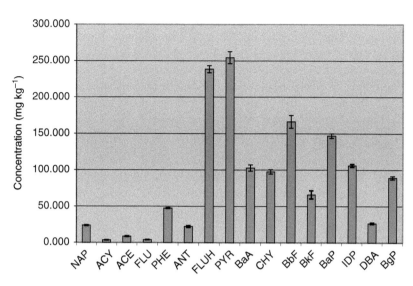

Figure 10.11 Determination of PAHs from a contaminated land soil using PLE with *in situ* clean-up using alumina (mean ± sd, n = 3).

Case Study C   Determination of Polycyclic Aromatic Hydrocarbons and Organophosphate Flame Retardants in Soils from a Public Open Space [8]

Background: This study has evaluated the occurrence of 18 polycyclic aromatic hydrocarbons (PAHs) and 11 organophosphorus flame retardants (OPFRs) in soils from a public open space. This was achieved using PLE and subsequent quantification by gas chromatography coupled to mass spectrometry (GC-MS) in selected ion monitoring (SIM) mode.

Experimental
Chemical Standards

Tris(1,3-dichloro-2-propyl) phosphate, TDCPP (95%); tetraethyl ethylene diphosphonate, TEEdP (97%); tris (2-ethylhexyl) phosphate, TEHP (97%); tri-m-cresyl phosphate, TCrP (95%); tri-n-butyl phosphate, TnBP (99%); triphenyl phosphate, TPhP (99%); triphenylphosphine oxide, TPPO (98.5%); tripropyl phosphate, TPrP (99%); tris (2-butoxyethyl) phosphate, TBOEP (94%); and tris (2-chloroethyl) phosphate, TCEP (97%), were all purchased from Sigma-Aldrich (Steinheim, Germany). Tri-iso-butyl phosphate, TiBP (95%), was purchased from Carbosynth Ltd (Compton, Berkshire, UK). The PAH Calibration Mix (16 compounds) (2000 µg ml^{-1} in methylene chloride was purchased from Restek (Bellefonte, PA, USA). Benzo(e)pyrene, B(e)P (10 µg ml^{-1}) in acetonitrile; benzo(j)fluoranthene, B(j)F (10 µg ml^{-1}) in hexane; and retene, Ret, (10 µg ml^{-1}) in hexane were purchased from Dr. Ehrenstorfer-LGC Standards (Augsburg, Germany).

Hydromatrix was purchased from Thermo Fisher (Hemel Hempstead, UK). Acetone (99.8%), dichloromethane (99.8%) and ethyl acetate (99.5%) were purchased from Fischer Scientific (Loughborough, UK). Mixed standard solutions of OPFRs and BPA were prepared each by weighting individuals and dissolving in ethyl acetate at approximately $1000\,\mu g\,ml^{-1}$ level. Mixed standard solution of PAHs was prepared in hexane at $1.0\,\mu g\,ml^{-1}$ level by diluting the different commercial standard solutions. All mixed standard solutions were subsequently diluted as necessary.

Soils from a Public Open Space

The public open space, POS (Figure 10.12), was based in the Walker area of Newcastle upon Tyne, UK. Historically, the site was a former Lead Works (from the 1860s to 1940) based adjacent to the river Tyne. Since the 1960s, the site has been maintained as a public recreational space. The site has level grassed open space, interspersed with overgrown steep grassed/shrub areas. The site is currently used for leisure and is interspersed with a network of tarmac paths, which cross the site allowing access to and from surrounding residential areas and the river boundary. Samples of soil were collected from 10 sample points on the site by digging a square hole of about $10\,cm^2$ from the topsoil. The grass on the top of the soil was removed. The topsoil collected was put inside a paper geochemical (Kraft) bag and labelled. During sampling, sample handling and sample preparation polyethylene gloves were worn. The soil samples were dried in the sample bags in an oven at a temperature of <40 °C for six days. The dried soil samples were gently disaggregated in a porcelain pestle and mortar and passed through a plastic sieve of mesh size 2 mm, followed by a sieve <250 μm and stored in sealed containers for subsequent analysis.

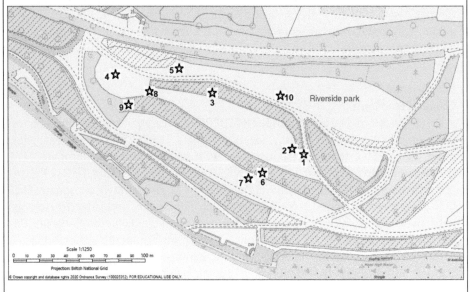

Figure 10.12 Schematic diagram of the public open space with sampling points.

Pressurized Liquid Extraction Procedure:

Samples were extracted using an ASE 200 Accelerated Solvent Extractor (Thermo Fisher, Sunnyvale, CA, USA) in 11 ml stainless steel vessels. A cellulose filter was placed in the bottom of the vessel and was filled until a quarter of the vessel volume of hydromatrix sorbent. Then, 1.0 g of sample, accurately weighed, was added to the vessel. Finally, the vessel was fully filled with hydromatrix and closed. Extraction was performed in a single cycle of a preheating of 5 minutes and 10 minutes static time at 100 °C and 2000 psi using acetone:dichloromethane (50:50, v/v) as extraction solvent. Flush volume of 50% and purge time of 60 seconds were programmed. A glass collector vial of 60 ml was used to collect the extract. The obtained extract was concentrated to dryness using a sample concentrator (Techne, DB-3, Dri-Block®, Essex, UK) at 35 °C and a N_2 stream of 10 psi. The residue was dissolved in 1.0 ml of ethyl acetate using a vortex and stored in 2 ml vial in a fridge at 4 °C until GC-MS analysis. Each sample was extracted in triplicate.

GC-MS System and Operation Conditions

Chromatographic separation was achieved using a TG-5MS column (30 m × 0.25 mm ID × 0.25 μm film thickness) from Thermo Scientific (Hemel Hempstead, UK). The chromatographic system consisted of a Trace 1300 gas chromatograph, TriPlus RSH with liquid sampling tool and an ISQ 7000 Single Quadrupole Mass Spectrometer (Thermo Scientific). Sample volume injection was 1.0 μl in splitless mode with a split flow of 30 ml min^{-1}, 1.0 minutes of splitless time and 10 ml min^{-1} of purge flow. The injector temperature was maintained at 300 °C. Helium was used as a carrier gas with a constant flow of 1.0 ml min^{-1}. The oven temperature program was 70 °C held for 1.0 minute, followed by an increase by 40 °C min^{-1} to 110 °C, and held 2.0 minutes. The temperature was increased to 170 °C by 5 °C min^{-1} and increased to 200 °C by 2.50 °C min^{-1} and held for 3.0 minutes. Finally, an increase to 310 °C by 5 °C min^{-1} was performed, and the temperature is held for 5.0 minutes with a total analysis time of 58 minutes. The temperatures of the source and MS transfer line were 300 °C and 280 °C, respectively. The MS was operated in selective-ion monitoring (SIM) mode using electron impact ionization (EI). Using these conditions, the separation of 18 PAHs and 11 OPFRs was assessed in a single chromatographic run in 58 minutes.

Results and Discussion

Method Validation

The calibration graphs for the PAHs and OPFRs identified in this work were constructed using data from three replicates of each standard solution. Calibration data for each pollutant are shown in Tables 10.7 and 10.8 for PAHs and OPFRs, respectively. All calibration graphs were linear with correlation coefficients (R^2) greater than 0.99. Limit of detection (LOD) and limit of quantitation (LOQ) were determined, using the slope of the calibration graph and the standard deviation of the intercept, based on the following equations: LOD = 3.3σ/s and LOQ = 10σ/s, where σ is standard deviation of intercept and s is the slope. Typical LOD data varied between 3 ng g^{-1} for TEHP to 322 ng g^{-1} for benzo[a]anthracene (BaA), with corresponding LOQ data of 9 ng g^{-1} and 975 ng g^{-1}, respectively. A procedural blank (total number >20) was included in all extraction

Table 10.7 Analytical figures of merit for analysis of compounds in soil from a public open space: PAHs.

Compound	Retention time (min)	Qualitative m/z	Quantitative m/z	Calibration range (ng ml⁻¹)	N° of data points	Calibration graph	R^2	Precision RSD (%) Standard solution[a]	Soil sample	LOD (ng g⁻¹)	LOQ (ng g⁻¹)
Acy	10.98	153, 76	152	0–250	5	Y = 1188.3x + 171 618	0.9943	12.6	4.9	132	399
Ace	12.37	76, 80	153	0–500	6	Y = 1946x − 12 786	0.9993	13.7	7.6	17	52
Fl	14.57	165, 139	166	0–500	6	Y = 1342.1x − 14 772	0.9979	11.6	14.5	50	153
Phe	19.06	160, 176	178	0–500	5	Y = 1278.7x − 15 257	0.9981	10.6	13	43	130
Ant	19.21	160, 176	178	0–500	6	Y = 960.4x − 20 679	0.9917	9.3	9.9	6	17
Ft	26.56	101, 106.5	202	0–500	5	Y = 984.8x − 23 588	0.9935	9.2	9.3	220	667
Pyr	28.07	101, 106.5	202	0–500	5	Y = 1124.9x − 26 961	0.9946	9.2	_b	29	89
Ret	31.80	234, 203.5	219	0–500	5	Y = 211.12x − 6652.7	0.9908	8.8	_b	31	94
BaA	38.06	113.5, 236	228	0–500	5	Y = 326.73x − 14 585	0.9924	8.5	4.2	322	975
Chry	38.29	113.5, 236	228	0–500	5	Y = 840.54x − 21 553	0.9919	8.1	9.7	142	432
BjF+BkF+BbF	44.01	101, 141	252	0–500	5	Y = 1366.5x − 61 960	0.9958	10.7	6.3	103	312
BeP	45.08	125, 132	252	0–500	5	Y = 578.01x − 24 037	0.9978	7.3	7.5	75	228
BaP	45.34	125, 132	252	0–500	5	Y = 804.64x − 37 335	0.9974	8.4	12	134	405
IP	49.82	276, 138	292	0–500	4	Y = 239.63x − 12 987	0.9936	6.3	11.2	16	49
DBahA	50.00	138, 126	278	0–500	5	Y = 394.3x − 16 402	0.9982	8.4	6.5	6	19
BghiP	50.67	228, 138	276	0–500	5	Y = 474.79x − 21 269	0.9918	8.8	13.9	18	54

Acenaphthene (Ace); acenaphthylene (Acy); anthracene (Ant); benzo(a)anthracene (BaA); benzo(a)pyrene (BaP); benzo(b)fluoranthene (BbF); benzo(a)pyrene (BeP); benzo(g,h,i)perylene (BghiP); benzo(j)fluoranthene (BjF); benzo(k)fluoranthene (BkF); chrysene (Chry); dibenzo(a,h)anthracene (DBahA); fluorene (Fl); fluoranthene (Ft); indeno(1,2,3-c,d)pyrene (IP); phenanthrene (Phe); retene (Ret); pyrene (Pyr).
[a] 100 µg l⁻¹.
[b] Not calculated (target compound concentration <LOD).

Table 10.8 Analytical figures of merit for analysis of compounds in soil from a public open space: OPFRs.

Compound	Retention time (min)	Qualitative m/z	Quantitative m/z	Calibration range (ng ml^{-1})	N° of data points	Calibration graph	R^2	Precision RSD (%) Standard solution[a]	Soil sample	LOD (ng g^{-1})	LOQ (ng g^{-1})
TPrP	9.96	182, 141	99	0–2000	8	Y = 1226.6x − 63137	0.9925	13.4	13.9	19	58
TiBP	13.02	155, 138.6	99	0–1000	7	Y = 928.57x − 22323	0.9965	11.4	15.3	225	74
TnBP	16.00	168, 157	99	0–1000	7	Y = 897.9x − 24569	0.9954	9.9	_b	15	44
TCEP	18.67	249, 205	143	0–1000	6	Y = 230.63x − 816.4	0.9962	9.1	11.9	59	178
TEEdP	21.49	173, 165	109	0–1000	7	Y = 123.51x − 8106.6	0.9950	3.8	11.5	9	18
TDCPP	35.64	209.5, 99	99	0–2000	5	Y = 280.76x − 36833	0.9922	6.2	4.3	5	16
TPhP	37.17	169, 215	326	0–1000	5	Y = 197.96x − 18386	0.9964	8	16.8	30	91
TBOEP	37.77	198.8, 125	101	0–1000	4	Y = 51.848x − 6243.5	0.9873	3.5	5.9	138	418
TEHP	39.07	113, 112	99	0–1000	6	Y = 389.46x − 20509	0.9940	3.8	16.7	3	8
TPPO	39.7	227, 199	152	0–2000	5	Y = 49.607x − 5377.3	0.9967	5.9	_b	31	95
TCrP	42.30	367.5, 261	165	0–1000	5	Y = 205.77x − 18393	0.9877	4.9	7.5	12	38

Tetraethyl ethylene diphosphonate (TEEdP); tri-m-cresyl phosphate (TCrP); triphenylphosphine oxide (TPPO); tris (2-butoxyethyl) phosphate (TBOEP); tris(2-chloroethyl) phosphate (TCEP); tri(2-ethylhexyl) phosphate (TEHP); tris(1,3-dichloro-2-propyl) phosphate (TDCPP); tri-iso-butyl phosphate (TiBP); tri-n-butyl phosphate (TnBP); triphenyl phosphate (TPhP); tripropyl phosphate (TPrP).
[a] 100 µg l^{-1}.
[b] Not calculated (target compound concentration <LOD).

batches, and the average value was subtracted from the samples. Control standards were injected at regular interval throughout the analysis. The results of precision and accuracy of the method are shown in Tables 10.7–10.9. All the 30 organic pollutants demonstrated good precision of (RSD ⩽ 16.7%). The precision of the data from the analytical standards varied between 3.5% for TBOEP and 13.7% for acenaphthene (Ace) (Table 10.7). In most cases, the precision of the analysed samples was slightly worse ranging from 4.2% for BaA to 16.7% for TEHP. The trueness (expressed as percent recovery) of the developed method was studied using spiked soil samples before and after PLE, at two concentration levels (100 or 200 µg l^{-1} for all target compounds) (Table 10.9). Typical recoveries pre-PLE ranged from 57–118% for TPrP and TCEP; post-PLE recoveries ranged from 87 to 137% for Ret and TnBP, respectively. Trueness of the method (PLE plus GC-MS quantification) was also assessed for 8 PAHs by analysing CRM 172 (sandy loam soil, AccuStandard Inc., New Haven, CT, USA). Concentrations found (Table 10.9) are in good agreement with the certified values after statistical evaluation by applying a t-test at 95% confidence level for two degrees of freedom. t_{cal} values for all PAHs (Table 10.9) are lower than the t_{tab} value of 4.30. These validation results indicated that the PLE-GC-MS method developed in this work was acceptable. Example chromatograms for all the compounds are shown in Figure 10.13, including an analytical standard and soil extracts.

Analysis of Pollutants in Public Open Space

A POS have been selected for consideration in this study using the developed multi-residue method of PLE-GC-MS. The site is used for recreational purposes so the exposure to humans can be assessed. It can be observed (Tables 10.10 and 10.11) that the major contaminants are the PAHs; the ΣPAHs across all sampling sites varies between 7.1 µg g^{-1} and 166 µg g^{-1}. Whereas the ΣOPFRs for the POS vary between 35 ng g^{-1} and 617 ng g^{-1}.

Table 10.9 Analytical recovery (pre- and post-PLE) for a soil sample and analysis of a soil certified reference material for PAHs (CRM 172).

	Analytical Recovery (%) (n = 3)		CRM172 (Sandy loam soil)		
Compound	Pre-PLE^a	Post-PLE^b	Certified value (ng g^{-1})	Found value (ng g^{-1}) (n = 3)	$|t_{exp}|^c$
HAPs					
Acy	86±7	96±8	55.6±18.1	-^{d}	
Ace	87±1	93±7	94.9±24.7	-^{d}	
Fl	97±1	90±4	66.4±11.2	-^{d}	
Phe	96±3	98±1	168±7.6	170.4±4.6	0.74
Ant	103±1	112±1	17.7±2.7	13.6±0.2	4.14
Ft	87±4	92±6	634±82.4	670.2±30.2	1.70

Table 10.9   (Continued)

Compound	Analytical Recovery (%) (n = 3) Pre-PLE[a]	Post-PLE[b]	CRM172 (Sandy loam soil) Certified value (ng g^{-1})	Found value (ng g^{-1}) (n = 3)	$|t_{exp}|$[c]
Pyr	95±3	123±1	86.5±13	73.9±4.8	3.71
Ret	95±3	87±4			
BaA	107±2	101±5	303±47.7	_d	
Chry	85±3	105±8	154±20.8	146.5±3.1	3.42
BjF+BkF+BbF	84±2	101±7	_e	_d	
BeP	96±1	109±9	_e	_d	
BaP	107±6	97±2	33.9±10.9	_d	
IP	116±5	111±7	150.7±30.5	151.3±4.9	0.17
DBahA	105±10	102±2	284±30.5	277.8±16.1	0.55
BghiP	96±1	110±5	452±81.2	392.4±21.2	4.00
OPFRs					
TPrP	57±1	108±5	Not certified		
TiBP	98±1	114±7			
TnBP	95±2	137±4			
TCEP	118±8	94±6			
TEEdP	90±11	114±4			
TDCPP	104±2	107±2			
TPhP	78±4	102±1			
TBOEP	116±2	87±3			
TEHP	82±1	96±7			
TPPO	82±4	93±2			
TCrP	88±2	95±4			

Acenaphthene (Ace); acenaphthylene (Acy); anthracene (Ant); benzo(a)anthracene (BaA); benzo(a)pyrene (BaP); benzo(b)fluoranthene (BbF); benzo(a)pyrene (BeP); benzo(g,h,i)perylene (BghiP); benzo(j)fluoranthene (BjF); benzo(k)fluoranthene (BkF); chrysene (Chry); dibenzo(a,h)anthracene (DBahA); fluorene (Fl); fluoranthene (Ft); indeno(1,2,3-c,d)pyrene (IP); phenanthrene (Phe); retene (Ret); tetraethyl ethylene diphosphonate (TEEdP); tri-m-cresyl phosphate (TCrP); triphenylphosphine oxide (TPPO); tris (2-butoxyethyl) phosphate (TBOEP); tris(2-chloroethyl) phosphate (TCEP); tri(2-ethylhexyl) phosphate (TEHP); tris(1,3-dichloro-2-propyl) phosphate (TDCPP); tri-iso-butyl phosphate (TiBP); tri-n-butyl phosphate (TnBP); triphenyl phosphate (TPhP); tripropyl phosphate (TPrP); pyrene (Pyr).

[a] Spiked 200 μg l^{-1} on soil.

[b] Spiked 100 μg l^{-1} on soil.

[c] t_{exp} calculated as follows: $t_{exp} = \left| []_{certified} - []_{found} \right| \times \dfrac{\sqrt{n}}{SD}$, $[]_{found}$ and SD are the mean and standard deviation values (n = 2) after PL10- GC-EI-MS and $[]_{certified}$ is the certified concentration.

[d] <LOQ.

[e] Not certified.

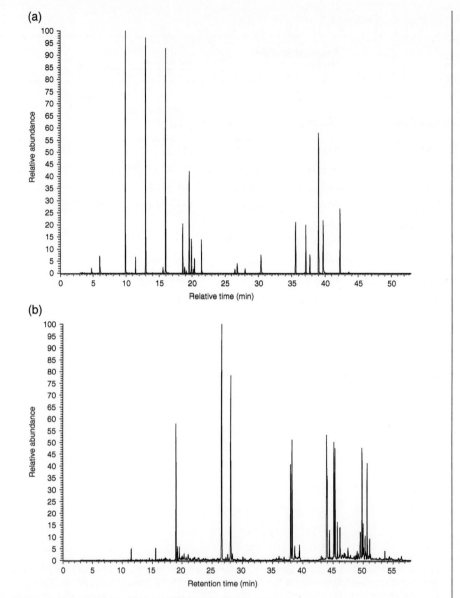

Figure 10.13 GC-MS chromatograms for a mixture of the target compounds. (a) A 200 ng ml^{-1} standard, and a PLE extract from (b) a soil sample from a public open space, using the optimized conditions.

Table 10.10 Analysis and characterization of soil at a public open space: PAHs.

	Public open space: sample identifier[a]									
	#1	#2	#3	#4	#5	#6	#7	#8	#9	#10
Compound	Concentration (ng g^{-1})±SD (n = 3)									
Acy	[b]	[b]	[b]	[b]	[b]	[b]	[b]	[b]	[b]	[b]
Ace	[b]	[b]	91.2 (87.2, 95.2)[c]	126 (112, 141)[c]	120±3.3	253 (300, 206)[c]	122±14	1380±201	[b]	[b]
Fl	[b]	[b]	[b]	[b]	[b]	[b]	[b]	1750±151	[b]	[b]
Phe	741±117	1250 (1250, 1250)[c]	7790 (7760, 7820)[c]	2705±486	3490±292	3940±462	4200±291	55 300±2920	2680±47	918±202
Ant	25.5±1.4	42.3±7.4	326 (388, 264)[c]	91.2±17	145±17	108±12	151±11	3460±198	97.7±20	25.8±4.3
Ft	2680±437	4580±272	35 900±11 200	8530±1230	10 500±873	6760±764	12 200±722	[b]	12 800±1190	3810±299
Pyr	390±72	661±27	5330 (4670, 6000)[c]	1220±159	1530±147	1050±95	1660±110	[b]	1830±170	527±40
Ret	[b]	[b]	[b]	529±342	288±13	593±11	325±5.2	[b]	293±41	[b]
BaA	1770±258	2910±37	28 100 (28 100, 28 100)[c]	5730±712	7860±638	5730±330	7580±495	[b]	7050 (6990, 7100)[c]	1780±163
Chry	339±60	497±3.7	4200±1160	864±89	1140±77	1030±41	1110±88	18 600±758	1140±209	[b]
BjF+BkF+BbF	[b]	[b]	3110 (3260, 2970)[c]	369±24	484±43	342±22	436±61	11 900±681	507 (495, 518)[c]	[b]
BeP	330±65	506±4.3	4930 (5090, 4780)[c]	883±103	1110±72	1020±61	1060±77	[b]	1220±140	282±9.3
BaP	[b]	[b]	3940 (3820, 4060)[c]	597±70	780±67	519±32	716±50	14 400±708	758 (752, 764)[c]	[b]

(Continued)

Table 10.10 (Continued)

	Public open space: sample identifier[a]									
	#1	#2	#3	#4	#5	#6	#7	#8	#9	#10
Compound	Concentration (ng g^{-1}) ± SD (n = 3)									
IP	512±98	764±13	12500 (12200, 12900)[c]	1410±156	1840±133	1350±158	1760±141	49100±2390	1930±396	403±51
DBahA	119±13	155±1.8	1780 (1970, 1580)[c]	268±28	333±24	317±32	324±21	9790±402	224 (243, 204)[c]	95.5±17
BghiP	224±39	331±5.5	2190 (2450, 1920)[c]	564±59	743±41	606±37	695±59	_[b]	816±100	183±18
ΣPAHs	7130 ±804[d]	11700 ±1380[d]	110000 ±10600[d]	23900 ±2370[d]	30400 ±3040[d]	23600 ±2078[d]	32300 ±3360[d]	166000 ±18900[d]	31300 ±3470[d]	8020 ±1150[d]
BaP$_{eq}$	718	1100	15500	2630	3400	2760	3260	31100	3320	646

Acenaphthene (Ace); acenaphthylene (Acy); anthracene (Ant); benzo(a)anthracene (BaA); benzo(a)pyrene (BaP); benzo(b)fluoranthene (BbF); benzo(a)pyrene (BeP); benzo(g,h,i)perylene (BghiP); benzo(j)fluoranthene (BjF); benzo(k)fluoranthene (BkF); chrysene (Chry); dibenzo(a,h)anthracene (DBahA); fluorene (Fl); fluoranthene (Ft); indeno(1,2,3-c,d)pyrene (IP); phenanthrene (Phe); retene (Ret); pyrene (Pyr).

[a] All public open space soil samples were determined on the <250 μm fraction.

[b] <LOQ.

[c] n = 2; mean (individual values).

[d] $SD_{sum} = \sqrt{\sum SD_i^2}$, SD_i is the SD of the PAHs$_i$.

Table 10.11 Analysis and characterization of soil at a public open space: OPFRs.

	Public open space: sample identifier[a]									
	#1	#2	#3	#4	#5	#6	#7	#8	#9	#10
Compound	Concentration (ng g^{-1}) ± SD (n = 3)									
TPrP	_b_	_b_	_b_	_b_	_b_	_b_	_b_	_b_	_b_	_b_
TiBP	_b_	_b_	_b_	_b_	_b_	_b_	_b_	256 ±24	_b_	_b_
TnBP	242 (227, 258)^c	_b_	_b_	_b_	_b_	75.1 (74.8, 75.4)^c	_b_	_b_	_b_	186 (158, 214)^c
TCEP	_b_	_b_	_b_	_b_	_b_	_b_	_b_	_b_	_b_	_b_
TEEdP	_b_	29.9 ±5.3	55.8 ±12	35.3 ±8.4	34.0 ±0.6	82.6 ±6.0	38.1 ±5.7	90.7 ±3.7	32.5 ±0.9	_b_
TDCPP	_b_	_b_	_b_	_b_	_b_	_b_	_b_	_b_	_b_	_b_
TPhP	_b_	_b_	175 ±30	_b_	_b_	106 ±1.5	_b_	_b_	_b_	_b_
TBOEP	_b_	_b_	_b_	_b_	_b_	_b_	_b_	_b_	_b_	_b_
TEHP	_b_	32.6 (32.9, 32.3)^c	107 ±14.4	_b_	28.1 (29.4, 26.7)^c	75.7 ±12	43.3 ±2.2	270 (272, 267)^c	48.7 ±0.6	66.4 ±1.3
TPPO	_b_	_b_	_b_	_b_	_b_	_b_	_b_	_b_	_b_	_b_
TCrP	_b_	_b_	_b_	_b_	_b_	_b_	_b_	_b_	_b_	_b_
ΣOPFR	242^d	62.5 ±1.4^d	338 ±49^d	35.3^d	62.1 ±3.0^d	339 ±13^d	81.4 ±2.6^d	617 ±81^d	81.2 ±8.1^d	252 ±60^d

Tetraethyl ethylene diphosphonate (TEEdP); tri-m-cresyl phosphate (TCrP); triphenylphosphine oxide (TPPO); tris (2-butoxyethyl) phosphate (TBOEP); tris(2-chloroethyl) phosphate (TCEP); tri(2-ethylhexyl) phosphate (TEHP); tris(1,3-dichloro-2-propyl) phosphate (TDCPP); tri-iso-butyl phosphate (TiBP); tri-n-butyl phosphate (TnBP); triphenyl phosphate (TPhP); tripropyl phosphate (TPrP).

[a] All public open space soil samples were determined on the <250 µm fraction.

[b] <LOQ.

[c] n = 2; mean (individual values).

[d] $SD_{sum} = \sqrt{\sum SD_i^2}$, SD_i is the SD of the OPFR$_i$.

10.8   Summary

A crossword of the key terms outlined in this chapter and Chapters 11–13 can be found in Appendix A2, with the solution in Appendix B2.

References

1 Lide, D.R. (1992–1993). *CRC Handbook of Chemistry and Physics*, 73e, 6–10. CRC Press Inc.
2 USEPA (2007), Test methods for Evaluating Solid Waste, Method 3545A – 1. USEPA, Washington DC.
3 Scott, W.C. and Dean, J.R. (2003). *J. Environ. Monit.* 5: 724–731.
4 Scott, W.C. and Dean, J.R. (2005). *J. Environ. Monit.* 7: 710–714.
5 Lorenzi, D., Cave, M., and Dean, J.R. (2010). *Environ. Geochem. Health* 32: 553–565.
6 Lorenzi, D., Entwistle, J., Cave, M., and Dean, J.R. (2011). *Chemosphere* 83: 970–977.
7 Lorenzi, D., Entwistle, J., Cave, M. et al. (2012). *Anal. Chim. Acta* 735: 54–61.
8 Sánchez-Piñero, J., Bowerbank, S.L., Moreda-Piñeiro, J. et al. (2020). *Environ. Pollut.* 266: 115372.

11

Microwave-Assisted Extraction

LEARNING OBJECTIVES

After completing this chapter, students should be able to:

- Understand the principles and instrumentation for microwave-assisted extraction.
- Comprehend the theoretical basis of microwave heating.
- Be able to identify and select appropriate operating conditions for microwave-assisted extraction.

11.1 Introduction

The use of microwaves in analytical science is not new. The first reported analytical use for microwave ovens was in 1975 for the digestion of samples for metal analysis [1], with the first use of microwaves for organic compound extraction some 10 years later [2]. The history of microwave ovens can be traced back to an accidental discovery. A scientist (Dr. Percy Spencer) working at Raytheon Corp., USA, on a radar-related project in the 1940s, noted some unusual heating effects when testing a new vacuum tube, a magnetron (Figure 11.1). His experiences noted that a candy bar in his pocket melted, as well as popping corn kernels becoming popcorn. Within a short time (by 1947), the first commercial microwave ovens ('radarange') for heating food appeared in the marketplace. Our focus in this chapter is on the application of microwave ovens to assist in the extraction of organic compounds from solid matrices.

11.2 Theoretical Considerations for MAE

Microwaves are high-frequency electromagnetic radiation with a typical wavelength of 1 mm to 1 m. All microwave ovens, both industrial and domestic, operate at a wavelength of 12.2 cm (or a frequency of 2.45 GHz) to prevent interference with radio transmissions. Microwaves themselves are fundamentally composed of two parts, an electric field component and a magnetic field component. These components are perpendicular to each other, and the direction of propagation (travel), and vary sinusoidally. The heating effect arises

Extraction Techniques for Environmental Analysis, First Edition. John R. Dean.
© 2022 John Wiley & Sons Ltd. Published 2022 by John Wiley & Sons Ltd.

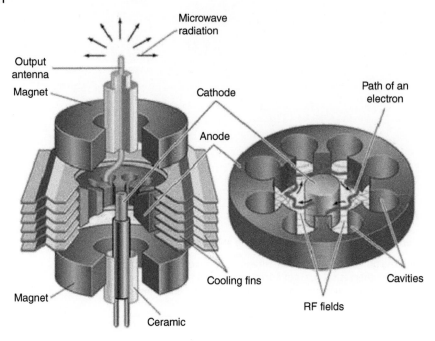

Output antenna

Magnet

Magnet

Ceramic

Microwave radiation

Cathode

Anode

Cooling fins

Path of an electron

Cavities

RF fields

Figure 11.1 A microwave heating source: Magnetron.

from the interaction of the electric field component with charged particles in the sample by dielectric polarization and one of its sub-component's, i.e. dipolar polarization.

In the case of dielectric polarization, charged particles or polar molecules are free to move, and this causes a current in the material. As a result, the polar molecules (or charged particles) reorientate themselves so that they are in-phase with the electric field. This dielectric polarization is composed of four components, each based upon the four different types of charged particles that are found in matter. These are electrons, nuclei, permanent dipoles, and charges at interfaces. The total dielectric polarization of a material is the sum of all four components.

$$\alpha_1 = \alpha_e + \alpha_a + \alpha_d + \alpha_i \tag{11.1}$$

where, α_1 is the total dielectric polarization, α_e the electronic polarization (polarization of electrons round the nuclei), α_a the atomic polarization (polarization of the nuclei), α_d the dipolar polarization (polarization of permanent dipoles in the material), and α_i the interfacial polarization (polarization of charges at the material interface).

The electric field of microwaves is in a state of flux, i.e. it is continually polarizing and depolarizing at 2450 million times per second. These frequent changes in the electric field of the microwaves cause similar changes in the dielectric polarization. Electronic (α_e) and atomic (α_a) polarization (and depolarization) occur more rapidly than the variation in the electric field. As a result, they have no effect on the heating of the polar molecule. Interfacial (α_i) polarization (also known as the Maxwell-Wagner effect) is strongly influenced by microwave frequencies, which in this case are fixed

(2.45 GHz). As a result, the most important component with respect to microwave heating is dipolar polarization (α_d).

Dipolar polarization (sometimes also referred to as orientation polarization) is the most significant heating mechanism. Realignment of the dipoles of the solvent molecules occurs with a rapidly changing electric field (i.e. 2.45×10^9 times per second). Each time a solvent molecule tries to realign itself within the electric field to retain itself in the same phase, it fails to do so (as a consequence of the rapidly changing electric field). This leads to vibration within the solvent molecules, which produces heat through frictional force.

11.2.1 Selecting an Organic Solvent for MAE

In practice, the choice of organic solvent for extraction is crucial to the success of this technique. In order for extraction to occur, the organic solvent must be able to absorb microwave radiation and pass it on in the form of heat to other molecules in the system (e.g. sample matrix). Therefore, only dielectric material or solvents with permanent dipoles get heated under microwave conditions. The efficiency with which different solvents heat up can be determined using the dissipation factor (tan δ). This is calculated as follows:

$$\tan\delta = \varepsilon''/\varepsilon' \tag{11.2}$$

where ε'' is a measure of the efficiency of conversion of microwave energy into heat (also known as the dielectric loss) and ε' is a measure of the ability of the solvent to absorb microwave energy (also known as the dielectric constant).

Table 11.1 shows dielectric constants and dissipation factors for some common solvents used for microwave extraction. In contrast to water, the organic solvent methanol will undergo less microwave absorption due to its lower ε' values (i.e. 32.63), but the overall

Table 11.1 Common organic solvents used in microwave-assisted extraction.

Solvent	Dielectric constant (ε') at 20 °C	Dissipation factor (tan δ) at 20 °C, 2.45 GHz	Boiling point °C	Closed-vessel temperature[a] (°C)
Acetone	20.7	0.054	56.2	164
Acetonitrile	37.5	0.062	81.6	194
Dichloromethane	8.93	0.042	39.8	140
Hexane	1.89	0.020	68.7	NH
Methanol	32.63	0.659	64.7	151
2-Propanol	19.9	0.799	82.4	145
Water	78.3	0.123	100.0	NA
Acetone/hexane (1:1, v/v)	NA	NA	52	156

[a] at 175 psig.
NA = not available.
NH = no heating in microwave.

heating with this solvent will remain higher (than water) due to its higher dissipation factor (tan δ is 0.659, for methanol). In contrast, a solvent like hexane remains transparent to microwave absorption (a low ε′ value, 1.89) and a low dissipation factor (tan δ value, 0.020) and undergoes no microwave heating.

11.2.2 Heating Methods

A reason for the reduced extraction time using a microwave oven can be attributed to the different heating methods employed by the microwave and conventional heating. Conventional heating (Figure 11.2) relies on an external heat source (i.e. an isomantle for organic solvents) to heat the external surface of the vessel (e.g. a round-bottomed flask). A finite period of time is required to heat the vessel before the heat is transferred to the solution. Thermal gradients are set up in the liquid due to convection currents. As a result, a temperature lag occurs in the heating process. Whereas in microwave heating (Figure 11.3) the interaction of the microwave energy with a polar solvent causes direct heating, without heating the vessel, hence temperature gradients are kept to a minimum. Therefore, the rate of heating using microwave radiation is faster than in a conventional method. In addition, energy is not lost due to the unnecessary heating of the vessel. Localized superheating can also occur. The heating profiles for water using a microwave oven and a conventional method have been compared (Figure 11.4). It can be seen (Figure 11.4) that the microwave heated water quickly reaches the boiling point of water (approximately 6–7 minutes), whereas conventionally heated water takes much longer (approximately 15 minutes).

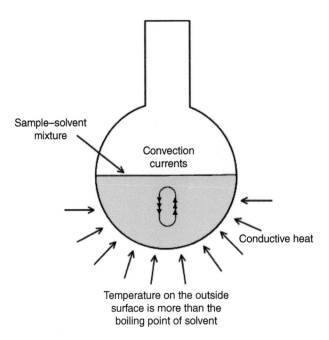

Figure 11.2 The fundamental basis of conventional heating.

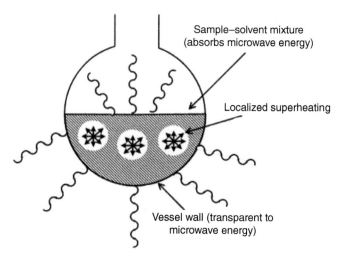

Figure 11.3 The fundamental basis of microwave heating.

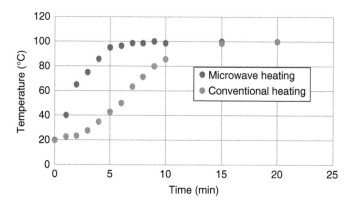

Figure 11.4 A comparison of heating methods: a heating profile comparison.

11.2.3 Calibration of a Microwave Instrument

Measurement of the available microwave power can be assessed by measuring the temperature rise in 1 kg of water exposed to microwave radiation for a fixed period of time. By producing a multiple point calibration, it is possible to relate power (in watts) to the partial power
setting of the microwave system. Typically for a 600 W microwave system, the following power
settings are measured: 100, 99, 98, 97, 95, 90, 80, 70, 60, 50, and 40%. This is done as follows:

- Equilibrate a large volume of water to room temperature (23 ± 2 °C). One kilogram of
  reagent is weighed ($1000.0 \, g \pm 0.1 \, g$) into a fluorocarbon beaker (or similar container that
  does not absorb microwave energy e.g. not glass) with lid. The initial temperature of the
  water should be 23 ± 2 °C measured to ± 0.05 °C.
- The covered beaker is then subjected to microwave radiation at the desired partial power
  settings for 2 minutes under normal conditions of operation, i.e. exhaust fan on.
- After the elapsed time, immediately remove the beaker, add a magnetic stirring bar and
  vigorously stir using a stirrer hot plate (without the heat on).

- Record the maximum temperature within the first 30 seconds to ± 0.05 °C.
- Using a new water sample for each subsequent measurement, determine three measurements at each power setting.

The apparent power absorbed by the sample (P), measured in Watts (W = J s^{-1}), can then be determined:

$$P = \frac{K \times C_p \times m \times \Delta T}{t}$$ (11.3)

where, K = the conversion factor for thermochemical calories s^{-1} to Watts (= 4.184), C_p = the heat capacity, thermal capacity or specific heat (cal g^{-1} °C^{-1}) of water, m = the mass of water sample (g), ΔT = the final temperature minus the initial temperature (°C), and t = the time in seconds (s).

Using the experimental conditions of 2 minutes and 1 kg of distilled water (heat capacity at 25 °C is 0.9997 cal g^{-1} °C^{-1}), the calibration equation simplifies to:

$$P = \Delta T \times 34.86$$ (11.4)

11.3 Instrumentation for MAE

The essential components of a pressurized MAE instrument (Figure 11.5) are:

- A microwave generator (magnetron).
- A wave guide for transmission of microwaves into the cavity.

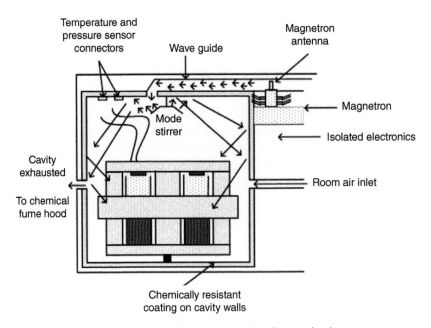

Figure 11.5 Schematic diagram of microwave-assisted extraction instrument.

- A resonant cavity (the inside of the 'oven').
- An electric power supply.
- A method of measuring and controlling temperature and pressure.
- An external vent connected to fume hood ducting (or similar).

In addition, consideration is required of the following:

- Selection of the extraction solvent or solvent mixture (e.g. hexane:acetone 1:1, v/v). Other solvent systems are also to be considered (Table 11.1).
- An appropriate extraction vessel. The commercial microwave instrument will operate with multiple extraction vessels (e.g. up to 40 extraction vessels) with direct temperature monitoring and control in at least a single reference extraction vessel. Each vessel has an exterior body made of microwave transparent material, such as polyether imide or tetrafloromethoxyl polymers, and an inert inner liner. The sample is placed inside the inner liner (e.g. made of Teflon fluoropolymer, perfluoroalkoxy, or glass). The extraction vessels are placed on a carousel, located within the microwave oven cavity.
- A safety system that if a solvent leaks from an extraction vessel(s), the solvent monitoring system will automatically shut off the magnetron but allow the exhaust fan to continue working venting the fumes into external ducting.
- A microwave energy output, typically, 1200 or 1800 W at 100% power with pressures between 50 and 1500 psi and temperatures between 100 and 300 °C.

11.4 A Typical Procedure for MAE

It is important that the extraction vessel is only partially filled to capacity to allow space within the vessel for gas expansion. An accurately weighed solid sample (e.g. 1–20 g) is placed in the vessel inner liner. Then, an appropriate polar solvent (or solvent mixture is added) (e.g. methanol, acetonitrile, dichloromethane, or acetone). Alternatively, manufacturers have developed polar stir bars (i.e. Weflon™ or Carboflow®), which will be heated by microwave radiation, thereby allowing non-polar solvents to be used. The sealed extraction vessels are placed on to the carousel and the extraction programme started. MAE can be performed under the following conditions:

- Organic solvent: hexane:acetone, 1:1, v/v.
- Pressure: 50–150 psi.
- Temperature: 100–115 °C.
- Extraction time (at temperature): 10–20 minutes.
- Cooling: to room temperature.

After extraction, the solvent (hexane:acetone, 1:1 v/v) extract will need to be filtered to remove excess sample matrix. The resultant solution can then be evaporated under a gentle stream of nitrogen gas to dryness (see Chapter 15) and re-constituted with an appropriate solvent prior to analysis or analysed directly.

11.5 Applications of MAE

Case Study A   Extraction of Polycyclic Aromatic Hydrocarbons Using Microwave-Assisted Extraction – A Comparison With Soxhlet Extraction [3–4]

Background: The extraction of polycyclic aromatic hydrocarbons (PAHs) from contaminated soil sites is of major environmental interest. This example compares the results, for extraction of PAHs from soils, using Soxhlet extraction and microwave-assisted extraction.

Experimental
Soil Preparation
The contaminated land soil was sampled from known sites and, after transportation to the laboratory, was air-dried. All large stones and extraneous material were removed by hand and the soils placed in a blender for 5 minutes. The powdered soils were then stored in air-tight containers until required.

Standards and Solvents
The sixteen PAHs studied were supplied (Sigma Chemicals Ltd., Poole, UK) at the $2000\,\mu g\ ml^{-1}$ concentration in benzene-dichloromethane, 1:1, v/v. The stock solution was serially diluted in dichloromethane as required. All solvents used (acetone, dichloromethane, and methanol) were HPLC grade and supplied by Rathburn Ltd. (Walkerburn, Scotland, UK).

Soxhlet Extraction Procedure
Soxhlet extractions were performed using 10 g portions of the soil to which was added 30 g of anhydrous sodium sulphate (Sigma Chemicals Ltd., Poole, UK). The mixture was transferred to a cellulose extraction thimble (Whatman, Maidstone, UK), covered with a loose wad of dichloromethane (DCM)-extracted cotton wool and inserted into the Soxhlet assembly. This apparatus was fitted with a 250 ml flask containing 100 ml of DCM. The assembly was heated and refluxed for 6 hours using an isomantle.

Microwave-Assisted Extraction Procedure
All microwave extractions were performed using a 1000 W MES 1000 microwave sample preparation unit (CEM, Buckingham, UK). The system consisted of a Teflon-coated microwave cavity in which a turntable, capable of holding 12 sample vessels, was placed. The outer body and cap of the vessels were made from microwave-transparent Ultem poly(ether imide). Inside each vessel was a PTFE liner into which the sample was placed (volume 100 ml) together with the extraction solvent. The cap of the vessel was also lined with PTFE and contained a PTFE rupture membrane pressure rated to 200 psi $(1\,psi = 6.89476 \times 103\,Pa)$. If the safety membrane were to break, solvent vapours could then escape through a small exhaust port and be carried via a PTFE tube, to a central well, which was air-sealed to prevent hot gases escaping into the cavity. A PTFE tube running from the centre of the well was connected to a solvent vapour detection system, which immediately stopped microwave heating on the breakage of the rupture membrane. One of the 12 vessels used contained a modified cap, which allowed an

optical fibre temperature sensor and a pressure sensing tube, connected to a pressure transducer, to be used in situ. The optical fibre was housed in a Pyrex tube, which protected it from solvent attack. This system allowed the temperature and pressure to be monitored during extraction. The MES 1000 software allowed temperature and pressure to be set in five separate heating stages. The instrument was then controlled by either the set temperature or pressure depending on which parameter reached its programmed set point first. A portion of soil (2 g) was weighed into a weighing boat and transferred into a PTFE vessel liner where 40 ml of dichloromethane (DCM) was added. New rupture membranes were fitted into each cap, which screwed onto the vessels. The vessels were then placed symmetrically on the microwave turntable together with the control containing the temperature and pressure sensory equipment. Each vessel was connected to the central well, the PTFE tubing, and extraction done using 40 ml of DCM at a constant temperature of 120 °C. The magnetron power was set at 50% for an extraction time of 20 minutes.

After the extraction was completed, the vessels were cooled until no residual pressure was observed. The solvent from each vessel was then filtered through a GF/A glass microbore filter (Whatman, Maidstone, UK) into a 50 ml calibrated flask and the remaining solid washed with solvent, which was also added prior to making up to volume with the same solvent. The extracts were then ready for analysis without further pre-concentration.

Analysis of Extracts by Gas Chromatography-Mass Spectrometry (GC-MS)

The analysis of all soil extracts was carried out on a HP 5890 Series 11+ GC fitted with a HP 5972A mass spectrometer and 7673 autosampler (Hewlett-Packard Ltd., Bracknell, Berkshire, UK). A 30 m × 0.25 mm i.d. × 0.25 μm film thickness DB-5 capillary column (J&W Scientific, supplied by Phase Separations Ltd., Clwyd, UK), operated with a constant flow rate of $1 \, ml \, min^{-1}$, was used to achieve separation using the following temperature programme. Following injection, the column was held at 85 °C for 3 minutes before commencing a $6 \, °C \, min^{-1}$ temperature programme to 300 °C. The column was held at the final temperature for 7 minutes. The split/splitless injector was held at 300 °C and operated in splitless mode with the split valve held closed for 1 minute following sample injection. The split flow was set at $40 \, ml \, min^{-1}$. The mass spectrometer transfer line carrying the column to the ion source was maintained at 270 °C. Electron impact (EI) ionization at 70 eV with an electron multiplier voltage set at 1500 V was used while operating in single-ion monitoring (SIM) mode throughout the chromatogram. A five-point calibration plot containing 0, 2, 5, 10, 20, and $50 \, μg \, ml^{-1}$ of each individual PAH and $20 \, μg \, ml^{-1}$ internal standard mix (3,6-dimethylphenanthrene and 6-ethylchrysene) was prepared and used to confirm the system linearity.

Results and Discussion

The results of successive extractions on soil sub-samples by each selected technique are shown in Table 11.2. Consistent results were obtained between Soxhlet extraction and microwave-assisted extraction. The additional advantage of MAE is the speed of extraction (20 minutes for multiple samples to be extracted, plus colling time) as compared to the 6 hours for Soxhlet (USA EPA Method 8100).

Table 11.2 A comparison of Soxhlet extraction with microwave extraction for the analysis of polycyclic aromatic hydrocarbons from contaminated soil (mg kg^{-1}).

Compound	Contaminated soil 1		Contaminated soil 2		Contaminated soil 3	
	Soxhlet[a]	MAE[b]	Soxhlet[c]	MAE[d]	Soxhlet[e]	MAE[f]
Naphthalene	4.2±0.9	5.9±0.7	6.0±1.1	10.1±0.4	12.1±0.6	13.5 (13.9, 13.0)
Acenaphthylene	2.6±0.5	3.0±0.2	0.7±0.5	1.3±0.6	1.8±0.1	2.8 (2.8, 2.8)
Acenaphthene	6.4±1.3	7.6±0.7	1.9±0.6	3.1±0.7	0.9±0.1	1.1 (1.1, 1.0)
Fluorene	8.6±1.4	8.9±0.9	2.5±0.6	4.4±0.9	0.8±0.1	0.9 (0.9, 0.8)
Phenanthrene	53.4±6.0	55.5±6.9	1.0±0.5	1.2±0.3	67.9±2.3	69.8 (70.5, 69.1)
Anthracene	13.6±1.1	15.5±2.8	2.3±0.8	2.4±1.1	1.9±0.7	3.5 (3.3, 3.6)
Fluoranthene	54.1±3.3	48.8±6.8	1.1±0.2	1.2±0.6	56.8±2.9	54.5 (54.3, 54.6)
Pyrene	43.0±3.1	38.0±5.5	1.0±0.3	1.4±0.6	34.3±0.5	35.9 (35.7, 36.0)
Benz(a)anthracene	25.3±1.9	20.6±2.5	2.9±1.0	3.6±0.8	10.9±0.1	11.4 (11.8, 11.0)
Chrysene	26.6±1.3	22.9±3.7	3.9±1.3	2.9±1.3	15.5±0.6	15.2 (15.5, 14.8)
Benzo[b]fluoranthene	15.1±2.9	14.4±3.2	2.8±0.7	3.8±2.3	13.4±0.5	12.3 (13.4, 11.2)
Benzo[k]fluoranthene	11.0±1.3	9.0±0.5	3.4±1.5	3.9±2.1	9.6±2.2	11.9 (10.6, 13.1)
Benzo[a]pyrene	15.3±2.5	12.8±1.4	4.1±2.4	5.8±2.7	2.1±0.1	2.0 (2.0, 2.0)
Indeno[1,2,3-cd]pyrene	7.2±2.0	6.5±0.9	6.0±4.4	7.4±4.5	2.8±0.2	2.2 (2.8, 1.6)
Dibenz[a,h]anthracene	3.4±2.0	3.3±0.7	12.9±11.6	8.7±1.7	1.8±0.1	1.7 (1.7, 1.7)
Benzo(ghi)perylene	7.6±1.7	7.3±0.6	5.6±3.6	7.1±2.6	10.0±0.4	9.2 (9.2, 9.2)
total	297.4±30.0	279.8±36.4	58.1±25.0	68.3±9.0	242.7±20.4	247.5±20.4

[a] n = 5 using the extraction solvent dichloromethane.
[b] n = 4 using the extraction solvent dichloromethane.
[c] n = 6 using the extraction solvent dichloromethane.
[d] n = 4 using the extraction solvent dichloromethane.
[e] n = 3 using the extraction solvent acetone.
[f] n = 2 using the extraction solvent dichloromethane; the individual concentrations are presented in brackets.

Case Study B   Extraction of Oligomers from Poly(ethylene terephthalate) by Microwave-Assisted Extraction – A Comparison with Soxhlet Extraction [5]

Background: Poly(ethylene terephthalate), PET, is a commercial polymer having a linear chain, which is produced in a polycondensation reaction, for instance, from ethylene glycol and terephthalic acid methyl ester. The main application of PET has been in the food packaging industry where its original use was in the production of soft drink bottles. However, recent developments in the consumer market have led to its widespread use for dual oven usable trays, where ready-made meals can be heated either by microwave or by conventional ovens. It has been found that low-molecular-weight oligomers can be extracted from the polymer during the cooking and storage of foodstuffs. This is of particular concern to the polymer manufacturer and the food retailer. This concern over plastic contact materials and issues relating to the recycling of waste plastics make it necessary to investigate the extractability of oligomers from PET.

Experimental
Materials and Reagents

Poly(ethylene terephthalate) film of 250 mm thickness was cut into small strips (approximately 0.5 × 5.0 cm). Solvents (xylene, dichloromethane, acetone, hexane, and tetrahydrofuran) of appropriate grade were used throughout the work. Cyclic trimer reference standard was supplied by ICI Research and Technology Centre, Wilton, UK.

Apparatus

Microwave-assisted extraction was performed using a commercial system MES 1000 (CEM, Buckingham, UK) operated at 950 W and 2.45 GHz. The system has capacity for 12 vessels to be extracted simultaneously. Samples were placed into 100 ml capacity lined vessels constructed of polyetherimide bodies and caps. Inside each vessel is an inner liner and cover with which the sample comes into contact. The inner liner container and cover are constructed of Teflon perfluoroalkoxy (PFA). The system allows both temperature and pressure measurements to be made in a single vessel. Each vessel is fitted with a rupture membrane that will fail if the pressure exceeds 200 psi (1 psi = 6894.76 Pa).

HPLC was performed using a gradient system P4000 connected to a UV/VIS detector UV1000 (Thermo Separations Products, Hemel Hempstead, Hertfordshire, UK). Samples and standards (10 ml) were injected into the chromatographic column. Separation was carried out using an ODS2 column (Phase Separations, Clwyd, UK) (25 cm × 4.6 mm), which was located in an oven maintained at 35 °C. The gradient mobile phase consisted of water–acetonitrile (65 + 35) for 15 minutes followed by 100% acetonitrile for 5 minutes at a flow rate of 1 ml min^{-1}. Detection was at 240 nm. Data acquisition was performed using Peak Simple Software (SRI Instruments, Torrance, CA, USA).

Procedure for Soxhlet Extraction

The Soxhlet apparatus was set up in the following manner. Three round-bottomed flasks were cleaned, rinsed in acetone, and left in an oven at 150 °C for 1 hour to remove any residue. After this, the flasks were left to cool to room temperature for 1 hour in a desiccator. Then, PET strips (15–20 g) were placed into a 30 × 100 mm cellulose extraction thimble and placed inside the Soxhlet apparatus. Xylene (190 ml) was placed in the round-bottomed flask and heated to reflux. The solvent was then boiled at this point for 24 hours. After extraction, the vessel was left to cool for 1 hour, the solvent was then removed by evaporation on a steam bath under a continuous stream of nitrogen. This procedure was also repeated using a different solvent system, i.e. dichloromethane.

Procedure for Microwave-Assisted Extraction

The PET film was prepared in the same manner as for the Soxhlet extraction. To each microwave vessel, 8 g of polymer strip was added together with 40 ml of dichloromethane solvent. Microwave-assisted extraction of total solvent-extractables from the PET film was done using operating conditions of 100% power, pressure at 150 psi, a 120-minute extraction time, a temperature of 120 °C, and dichloromethane as the solvent. Temperatures more than 125 °C were discontinued as they led to polymer fusion in the extraction vessel.

Chromatographic Analysis
The extracted residue was dissolved in tetrahydrofuran (THF) and made up to 50 ml in a calibrated flask. A portion (1 ml) of this solution was then further diluted to 50 ml. A series of calibration standards were prepared (0–80 µg ml^{-1}) and the extracts analysed by external calibration. The correlation coefficients for the calibration graphs were >0.9997.

Results and Discussion
A comparison of Soxhlet extraction and MAE has been made using a variety of solvent systems. Traditionally, xylene has been the solvent of choice due to its capacity to extract solvent extractables and oligomers. However, xylene has a relatively high boiling point (137–142 °C), and this can extend the length of time required to prepare the sample for analysis. Soxhlet extraction of PET film to recover total solvent-extractables was done using xylene as the solvent (for 24 hours). In addition, the process was repeated using dichloromethane as the solvent. The results (Table 11.3) show the effectiveness of xylene to recover the total solvent-extractables.

The results by MAE are shown in Table 11.3. Comparable data, using MAE with dichloromethane, was obtained to that using Soxhlet extraction with xylene. In addition, to determining the total solvent-extractables from the PET film, it was also possible to determine the amount of cyclic trimer present using HPLC (Table 11.3). An example chromatogram is shown in Figure 11.6.

Table 11.3 A comparison of Soxhlet extraction with microwave extraction for the analysis of total solvent-extractables and cyclic trimer from poly(ethylene terephthalate) film (% m/m).

Extraction technique	Solvent	Total solvent-extractables (%, m/m)	Number of repeats
Soxhlet extraction	Xylene	1.24±0.16	11
	Dichloromethane	0.66±0.10	12
MAE	Dichloromethane	1.16±0.06	12
		Cyclic trimer (%, m/m)	
MAE	Dichloromethane	57.2±3.7	12

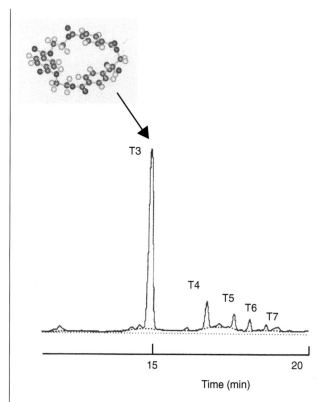

Figure 11.6 High-performance liquid chromatographic analysis of cyclic trimer (T3) in poly(ethylene terephthalate), PET.

11.6 Summary

A crossword of the key terms outlined in this chapter and Chapters 10, 12 and 13 can be found in Appendix A2, with the solution in Appendix B2.

References

1 Abu-Samra, A., Morris, J.S., and Koirtyohann, S.R. (1975). *Anal. Chem.* 47: 1475–1477.
2 Ganzler, K., Salgo, A., and Valko, K. (1986). *J. Chromatogra. A* 371: 299–306.
3 Barnabas, I.J., Dean, J.R., Fowlis, I.A., and Owen, S.P. (1995). *Analyst* 120: 1897–1904.
4 Dean, J.R., Barnabas, I.J., and Fowlis, I.A. (1995). *Anal. Proc. including Anal. Comm.* 32: 305–308.
5 Costley, C.T., Dean, J.R., Newton, I., and Carroll, J. (1997). *Anal. Comm.* 34: 89–91.

12

Matrix Solid-Phase Dispersion

LEARNING OBJECTIVES

After completing this chapter, students should be able to:

- Understand the principle of matrix solid-phase dispersion.
- Be aware of the practical considerations necessary to perform matrix solid-phase dispersion.
- Be able to identify and select appropriate operating conditions for MSPD.

12.1 Introduction

Matrix solid-phase dispersion (MSPD) is used for the extraction and fractionation of solid, semi-solid, or viscous biological samples. The process of MSPD is analogous to solid-phase extraction (SPE), as described in Chapter 4, except that the sample, in this case, is not a liquid. But while MSPD is similar in appearance to SPE, its performance and function are different. Essentially, MSPD differs in the following respects:

1) The sample is dissipated, by mixing with the support material over a large surface area.
2) The sample is homogeneously distributed through the column.

12.2 Practical Considerations for MSPD

The concept of MSPD is that a sample is ground with a support material, e.g. octadecylsilane (ODS or C18), silica, alumina, or florisil in a glass or agate mortar with the pestle for a short time, e.g. 30 seconds. In selecting the support material, the particle size is important, typically, 40–100 μm. This is so that restricted flow does not occur through the MSPD column if smaller particle sizes (3–10 μm) had been used. The mechanical grinding of the sample, in the mortar and pestle, with the support is critical to the effectiveness of MSPD. The abrasiveness of the support leads to shearing and disruption of the sample matrix, which in turn results in a large surface area for solvent interaction. After grinding, the blended sample mixture is then quantitatively transferred to an empty SPE cartridge fitted with a frit. By addition of single, or multiple solvents, it is then possible to perform clean-up and or (selective) elution of compounds (Figure 12.1).

Extraction Techniques for Environmental Analysis, First Edition. John R. Dean.
© 2022 John Wiley & Sons Ltd. Published 2022 by John Wiley & Sons Ltd.

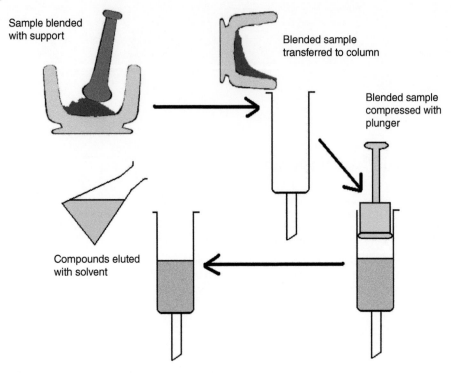

Sample blended
with support

Blended sample
transferred to column

Blended sample
compressed with
plunger

Compounds eluted
with solvent

Figure 12.1 Schematic diagram of apparatus for matrix solid-phase dispersion.

12.3 Optimization of MSPD

In considering the optimization of MSPD, it is important to consider the following:

- Selection of support material and its effectiveness, e.g. use of end-capped or non-end-capped ODS, with different carbon loadings (i.e. 10–20%), alumina, florisil, or silica.
- Ratio of sample to support material. The ratio of sample to sorbent can be varied between 1 : 1 and 1 : 4 w/w. For example, 0.5 g of sample to 2.0 g of C18 (1 : 4 w/w).
- Addition of acids or bases, which may affect clean-up and elution of compound(s).
- Selection of solvent(s) for clean-up, i.e. removal of extraneous material, e.g. fats.
- Selection of solvent(s) for elution of compound(s).
- Elution volume, i.e. for a 0.5 g sample mixed with 2.0 g of support material, then the target compounds typically elute in the first 4 ml of solvent.
- Influence of the sample matrix, i.e. the different properties of the sample will influence the recovery of target compounds.
- Requirement for additional clean-up procedures, e.g. alumina SPE, prior to analysis.

A typical procedure for performing matrix solid-phase dispersion extraction is shown in Figure 12.2. To complement the MSPD approach also refer to the solid-phase extraction method development in Section 4.4.

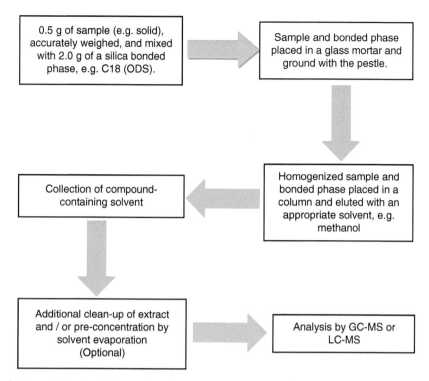

Figure 12.2 Typical procedure for matrix solid-phase dispersion.

12.4 Application of MSPD

Case Study A   Identification and Extraction of Alcohol Ethoxylated Non-Ionic Surfactants from Fish Samples

Background: The term 'surfactant' is a derivation from the phrase 'surface active agent'. This type of compound incorporates both hydrophobic and hydrophilic character into its structure. The most common structure being a linear fatty type of molecule in which one end is oil soluble and the other end water soluble. Surfactants are classified as 'anionic', 'cationic', or 'non-ionic' based on the nature of their ionic charges in solution. This classification applies to the hydrophilic end of the molecule, as the hydrophobic end is always non-ionic. Therefore 'anionic' refers to negatively charged surfactants, 'cationic' refers to positively charged surfactants, while 'non-ionic' refers to those that are uncharged in solution. Table 12.1 shows the chemical structure of some typical surfactants. Of specific concern in this case study are non-ionic surfactants, which are formed of adducts of long-chain alcohols or alkylphenols with a number of ethylene oxide (EO) units. The generic formula for an alcohol ethoxylate (AE) is:

$$RCH_2O(CH_2CH_2O)_n H$$

Table 12.1 Chemical structures of common surfactants.

Anionic, e.g. Linear alkyl benzene sulphonate (LAS)

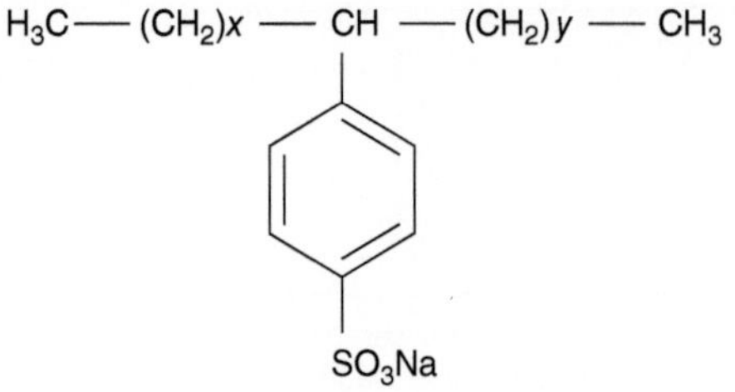

H_3C——$(CH_2)x$——CH——$(CH_2)y$——CH_3

SO_3Na

where $x + y = 7$–10 carbon chain length.

Non-ionic, e.g. alkylphenol ethoxylates (APEs)

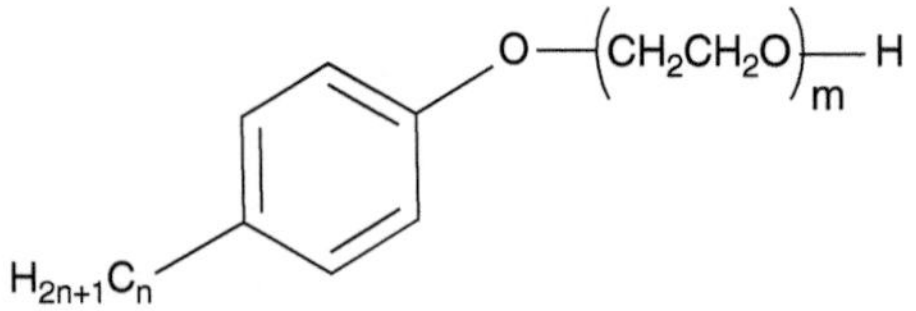

O——$(CH_2CH_2O)_m$——H

$H_{2n+1}C_n$

where $n = 8$ or 9, and $m = 1$–20.

Non-ionic, e.g. alkyl ethoxylates (AEs)

C_nH_{2n+1}-O-$(CH_2CH_2O)_m$-H

Where $n = 12$–18 and $m = 0$–20. In domestic detergents, $m =$ usually averages 7–12.

Cationic, e.g. alkyl quarternary

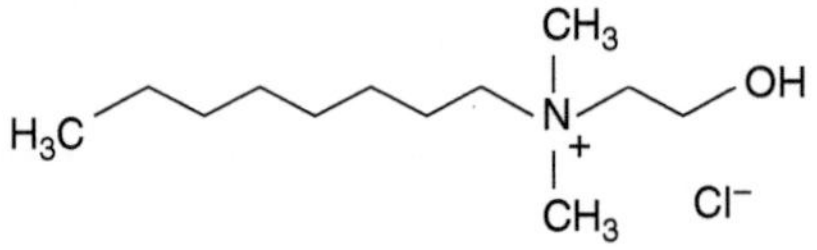

---

where R = alkyl chain.

There most common use is in household detergents/cleaners, with disposal in household wastewater which leads to low levels of AEs in the environment.

Chemicals and Reagents

A commercial alcohol ethoxylate was used: Lutensol (a linear AE with C13 and C15 alkyl chain and average of 7 EO) supplied by Unilever Ltd, Port Sunlight, UK. Acetone, acetonitrile (far UV grade), cyclohexane, dichloromethane, ethyl acetate, heptane, hexane, isopropyl alcohol, and methanol (all HPLC grade) were obtained from Fisher Scientific, Leicestershire, UK. 1,2-Dichloroethane (HPLC grade), heptane, hydrobromic acid (33% in glacial acetic acid), isopropyl alcohol, palmitic acid myristyl ester, phenyl isocyanate (98+ %), 1,1,1-trichloroethane, and undecanol were obtained from Sigma-Aldrich Chemical Co. Ltd, Dorset, UK. Octadecyl silica (ODS) was obtained from Varian sample preparation products, Varian Medical Systems Ltd, Sussex, UK.

Analysis

High-performance liquid chromatography (HPLC) was done using a Thermal Separations (Thermo Separation Products, Staffs., UK) system consisting of a SpectraSystem P4000 pump connected to a fluorescence detector, FD (FL2000, Thermal Separations). An injection volume of 20 μl was used to introduce samples and standards into the system. Normal-phase-HPLC-FD separation was achieved using a Genesis NH_2, 4 μm, 1 cm guard column (Jones Chromatography Ltd, Hengoed, UK) coupled to a Genesis NH_2, 250 mm × 4.6 mm, 4 μm i.d. analytical column (Jones Chromatography Ltd) at a flow rate of 1 ml min^{-1}. Fluorescence spectra were obtained at an excitation wavelength of 238 nm and an emission wavelength of 300 nm. The mobile phase consisted of solvent A, hexane, and solvent B, isopropyl alcohol, with a linear gradient of 98% A and 2% B held for 5 minutes. Then 98% A to 50% A in 20 minutes, and then to 20% A in 5 minutes. Finally, the gradient was held at 20% A for 5 minutes. Solvent equilibration between each run took 15 minutes.

Sample extracts and standards for NP-HPLC-FD were derivatized to allow fluorescence detection. The relevant amount of standard (or sample extract) was placed in a glass vial with 500 μl of acetonitrile and 20 μl of phenyl isocyanate (PIC) added. The vial was then capped and mixed thoroughly on a vortex mixer. The vial was then heated in an oven, at 60 °C, for 1 hour. After cooling to room temperature, the underivatized components were removed, under a gentle stream of N_2, at 100 °C for 10 minutes. The dried residue was reconstituted in 1 ml hexane/1,2-dichloroethane (90 : 10, v/v). The derivatized standard (or sample extract) was then analysed using NP-HPLC-FD. Figure 12.3 illustrates the reaction scheme to generate a fluorescent product for detection.

Gas chromatography-mass spectrometry (GC-MS) was carried out on a Shimadzu (GC-MS, QP5000 Shimadzu, Milton Keynes, UK). The column used was a DB5-ms (J & W Scientific, Folson, California, USA) with dimensions of length 30 m × 0.25 mm i.d. × 0.25 μm film thickness. The temperature program used for analysis was oven: 70 °C for 2 minutes, then 10 °C min^{-1} to 250 °C, with a final hold time of 20 minutes. The injection port and detector temperatures were set at 230 °C, and the interface temperature was set at 240 °C. The helium flow rate was set at 1 ml min^{-1}. The GC-MSD was operated in selected-ion monitoring mode with a splitless injection volume of 1.0 μl. The EI MS detector scan range was 45–425 m/z. Standards (and sample extracts) were derivatized to allow analysis of the AEs; the AEs were derivatized into alkyl bromides

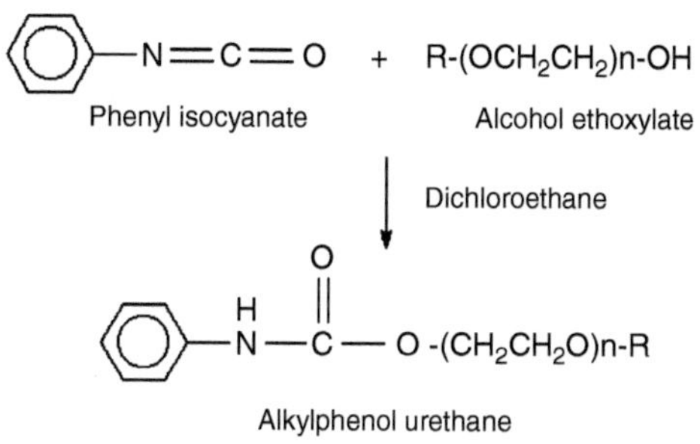

Figure 12.3 Derivatization of alcohol ethoxylates, using phenyl isocyanate followed by normal-phase high-performance liquid chromatography with fluorescence detection.

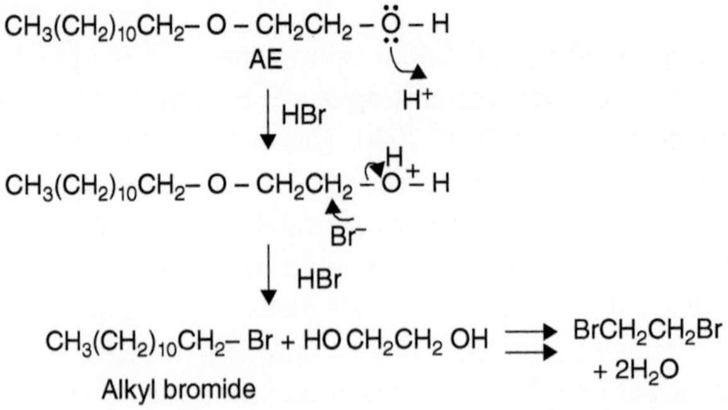

$$CH_3(CH_2)_{10}CH_2-O-CH_2CH_2-\ddot{O}-H$$
$$AE \qquad H^+$$

$\downarrow$ HBr

$$CH_3(CH_2)_{10}CH_2-O-CH_2CH_2-\overset{H}{\underset{Br^-}{\overset{+}{O}}}H$$

$\downarrow$ HBr

$$CH_3(CH_2)_{10}CH_2-Br + HO\,CH_2CH_2\,OH \longrightarrow BrCH_2CH_2Br$$
$$\text{Alkyl bromide} \qquad\qquad\qquad + 2H_2O$$

Figure 12.4 Derivatization of alcohol ethoxylates (e.g. C12EO1), to increase their volatility, using HBr fission followed by gas chromatography-mass spectrometry detection.

using HBr fission (Figure 12.4). The key identifier ions used for alkyl bromides were m/z 135 and 137.

The following procedure was used for the derivatization of AEs via HBr fission:

1) The relevant amount of standard (or sample extract) in 500 μl of ethyl acetate was dispensed into a Chromacol screw capped vial. An internal standard (undecanol) was then added.
2) Then, 0.5 ml of HBr (33% in glacial acetic acid) was added to the mixture and the vial was capped tightly. [Note: if acetonitrile is present (as a solvent) a white precipitate consistent with CH_3CNHBr is formed. This does not appear to affect the reaction.]
3) The sample vial (Chromacol screw cap vial) was placed in a heating block (Reacti-Therm Heating Module, Pierce and Warriner Ltd, Cheshire, UK.), pre-heated to $100-105\ °C$ for 4 hours.
4) The vial was then removed from the heating block and left to cool to ambient temperature, or in a refrigerator, before opening the vial.
5) The vial was uncapped and 1,1,1-trichloroethane (500 μl) added. The vial was then re-capped and vortex mixed for 5 seconds.
6) Then, 4 ml of NaOH (2 M) was then added to the vial, vortex mixed for 10 seconds, the lower organic layer was allowed to settle, then removed to a glass vial.
7) The sample was extracted twice more (from stage 5) with 500 μl aliquots of 1,1,1-trichloroethane.
8) Finally, the combined extracts were made up to 1.5 ml.

Procedure for Matrix Solid-Phase Dispersion
Homogenized rainbow trout tissue (obtained from a local retail outlet) was weighed and placed into a mortar, 1 ml of methanol per 1 g of fish was added and 15 μl of Lutensol (1000 μg ml^{-1} in ethyl acetate) was spiked onto the fish, the solvent was allowed to evaporate. Then, ODS was added (4 g per 1 g of fish tissue) and the mixture was ground with a pestle until a free-flowing powder was obtained. The mixture was placed into an empty SPE cartridge with a filter place underneath and on top of the matrix solid-phase dispersion extract (Figure 12.5).

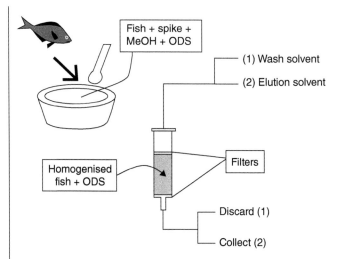

Figure 12.5 Matrix solid-phase dispersion extraction of fish tissue.

The MSPD column was washed with 20 ml of hexane, which was discarded. The collected eluents were taken to 1 m at 60 °C under a gentle stream of nitrogen, and then to dryness under nitrogen and in ice.

Extract Clean-Up

The collected MSPD fish extracts were then subjected to a clean-up using deactivated Alumina-N solid-phase extraction cartridges with a SPE vacuum manifold (Waters Ltd, Hertfordshire, UK). The first Al-N cartridge was conditioned with 20 ml of DCM:MeOH (100 : 5, v/v) at a flow rate of $2-3$ ml min^{-1}. The dried residue from the MSPD step was re-solvated in 5 ml DCM:MeOH (100 : 5, v/v), this was passed through the Al-N cartridge and collected in a glass vial. The sample container was washed with another 2×5 ml DCM:MeOH (100 : 5, v/v), and these were also passed through the cartridge, which was then washed with a further 5 ml of DCM:MeOH (100 : 5, v/v). All eluents were collected, taken to 1 ml at 60 °C, and to dryness under ice, both under a gentle stream of nitrogen.

The second Al-N cartridge was conditioned with 20 ml cyclohexane, at a flow rate $2-3$ of ml min^{-1}. Then, the dried extract was re-solvated in 3 ml cyclohexane, using a sonic bath. This solution was then transferred to the Al-N cartridge. The vial containing the extract was rinsed with 2×5 ml cyclohexane, and then transferred to the cartridge. The eluate was discarded and the AEs eluted with 20 ml DCM:MeOH (100 : 5, v/v). This extract was taken to 1 ml at 60 °C and to dryness under ice, both under nitrogen.

Results and Discussion
Initial Analysis of Fish Tissue Extracts Using NP-HPLC-FD

Preliminary analysis using NP-HPLC-FD of the fish extracts, using a previously developed method for separation of AEs based on their EO unit separation, proved unsuccessful. The resultant chromatogram was complex indicating significant interferences on AE detection due to elution of co-extractives from the sample matrix (Figure 12.6). An alternative approach was sought to identify the AE, in fish tissue, based on

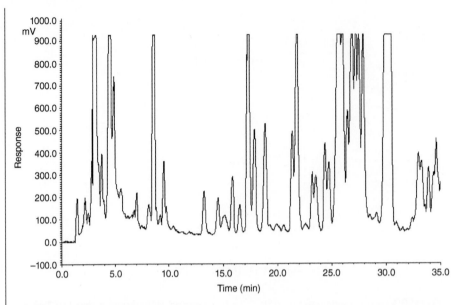

Figure 12.6 Normal-phase high-performance liquid chromatogram of fish tissue extracted using matrix solid-phase dispersion.

GC-MS. This involved derivatization of the AEs into alkyl bromides, using HBr fission (Figure 12.4).

Analysis of Fish Tissue Extracts Using HBr Derivatization and GC-MS Detection
External calibration was done using standard solutions of Lutensol (C13 and C15, with an average EO7) over the concentration range (0, 1, 2, 5, 8, 30, and 50 µg mL^{-1}). An internal standard, undecanol, was used to account for losses from derivatization or sample preparation, and any changes in GC-MS conditions. The linearity of the derivatized alkyl bromides was confirmed with a correlation coefficient of $R^2 = 0.9973$. A chromatogram of a Lutensol standard is shown in Figure 12.7a.

To assess the effectiveness of the elution, for MSPD of the fish tissue, three solvent systems were evaluated: (i) 20 ml of methanol followed by 20 ml of acetonitrile, (ii) 20 ml of a mixture of dichloromethane:acetonitrile (1 : 1, v/v) followed by 20 ml of methanol, and (iii) 20 ml of a mixture of dichloromethane:acetone (1 : 1, v/v). The collected eluents were taken to 1 ml at 60 °C under a gentle stream of nitrogen and to dryness under ice. Sample fish extracts were then derivatized, using HBr, and analysed by GC-MS. The results (Table 12.2) indicate that the best recoveries for Lutensol were obtained using a solvent mixture of dichloromethane/acetone (1 : 1, v/v), a recovery of 94% with 4.1% RSD. This approach is used for subsequent investigation. A chromatogram (Figure 12.7b) shows the analysis of a MSPD extract of fish tissue, with subsequent clean-up using deactivated alumina-N. In addition, a major peak was identified at 19.39 minutes. Subsequent investigation and analysis identified this as a wax ester, from the fish tissue, as the HBr derivatized form of palmitic acid myristyl ester, C16Br (Figure 12.7c).

Table 12.2 Recovery of Lutensol using matrix solid-phase dispersion extraction: Solvent optimization from fish tissue[a] (n = 3).

Solvent system	% Recovery	% RSD
Methanol/acetonitrile	76.7	13.0
Dichloromethane/acetonitrile	82.6	3.5
Dichloromethane/acetone	93.8[b]	4.1

[a] Spiked with 15 μg Lutensol/ g^{-1} fish tissue.
[b] n = 2.

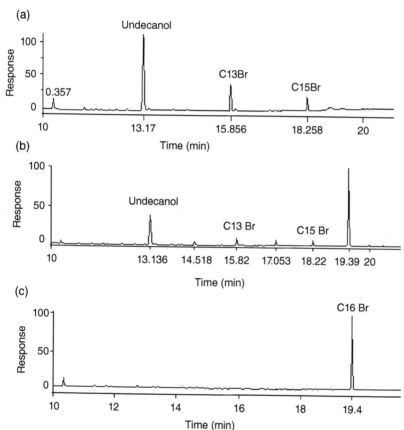

Figure 12.7 Gas chromatogram of (a) Lutensol standard (10 μg ml^{-1}), (b) MSPD extracted fish tissue (spiked with 15 μg Lutensol g^{-1} fish tissue), and (c) a standard of palmitic acid myristyl ester. [Note: post HBr fission derivatization.]

Using the optimized method for the isolation, identification, and quantitation of AEs from fish tissue using MSPD, with alumina-N clean-up, followed by HBr fission derivatization prior to GC-MS was applied to the analysis of rainbow trout tissue. The recovery of AEs from five rainbow trout, which were spiked with 15 μg of Lutensol, are shown in

Table 12.3 Recovery of Lutensol using matrix solid-phase dispersion extraction: Rainbow trout tissue[a] (n = 3).

Rainbow trout	% Recovery	% RSD
A	102.5	4.9
B	100.3	2.9
C	98.4	3.1
D	92.7	5.3
E	95.5	3.0
Average	97.9	4.0

[a] Spiked with $15\,\mu g$ Lutensol/ g^{-1} fish tissue.

Table 12.3. Good recoveries were obtained from each individual fish extract with precision, in all cases <5% RSD. Individual recoveries, across the 5 rainbow trout, ranged from 92.7 to 102.5%, with an overall averaged recovery of 97.9%.

It was concluded that the developed method, using MSPD with post-extract clean-up, followed by GC-MS of derivatized extracts, allowed quantitative analysis of AEs, based on their alkyl chain length. The method could be used to investigate the bioaccumulation of AEs in aquatic organisms. However, further evaluation would be required to different AE alkyl chain lengths from fish tissue, due to the potential impact of wax ester interference.

12.5 Summary

A crossword of the key terms outlined in this chapter and Chapters 10–11 and 13 can be found in Appendix A2, with the solution in Appendix B2.

13

Supercritical Fluid Extraction

LEARNING OBJECTIVES

After completing this chapter, students should be able to:

- Understand the principles and instrumentation for supercritical fluid extraction.
- Be able to interpret a phase diagram for a supercritical fluid.
- Be able to identify and select appropriate operating conditions for supercritical fluid extraction.

13.1   Introduction

Supercritical fluid extraction is an extraction technique that uses pressure and temperature to effect recovery of organic compounds from solid, semi-solid samples. The discovery of the supercritical phase is attributed to Baron Cagniard de la Tour in 1822 [1], who observed that the boundary between a gas and a liquid disappeared, for certain substances, when their temperature was increased in a sealed glass container. However, while some work was done on supercritical fluids, it largely remained ignored (at least in the context of analytical science) until the 1990s [2]. In the food industry, the awakening was earlier with a patent filed in 1964 that demonstrated the decaffeination of coffee using supercritical carbon dioxide. The use of supercritical fluids in the laboratory was initially focused on their use in chromatography, particularly capillary column supercritical fluid chromatography (SFC), and most recently as packed column SFC. The use of SFE for extraction was commercialized in the mid-1980s. The major advantage of SFE is the diversity of properties that it can exhibit. These include as follows:

- Variable solvating power that provides properties intermediate between gases and liquids, so that extraction can exhibit gas-like or liquid-like effects.
- High diffusivity that allows penetration of solid matrices and mass transfer.
- Low viscosity that provides good flow characteristics and mass transfer.
- Minimal surface tension that allows the supercritical fluid to penetrate within low porosity matrices.

These properties of a supercritical fluid allow selective extraction of organic compounds from sample matrices.

Extraction Techniques for Environmental Analysis, First Edition. John R. Dean.
© 2022 John Wiley & Sons Ltd. Published 2022 by John Wiley & Sons Ltd.

13.2 Theoretical Considerations for SFE

A phase diagram for a pure substance is shown in Figure 13.1. In this diagram can be seen the regions where a substance occurs, because of temperature or pressure changes, as a single phase, i.e. solid, liquid, or gas. The divisions between these regions are bounded by curves indicating the co-existence of two phases, e.g. solid–gas corresponding to sublimation; solid–liquid corresponding to melting; and finally, liquid–gas corresponding to vaporization. The three curves intersect at the triple point where the three phases co-exist in equilibrium. At the critical point, designated by both a critical temperature and a critical pressure, no liquefaction will take place on raising the pressure and no gas will be formed on increasing the temperature. It is this defined region, which is the supercritical region. Some critical properties for a range of substances are shown in Table 13.1; however, carbon dioxide is the main focus in analytical science.

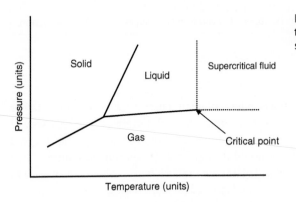

Figure 13.1 Schematic diagram of a typical phase diagram for a pure substance.

Table 13.1 Critical properties of selected substances.

Substances	Critical temperature (°C)	Critical pressure	
		Atm	psi
Ammonia	132.4	115.0	1646.2
Carbon dioxide	31.1	74.8	1070.4
Chlorodifluoromethane	96.3	50.3	720.8
Ethane	32.4	49.5	707.8
Methanol	240.1	82.0	1173.4
Nitrous oxide	36.6	73.4	1050.1
Water	374.4	224.1	3208.2
Xenon	16.7	59.2	847.0

13.3   Supercritical CO_2

The most common supercritical fluid is CO_2, which has a critical temperature of 31.1 °C and critical pressure of 1070 psi (or 74.8 atm). The main advantages that CO_2 has as a super-critical fluid are that it has:

- Moderate critical pressure (74.8 bar).
- Low critical temperature (31.1 °C).
- Low toxicity and reactivity.
- High purity at low cost.
- Useful for extractions at temperatures <150 °C.
- Ideal for extraction of thermally labile compounds.
- Ideal extractant for non-polar species, e.g. alkanes.
- Reasonably good extractant for moderately polar species, e.g. PAHs, PCBs.
- Can directly vent to the atmosphere.
- Little opportunity for chemical change in the absence of light and air.

The non-polar aspect of CO_2 means that in practice to extract polar organic compounds a co-solvent (or modifier) is required to be added, e.g. 10%, v/v methanol. It is possible to add the modifier in several ways including:

- Spiking of the organic solvent directly on the sample in the extraction vessel.
- Purchase of pre-mixed cylinders, e.g. 10% methanol modified CO_2.
- Addition of a second pump that allows in-line mixing of CO_2 and the organic solvent, prior to the extraction vessel.

13.4   Instrumentation for SFE

The essential components of an SFE instrument are:

- A cylinder of high purity CO_2. The CO_2 is supplied in a cylinder fitted with a dip tube, which allows liquefied CO_2 to be pumped by a reciprocating (or syringe) pump.
- A reciprocating piston pump (to deliver the CO_2), capable of operating between 0.5 and 10 ml min^{-1}. To allow pumping of the liquefied CO_2, without cavitation, requires the pump head to be cooled. This is achieved using a jacketed pump-head, which is either cooled via an ethylene-glycol mixture pumped using a re-circulating water bath or a peltier device.
- An organic solvent supply system with an additional reciprocating piston pump (e.g. for a co-solvent, i.e. methanol), capable of operating between 0.1 and 10 ml min^{-1}. The second pump does not require any pump head cooling as the CO_2 and modifier are mixed using a T-piece.
- An extraction vessel. This should be made of stainless steel and be capable of withstanding high pressures (up to 10 000 psi) safely.
- An oven capable of precise temperature control (in the range 31–250 °C).
- A method for creating pressure within the system capable of operating up to a maximum pressure of 400 bar (5801 psi). Pressure can be established within the

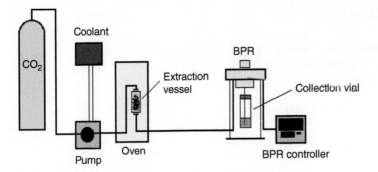

BPR = back pressure regulator

Figure 13.2 Schematic diagram of a supercritical fluid extraction system.

extraction vessel using a variable (mechanical or electronically controlled) restrictor. The variable restrictor (e.g. back-pressure regulator) allows a constant, operator-selected flow rate whose pre-selected pressure is maintained by the size of the variable orifice.

- An extract collection system that allows the escaping CO_2 to vent (as a gas). As a result of adiabatic expansion of the CO_2 upon exiting the restrictor, a build-up of ice is common unless the restrictor is heated. Sample extracts are collected in a vial prior to subsequent analysis as follows:
  - In an open vial containing organic solvent. If necessary, the co-solvent can be evaporated under a gentle stream of nitrogen gas to dryness (see Chapter 15) and re-constituted with an appropriate solvent prior to analysis or directly analysed without concentration.
  - In a sealed vial containing solvent but with the addition of a solid-phase extraction cartridge (see Chapter 4) through which CO_2 can escape but retains any organic compounds.
  - Directly on to a solid-phase extraction cartridge (see Chapter 4) through which CO_2 can escape but retains any organic compounds.
  - A schematic diagram of a basic SFE system is shown in Figure 13.2.

13.5 A Typical Procedure for SFE

The typical operating procedure for SFE is shown in Figure 13.3. It is important that the extraction vessel is full to its capacity as this allows maximum solvent–matrix–compound interaction. Initially, a filter paper is placed inside the extraction vessel and seated in the base. This prevents small particles exiting the extraction vessel and blocking or restricting the flow. Then, alumina (e.g. 2 g), or a similar absorbent (see Table 13.2), is added into the extraction vessel on top of the filter paper to allow *in situ* clean-up. This is followed by a drying agent to remove excess water; typically, anhydrous sodium sulphate (e.g. 0.5 g). Then, the sample is added (e.g. 2 g, accurately weighed), which is mixed with a similar quantity of high-purity diatomaceous earth (hydromatrix). Any residual space in the

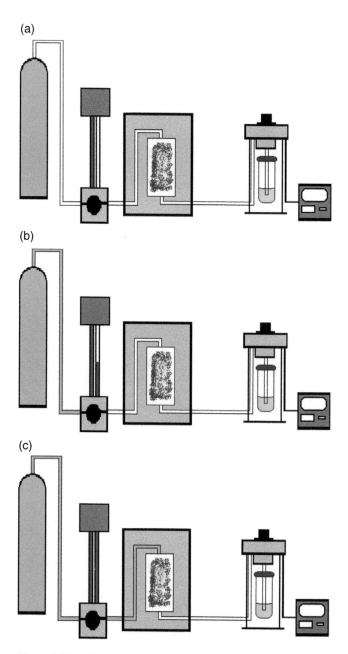

Figure 13.3 Operation of a supercritical fluid extraction system: (a) system set-up established with sample pre-loaded, (b) pump head cooled using a recirculating water bath and liquid carbon dioxide starts to flow, (c) supercritical carbon dioxide fills the extraction vessel, (d) extracted compounds are recovered from the sample matrix and recovered in collection solvent, (e) all extracted compounds are recovered from sample matrix, and (f) extraction complete.

(d)

(e)

(f)

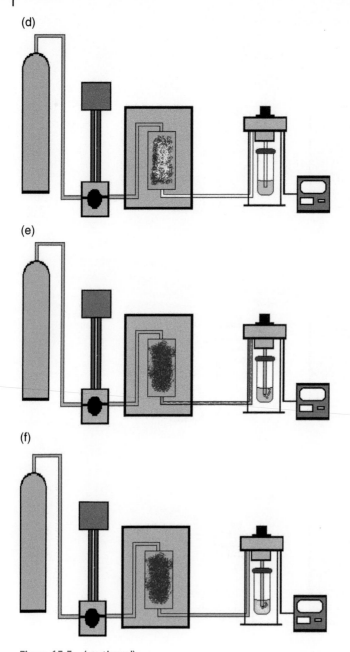

Figure 13.3 (continued)

extraction vessel is filled with more hydromatrix, and finally another filter paper is placed on top prior to extraction vessel closure.

The sealed extraction vessel is placed in the oven, and connected in series, and the extraction programme started. In selecting SFE operating conditions, the following should be considered:

Table 13.2 *In situ* extraction vessel absorbents and their use.

Absorbent	Potential interference	Compound under investigation
Carbon	Organics	Non-polar compounds, dioxins
Copper	Elemental sulphur	Multiresidue pesticides
Ion-exchange resins	Organics, metals, and ionic interferences	Anions, cations, arsenic speciation
C18	Organics, lipids	Non-polar compounds
Acid-impregnated silica gel	Lipids and oils	PCB and bromine flame retardants
Alumina[b]	Lipids, petroleum waste	Amines, perchlorates, and PCBs
Florisil[a]	Oils, lipids, and waxes	Pesticides and aromatics

[a] Florisil™ is magnesium silicate with basic properties and allows selective elution of compounds based on elution strength.
[b] Alumina is a highly porous and granular form of aluminium oxide, which is available in 3 pH ranges (basic, neutral, and acidic).

- Extraction temperature.
  - For thermolabile compounds, the temperature should be within the range 35–60 °C, i.e. close to the critical point but not so high temperature that a compound degradation might occur.
  - For non-thermally labile compounds, the temperature can exceed 60 °C (up to 200 °C).
- Extraction pressure.
  - The higher the pressure, the larger is the solvating power (often described in terms of CO_2 density, which can vary between 0.15 and 1.0 g ml^{-1}) and the smaller is the extraction selectivity.
- Flow rate of liquid CO_2.
  - A typical flow rate of 1 ml min^{-1} is used.
- Extraction time.
  - Often a compromise between obtaining a good recovery and the duration of the process. Typical extraction times may range from 30 to 60 minutes.
- Sample matrix particle size.
  - The smaller the uniform particle size, the more likely that efficient extraction takes place. However, a very small sample particle size can lead to channelling in the sample extraction vessel (leading to poor CO_2 to analyte interaction and consequently poorer extraction efficiency). Sample particle sizes in the range 0.25–2.0 mm are often used.
- Addition of a modifier.
  - The lack of a permanent dipole in CO_2 means that polar compounds will often have poor recoveries. This situation is often addressed by the addition of a polar organic solvent modifier, typically 5 or 10%, v/v methanol.

Recommended initial SFE operating conditions:

- Supercritical CO_2 will generally solvate GC-able compounds under extraction conditions of pressure, 400 atm, and a temperature of 50 °C.

- For fairly polar or compounds with high molecular masses, the addition of an organic modifier (10%, v/v methanol) may be necessary with a subsequent increase in temperature to 70 °C.
- For ionic compounds, the addition of an ion-pairing reagent may be beneficial.

13.6 Application of SFE

Case Study A   Extraction of Polycyclic Aromatic Hydrocarbons Using Supercritical Fluid Extraction – A Comparison with Soxhlet Extraction

Background: The extraction of polycyclic aromatic hydrocarbons (PAHs) from contaminated soil sites is of major environmental interest. This example compares the results, for extraction of PAHs from soils, using Soxhlet extraction and supercritical fluid extraction.

Experimental
Soil Preparation
Coal-derived contaminated land soil (organic carbon content, 10.2%) was sampled from known sites and transported to the laboratory where it was air-dried for 24 hours and then sieved through a 2 mm sieve. The fine powdered soils were then stored in air-tight containers until required.

Standards and Solvents
A standard PAH mixture was supplied by Chem Service, West Chester, PA, USA (PAH mixture 610/525/550). Two internal standards (3,6-dimethyl phenanthrene and 6-ethyl chrysene) were purchased from Lancaster Chemicals (Lancashire, UK). All solvents (acetone, dichloromethane, methanol) used were of analytical grade (Merck, Poole, Dorset, UK).

Soxhlet Extraction Procedure
Soxhlet extractions were performed using 10 g portions of the soil to which was added 30 g of anhydrous sodium sulphate (Sigma Chemicals Ltd., Poole, UK). The mixture was transferred to a cellulose extraction thimble (Whatman, Maidstone, UK) and inserted into a Soxhlet assembly fitted with a 250 ml flask. A 150 ml of dichloromethane was added and the whole assembly heated and refluxed for 24 hours using an isomantle. The extracts were concentrated to 10 ml using a rotary evaporator and then diluted twofold before the addition of the internal standards

Supercritical Fluid Extraction Procedure
All extractions were carried out using a Jasco SFE system (Mettler-Toledo, Halstead, Essex, UK) consisting of dual pumps for carbon dioxide and modifier addition and fitted with a back pressure regulator. Into each sample vessel was placed 1 g of soil. After equilibration, the sample was extracted by use of the following conditions: pressure, 250 kg cm^{-2}; temperature, 70 °C; 30 minutes of dynamic extraction time preceded by a 5-minute static period; flow rate, 2 ml min^{-1}; and a 20% concentration of methanol. Two portions of the 1 g extracts were combined and then concentrated to 5 ml using a rotary evaporator before the addition of internal standards.

Analysis of Extracts by Gas Chromatography-Flame-Ionization Detector (GC-FID)

Gas chromatographic separation and identification of the PAHs was performed on a Carlo Erba HRGC 5300 Mega Series (Fisons, Crawley, Surrey, UK) with on-column injection and flame-ionization detection. A 30 m × 0.32 mm I.D., 0.1 mm film thickness DB-5 HT capillary column (J&W Scientific, Phase Separations, Clwyd, UK) was used to achieve separation with the following temperature programme: initial column temperature, 50 °C; hold for 2 minutes; increase at 15 °C minutes to 90 C; hold for 2 minutes; increase at 6 °C minutes to 300 °C; hold for 8 minutes. The detector temperature was set at 290 °C. PAH quantitation was carried out using a five-point calibration plot containing 0, 20, 30, 40, and 50 mg ml^{-1} PAH standard mixture and 25 mg ml^{-1} internal standards (3,6-dimethyl phenanthrene and 6-ethyl chrysene). Correlation coefficients > 0.9997 were obtained. No sample clean-up was done on the extracts prior to analysis.

Results and Discussion

The results of successive extractions on soil sub-samples by each selected technique are shown in Table 13.3. Consistent results were obtained between Soxhlet extraction and supercritical fluid extraction. Similar individual and overall PAH recoveries were obtained by both extraction techniques. The additional advantage of SFE is the speed of extraction (60 minutes for the dual samples to be extracted) as compared to the 24 hours for Soxhlet.

Table 13.3 A comparison of Soxhlet extraction with 20% methanol-modified supercritical fluid extraction for the analysis of polycyclic aromatic hydrocarbons from contaminated soil (mean ± sd, mg kg^{-1}).

Compound	Contaminated soil	
	Soxhlet[a]	SFE[b]
Naphthalene	214 ± 30	193 ± 19
Acenaphthylene	30 ± 5	25 ± 2
Acenaphthene	56 ± 6	48 ± 6
Fluorene	102 ± 6	107 ± 5
Phenanthrene	291 ± 20	311 ± 12
Anthracene	82 ± 4	73 ± 4
Fluoranthene	219 ± 11	223 ± 11
Pyrene	181 ± 18	156 ± 11
Benz(a)anthracene	87 ± 12	92 ± 8
Chrysene	49 ± 11	59 ± 5
Benzo[b]fluoranthene + Benzo[k]fluoranthene	139 ± 15	89 ± 9
Benzo[a]pyrene	39 ± 9	45 ± 5
Indeno[1,2,3-cd]pyrene	76 ± 8	75 ± 9
Dibenz[a,h]anthracene	ND	ND
Benzo(ghi)perylene	58 ± 5	48 ± 2
Total	1623 ± 80	1544 ± 82

[a] n = 6 using the extraction solvent dichloromethane.
[b] n = 6 using 20% methanol-modified supercritical CO_2.

13.7   Summary

A crossword of the key terms outlined in this chapter and Chapters 10–12 can be found in Appendix A2, with the solution in Appendix B2.

References

1 de la Tour, C. (1822). *Annales de Chimie* 21: 127–132.
2 Dean, J.R. (ed.) (1993). *Applications of Supercritical Fluids in Industrial Analysis*. Glasgow: Blackie Academic and Professional.

Section E

Extraction of Gaseous Samples

14

Air Sampling

LEARNING OBJECTIVES

After completing this chapter, students should be able to:

- Understand the principles of air sampling.
- Be aware of the equipment for active and passive sampling of air.
- Calculate the concentration of a compound in air analysis.
- Calculate the workplace exposure limit for a compound.

14.1 Introduction

Of major concern in air sampling is the presence of either particulate matter (PM) or chemical vapours. The former is characterized based on the size of airborne particulates. Particulate matter with a diameter of 10 µm or less, is referred to as PM_{10}, and particulates with a diameter of 2.5 µm as $PM_{2.5}$ are known to have a significant impact on human health and the environment. The particles are a complex but stable suspension of liquid droplets (e.g. smog) and solid particles in the atmosphere, whereas the presence of chemical vapours, which are volatile (hence are classified as volatile organic compounds, VOCs), can be prevalent in certain environments (e.g. exposure during filling a vehicle with fuel (petrol) in a garage).

The presence of these pollutants in the air can be due to a range of natural (biogenic) or man-made (anthropogenic) processes. Anthropogenic sources are characterized by fossil fuel emissions (e.g. from coal, gas and oil fired power plants, as well as industrial, commercial and institutional sources such as heaters and boilers), other industrial processes (e.g. chemical production, petroleum refining, metals production, and other processes other than fuel combustion), on-road vehicles (e.g. cars, trucks, buses, and motorcycles), and non-road vehicles and engines (e.g. farm and construction equipment, lawnmowers, chainsaws, boats, ships, aircraft). In the case of biogenic source emissions, the main sources are pollen, fungal spores, microorganisms, and viruses, as well vegetation for VOCs. A range of compounds are present in the air (Table 14.1) and include natural or biogenic VOCs (i.e. isoprene, terpenes, alkanes, alkenes, alcohols, esters, carbonyls, and acids), man-made anthropogenic VOCs (i.e. BTEX (benzene, toluene, ethylbenzene, and xylene's), and PAHs (polycyclic aromatic hydrocarbons).

Extraction Techniques for Environmental Analysis, First Edition. John R. Dean.
© 2022 John Wiley & Sons Ltd. Published 2022 by John Wiley & Sons Ltd.

Table 14.1 Typical volatile organic compounds monitored in the atmosphere.

1,1,1,2-Tetrachloroethane	1,2-Dichloropropane	Carbon tetrachloride	m,p-Xylene
1,1,1Ttrichloroethane	1,3,5-Trimethylbenzene	Chlorobenzene	Naphthalene
1,1,2,2-Tetrachloroethane	1,3-Dichlorobenzene	Chloroform	n-Butylbenzene
1,1,2-Trichloroethane	1,3-Dichloropropane	Cis,trans 1,3-dichloropropene	n-Heptane
1,1-Dichloroethane	1,4-Dichlorobenzene	Cis,trans-1,2-dichloroethylene	n-Hexane
1,1-Dichloroethylene	1-Pentene	Dibromochloromethane	n-Octane
1,1-Dichloropropene	2,2-Dichloropropane	Dibromomethane	n-Pentane
1,2,3-Trichlorobenzene	2-Chlorotoluene	Dichloromethane	n-Propylbenzene
1,2,3-Trichloropropane	2-Cis,trans-pentene	Ethylbenzene	o-Xylene
1,2,3-Trimethylbenzene	4-Chlorotoluene	Hexachlorobutadiene	p-Isopropyltoluene
1,2,4-Trichlorobenzene	Benzene	i-hexene	Sec-tert-butylbenzene
1,2-Dibromo-3-chloropropane	Bromobenzene	i-octane	Styrene
1,2-Dibromoethane	Bromochloromethane	i-pentane	Tetrachloroethene
1,2-Dichlorobenzene	Bromodichloromethane	Isoprene	Toluene
1,2-Dichloroethane	Bromoform	Isopropylbenzene	Trichloroethylene

14.2 Techniques Used for Air Sampling

A range of techniques are used to sample and pre-concentrate VOCs in air samples and include:

- whole air collection in containers, or
- enrichment into solid sorbents: with active or passive sampling.

14.2.1 Whole Air Collection

This is the simplest approach for collecting air samples and uses bags (Figure 14.1) or canisters. Samples are analysed either by direct injection into a GC using a gas-tight syringe or more often the air within the container needs to be pre-concentrated to allow measurement of VOCs; this can be done using, for example, a SPME device (see Chapter 5).

The most common containers for collecting the whole air samples are plastic bags (e.g. Tedlar, Teflon, or aluminized Tedlar) (Figure 14.1) and stainless steel containers. The plastic bags are available in a range of sizes, from 500 ml to 100 l and can be re-used provided they are cleaned-out; cleaning takes place by repeatedly filling the bag with pure N_2 and evacuating with a slight negative pressure. Samples collected in plastic bags should be analysed within 24–48 hours to prevent losses. Stainless steel containers should be pre-treated to prevent internal surface reactivity by either a chrome-nickel oxide (Summa® passivation) or chemically bonding a fused silica layer to the inner surface.

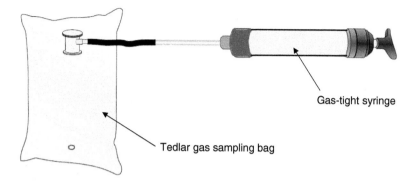

Gas-tight syringe

Tedlar gas sampling bag

Figure 14.1 Tedlar bag being sampled by a gas-tight syringe.

14.2.2 Enrichment Onto Solid Sorbents

14.2.2.1 Active Methods
In this approach, a defined volume of an air sample is pumped through a solid adsorbent (or mixture of adsorbents), located within a tube, where the VOCs are retained (Figure 2.15). The tube typically has dimensions of 3.5 in with a ¼ in external diameter capable of sampling air at flow rates ranging from 10 to 200 ml min^{-1}. Stainless steel tubes are manufactured specifically for thermal desorption. Typical adsorbents used for this approach are:

- Porous organic polymers, such as Tenax™, Chromosorb, and Porapak
- Graphitized carbon blacks, such as Carbotrap and Carbograph
- Carbon molecular sieves, such as carbosieve and carboxen
- Active charcoal. [Note: Tenax is a polymer based on 2,6-diphenyl-p-phenylene oxide.]

It may be necessary to cryogenically cool the trap during sampling to retain the VOCs. Loss of trap efficiency can result from the presence of ozone and humidity, the former can lead to loss of VOCs, particularly unsaturated compounds, by reaction. This can be prevented by the inclusion of a moisture trap attached to the sampling tube. This is particularly important when using activated carbon as the adsorbent.

14.2.2.2 Passive Methods
The determination of VOCs by passive samplers relies on the diffusion of the compounds from the air to the inside of the sampling device. At that point, the VOCs are either trapped on the surface or within a trapping medium. The process can be described using Fick's first law of diffusion, which can be represented as follows:

$$m/(t \times A) = D(C_a - C_f)/L \tag{14.1}$$

where m = mass of substance that diffuses (µg), t = sampling interval (s), A = cross-sectional area of the diffusion path (cm^2), D = diffusion coefficient for the substance in air (cm^2 s^{-1}), C_a = concentration of substance in air (µg cm^{-3}), C_f = concentration of substance above the sorbent, and L = diffusion path length (cm). If it is assumed that adsorbent acts as a 'zero-sink' for the substance, then $C_f = 0$, and then the equation can be simplified to:

$$m/(t \times C_a) = D \times A/L \tag{14.2}$$

The term '$m/(t \times C_a)$' is often called the uptake or sampling rate (R_s); in principle, it is constant for a compound and a type of sampling device and so once determined can be used to determine the concentration of the substance in the air (C_a), from a measured mass of substance. The Eq. (14.2) is often further simplified to:

$$R_s = D \times A/L \tag{14.3}$$

To determine the diffusion coefficient, R_s:

1) Use the published theoretical values of diffusion coefficients [1].
2) Experimentally determine the uptake rate coefficients based on the exposure of the sampler to a standard gas mixture in a chamber [2].
3) Calculate the diffusion coefficient using Eq. (14.4) [3].

$$D = 10^{-3} \left[T^{1.75} \left(\{1/m_{air}\} + \{1/m\} \right)^{1/2} \right] / P \left[V_{air}^{1/3} + V^{1/3} \right]^2 \tag{14.4}$$

where T = absolute temperature (K), m_{air} = average molecular mass of air (28.97 g mol^{-1}), m = molecular mass of the compound (g mol^{-1}), P = gas phase pressure (atm), V_{air} = average molar volume of gases in air (~20.1 cm^3 mol^{-1}), and V = molar volume of the compound (cm^3 mol^{-1}).

Once the diffusion coefficient is determined, use Eq. (14.3) to determine the uptake or sampling rate for the compound by measuring the cross-sectional area of the diffusion path and its diffusion path length. A range of devices have been used as passive samplers but are largely based on either tubes or boxes (badges):

- Badge-type samplers: characterized by a shorter diffusion path and a greater cross-sectional area resulting in higher uptake rates (Figure 14.2a).
- Tube-type samplers: characterized by a long, axial diffusion path and a low cross-sectional area resulting in relatively low sampling rates (Figure 14.2b).

14.3 Thermal Desorption

The determination of VOCs in the atmosphere can be done by trapping them on a solid support material (e.g. Tenax™ or Chromosorb™) and then using thermal desorption directly into a gas chromatograph (Figure 14.3) for subsequent analysis. Sampling of VOCs can be done actively or passively. In active (pumped) sampling, air is drawn through a sorbent (e.g. Tenax™ or Chromosorb™) located within a stainless steel tube (e.g. 89 × 6.4 mm outside diameter × 5 mm internal diameter) at a fixed flow rate (e.g. 20–10 ml min^{-1}) for a period. To allow maximum sampling efficiency, the safe sampling volume (SSV) should not be exceeded and that the appropriate desorption temperatures are applied (see Table 14.2).

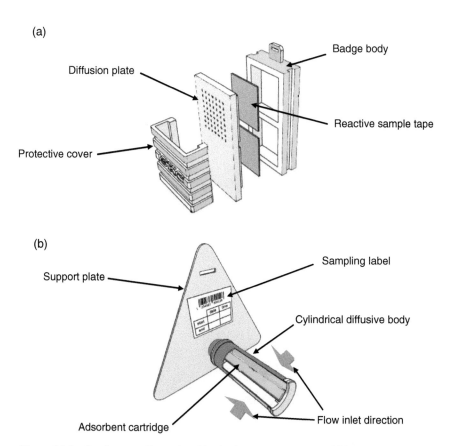

Figure 14.2 Passive sampling using (a) a badge-type sampler and (b) a tube-type sampler.

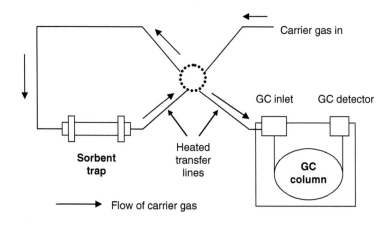

Procedure:
- A pre-loaded sorbent trap is placed within the thermal desorption unit.
- The sorbent trap is heated (approximately 150–250 °C) in a flowing carrier gas stream.
- Compounds thermally desorbed by the rapid heating or the trap and passage of a carrier gas stream.
- Compounds transferred, via heated transfer lines, to the injection port of the GC for analysis.

Figure 14.3 Schematic diagram of the thermal desorption process and its operating procedure.

Table 14.2 Examples of safe sampling volumes for thermal desorption-GC using different sorbents. Data from [4].

Compound	Boiling point (°C)	Vapour pressure kPa (25 °C)	Sorbent: Tenax TA (200 mg) at 20 °C				Sorbent: Chromosorb 106 (300 mg) at 20 °C			
			Retention volume (l)	SSV (l)[a]	SSV per g (l g^{-1})	Desorption temperature (°C)	Retention volume (l)	SSV[a] (l)	SSV per g (l g^{-1})	Desorption temperature (°C)
Benzene	80	10.1	12.5	6.2	31	120	53	26	87	160
Toluene	111	2.9	76	38	90	140	165	770	270	200
Ethylbenzene	136	0.93	360	180	900	145	730	360	1200	250
Hexane	69	16	6.4	3.2	16	110	60	30	100	140
Heptane	98	4.7	34	17	85	130	325	160	530	180
Octane	125	1.4	160	80	390	140	2076	1000	3300	200
Nonane	151	-	1400	700	3500	150	14000	7000	23000	220
Decane	174	-	4200	2100	10000	160	74000	37000	120000	250

[a] SSV = safe sampling volumes.

This could be done as part of an occupational exposure assessment where an individual would wear the sorbent tube close to their airway (within 200 mm of their nose and mouth, i.e. in the breathing zone) and have a portable sampling pump, connected to the sorbent tube, and drawing air through the sorbent. This will be done at a sampling rate of 50–100 ml min^{-1} and with a sampled volume of air of 1–10 l. The portable sampling pump would be located on their person (e.g. in a pocket). Whereas in passive (or diffusive) trapping, a sorbent, e.g. tube, filter paper, or solid-phase microextraction (SPME) fibre (see Chapter 5), is left exposed to the air for a period and the concentration of VOCs on the sorbent calculated. For passive (diffusive) sampling using a sorbent tube, the minimum sampling time will be from a few minutes (e.g. 30 minutes) and up to 8 hours. An effective way to desorb the VOCs from the sorbent is via thermal desorption; other approaches include solvent extraction, e.g. Soxhlet extraction (see Chapter 9).

In active (pumped) sampling, record the time, temperature, flow rate, barometric pressure when the pump is switched on and off. Whereas, in passive (diffusive) sampling record the time, temperature and barometric pressure at the start and end of the sampling period. The sampling sorbent tubes should be securely capped at the end of the active or passive sampling and transported back to the laboratory for analysis.

The term thermal desorption (TD) is used to refer to the removal of VOCs from a solid sorbent by heat and then their introduction, via a heated transfer line, and carrier gas flow into the sample injection port of a gas chromatograph (see Section 16.3). A schematic diagram of a complete TD-GC set-up is shown in Figure 14.3. The important parameters in TD-GC are the temperature of the trap (during the flow rate desorption process) and the flow rate of the carrier gas. In method development for TD-GC, it will be necessary for both temperature and carrier gas flow rate to be optimized for target VOCs. The sample containing sorbent trap is placed inside the thermal desorption unit (TDU). The trap is then desorbed using a purge flow rate (e.g. 20–50 ml min^{-1}) and temperature (e.g. 200–320 °C) for a time (e.g. 3–10 minutes). The desorbed compounds are transferred through a heated transfer line (e.g. 200 °C) directly to the injection port of the GC. After suitable calibration of the TDU-GC instrument, it is possible to calculate the concentration of the detected VOCs [4].

For active (pumped) sampling of air, the concentration of VOC (C), in units of mg m^{-3}, can be calculated as follows:

$$C = (Ms - Mb) / Vs \tag{14.5}$$

where Ms = mass of analyte in sample (µg), Mb = mass of analyte in blanks (µg), and Vs = air volume of sample (l).

If it is necessary to calculate the concentration of the VOC (mg m^{-3}) under specific atmospheric conditions and temperature (C_{corr}), then the following correction can be made:

$$C_{corr} = (C \times T \times 101) / M_w \times 298 \times P \tag{14.6}$$

where T = temperature of sampled air (K), Mw = molecular weight of VOC (g mol^{-1}), and P = barometric pressure of sampled air (kPa).

It is also possible to express the concentration of the VOC as a volume fraction in air (C'), in units of ppm.

$$C' = (C \times 24.5 \times T \times 101) / (M_w \times 298 \times P) \tag{14.7}$$

where 24.5 = molar volume of an ideal gas at 298 K (l mol^{-1}).

For passive (diffusive) sampling of air, the concentration of VOC (C), in units of mg m^{-3}, can be calculated as follows:

$$C = (Ms - Mb)/(U' \times t \times 1000) \tag{14.8}$$

where Ms = mass of analyte in sample (μg), Mb = mass of analyte in blanks (μg), U' = uptake rate (cm^3 min^{-1}), and t = sampling period (minutes).

If it is necessary to calculate the concentration of the VOC (mg m^{-3}) under specific atmospheric conditions and temperature (C$_{corr}$), then the following correction can be made:

$$C_{corr} = (C \times T \times 101)/(298 \times P) \tag{14.9}$$

It is also possible to express the concentration of the VOC as a volume fraction in air (C'), in units of ppm.

$$C' = (Ms - Mb)/U \times t \times 1000) \tag{14.10}$$

where U = uptake rate (ng^{-1} ppm^{-1} min^{-1}).

Note, uptake rates in cm^3 min^{-1} (U') and ng^{-1} ppm^{-1} min^{-1} (U) are related as follows:

$$U' = U \times (24.5 \times T \times 101)/(M_w \times 298 \times P) \tag{14.11}$$

Examples of WELs are shown in Table 14.3 [5]. It is also possible to convert WELs given in ppm to mg m^{-3} as follows:

$$WEL(mg m^{-3}) = (WEL(ppm) \times Mw)/24.05526 \tag{14.12}$$

where 24.05 526 = molar volume of an ideal gas at 293 K and 1 atmosphere pressure (760 mm Hg, 101 325 Pa or 1.01 325 bar) (1 mol^{-1}).

Table 14.3 Example workplace exposure limits (WELs) for selected VOCs. Data from [5].[e]

		Workplace exposure limit			
		Long term[a]		Short term[b]	
Compound	CAS number	ppm	mg m^{-3}	Ppm	mg m^{-3}
Benzene[c,d]	71-43-2	1	3.25		
Toluene[d]	100-88-3	50	191	100	384
Ethylbenzene[d]	100-41-4	100	441	125	552
Hexane	110-54-3	20	72		
Heptane	142-82-5	500	2085		

[a] Long-term exposure limit (8-hour time weighted average reference period).
[b] Short-term exposure limit (15 minutes reference period).
[c] Capable of causing cancer and/or heritable genetic damage.
[d] Can be absorbed through the skin; dermal absorption will lead to systemic toxicity.
[e] A total of 426 substances are currently listed [5] with WELs; they include organic and inorganic chemicals, fumes, dusts and fibres).

14.4 Workplace Exposure Limits

Substances that are assigned a workplace exposure limits (WEL) are subject to the requirement of COSHH (see Section 1.5). These regulations require employers to prevent or control exposure to the hazardous substances. Under COSHH, the term 'control' is defined as adequate only if (i) the principles of good control procedures are applied, (ii) any WEL is not exceeded, and (iii) exposure to asthmagens, carcinogens, and mutagens are reduced to as low as is reasonably practicable. Workplace exposure limits are used in the United Kingdom to assess occupational exposure and are used to assess the risk to workers from hazardous substances in the air. The WELs are used as a time weighted average (TWA), using a long-term time period (8 hours) and a short-term time period (15 minutes). Some examples of the WELs are shown in Table 14.3 [5]. The term '8 hours reference period' relates to the procedure whereby the occupational exposure in any 24 hours period are treated as equivalent to a single uniform exposure for 8 hours, i.e. the 8-hour time weighted average exposure. The 8-hour TWA can be represented as:

$$8-\text{hour TWA} = \left(C_1T_1 + C_2T_2 + C_3T_3 + C_4T_4 + C_5T_5 + C_6T_6 + C_7T_7 + C_8T_8\right)/8 \quad (14.13)$$

where C is the occupational exposure in each hour and T is the associated exposure time in hours in any 24 hours period.

Whereas a short-term exposure limit is used to help prevent effects, e.g. eye irritation, which may occur following exposure to a substance for a few minutes.

14.5 Biological Monitoring

Sometimes air monitoring may not provide a reliable indication of exposure to a substance, so a useful complementary approach is biological monitoring. This involves the measurement and assessment of the hazardous substances (or their metabolites) in tissue, secretions, excreta or expired air, in exposed workers. Examples of biological monitoring guidance values (BMGVs) are shown in Table 14.4. If a BMGV is exceeded, it is an indication that further investigation is required into current control measures, and work practices

Table 14.4 Example biological monitoring guidance values (BMGVs) for selected substances. Data from [5].

Substance	BMGV	Sampling time
Dichloromethane	30 ppm carbon monoxide in end-tidal breath	Post-shift
Lindane	35 nmol l^{-1} (10 µg l^{-1}) of lindane in whole blood (equivalent to 70 nmol l^{-1} of lindane in plasma)	Randon
Polycyclic aromatic hydrocarbons (PAHs)	4 µmol 1-hydroxypyrene/mol creatinine in urine	Post-shift
Xylene (o-, m-, p- or mixed isomers)	650 mmol methyl hippuric acid/mol creatinine in urine	Post-shift

are required. [Note; BMGVs are not an alternate or replacement for airborne occupational exposure limits.]

14.6 Particulate Matter

A variety of methods are available to monitor airborne particulate matter including a filter-based gravimetric sampler, a tapered element oscillating microbalance (TEOM), β-attenuation monitor, or optical analysers. One such filter-based gravimetric approach uses an ultra-high volume particulate sampler (Figure 14.4) fitted with an impactor cutoff of 10 μm with a flow rate of 2001 min^{-1}. A Munktell quartz-microfibre filter is inserted within the sampler and operated as indicated for up to 100 hours. Then, the filter is removed and analysed for the composition of the collected PM_{10} fraction. Schematic diagrams of scanning electron micrographs of a quartz-filter pre- and post-sampling for PM10 in a city centre in NE England are shown in Figure 14.5. Small particulates are clearly visible in Figure 14.5b, as compared to Figure 14.5a. For chromatographic analysis, an appropriate extraction technique would need to be used to recover the organic compounds, alternatively, and for metal analysis, the filter could be analysed using X-ray fluorescence spectrometry for metal contaminants.

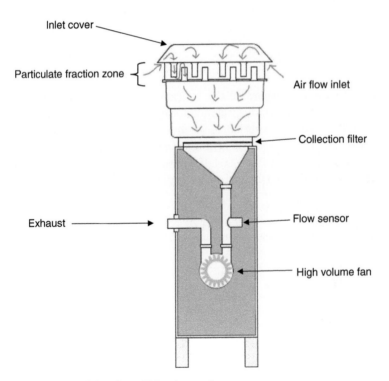

Figure 14.4 High-volume PM_{10} air sampler.

(a)

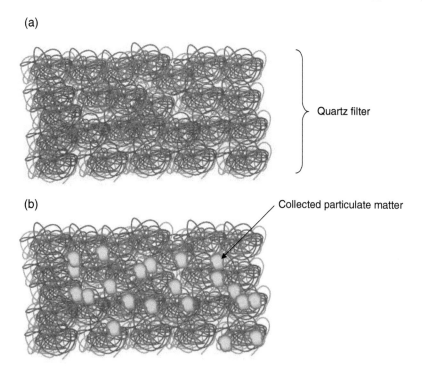

Quartz filter

(b)

Collected particulate matter

Figure 14.5 Schematic diagram of scanning electron microscope (SEM) images of a quartz filter: (a) pre-airborne sample collection and (b) post-airborne sample collection.

14.7 Application of Air Sampling

Case Study A   Calculate the Workplace Exposure Limit for an 8-Hour Exposure

Background: Workplace exposure limits are used to assess the risk to humans from exposure to hazardous substances. The 8-hour WEL is considered as it reflects the working day.

Activity: A scientist works for 6 hours 5 minutes on a process in which they are exposed to a substance hazardous to their health. The average exposure during that period is measured as $0.15\,\text{mg m}^{-3}$, what is the 8-hour TWA?

The 8-hour TWA = 6 hours 5 minutes (6.08 hours) at $0.15\,\text{mg m}^{-3}$ and 1 hour 55 minutes (1.92 hours) at $0\,\text{mg m}^{-3}$.

$$8-\text{hour TWA} = (0.15 \times 6.08) + (0 + 1.92)/8 = 0.114\,\text{mgm}^{-3} \tag{14.14}$$

This calculated 8-hour TWA is then compared to the WEL (e.g. Table 14.3) for the specific substance.

14.8   Summary

A word search of the key terms outlined in this chapter can be found in Appendix A2, with the solution in Appendix B2.

References

1 Lide, D.R. (2005). *CRC Handbook of Chemistry and Physics*, 86e. CRC Press.
2 Partyka, M., Zabiegala, B., Namiesnik, J., and Przyjazny, A. (2007). *Critical Reviews in Analytical Chemistry* 37: 51–78.
3 Schwarzenbach, R.P., Gschwend, P.M., and Imboden, D.M. (1993). *Environmental Organic Chemistry*. New York: Wiley-VCH.
4 Methods for the Determination of Hazardous Substances (MDHS) 104, Health and Safety Executive (2016). Volatile organic compounds in air. http://www.hse.gov.uk/pubns/mdhs/index.htm (accessed 18 October 2021).
5 EH40/2005 (2020). Workplace exposure limits, 4e. Health and Safety Executive, TSO (The Stationery Office), Norwich, UK (ISBN: 978 07176 67338).

Section F

Post-Extraction

15

Pre-Concentration and Associated Sample Extract Procedures

LEARNING OBJECTIVES

After completing this chapter, students should be able to:

- Understanding the principle of why pre-concentration of sample extracts is necessary.
- Be aware of the options for sample extract pre-concentration.
- Be aware of the sample extract clean-up options.

15.1 Introduction

In environmental organic analysis, it is often necessary to either pre-concentrate and/or clean-up the organic extract (post-extraction). Frequently, the extraction method (irrespective of whether it is from a solid or liquid sample) has not concentrated the organic solvent containing extract sufficiently for trace analysis. In these situations, further treatment of the organic solvent extract is required, which involves evaporation of the organic solvent without loss of the organic compounds prior to analysis. Alternatively, the extract (whether concentrated or not) is insufficiently 'clean' to undergo the analysis, e.g. chromatography. The term 'clean' in this context means that the extract still contains matrix components, which could either interfere with the analysis or cause deterioration in the performance of the analytical technique. An obvious issue is contamination of the chromatographic column leading to quantitation problems, e.g. interfering peaks.

15.2 Solvent Evaporation Techniques

Pre-concentration is concerned with the reduction, in volume, of a larger organic solvent extract into a smaller extract. A range of solvent evaporation techniques are available:

- Needle evaporator
- Automated evaporator (TurboVap)
- Rotary evaporator
- Kuderna–Danish evaporative concentrator

Extraction Techniques for Environmental Analysis, First Edition. John R. Dean.
© 2022 John Wiley & Sons Ltd. Published 2022 by John Wiley & Sons Ltd.

In all cases, the evaporation method is slow with the risk of contamination from the solvent, glassware, and the gas used for evaporation high.

15.2.1 Needle Evaporation

The sample extract (e.g. in a 2 ml vial) is placed in an aluminium block and heated (Figure 15.1a). Then, the needle manifold (Figure 15.1b) is lowered over the top of the vial to within 1 cm of the surface of the extract. Then a gentle flow of nitrogen gas is applied to the surface of the extract (within the vial) via the needle. It is necessary to check the evaporation frequently to prevent the extract 'going dry'. Multiple sample extracts, in a range of vial volumes, can be accommodated depending on the capacity of the heating block and number of needles operational.

15.2.2 Automated Evaporator (TurboVap)

A flow of nitrogen (15 psi) is directed onto the sample extract container (e.g. 20–250 ml), which is in a temperature-controlled water bath (e.g. 40 °C). The sample extract, within the glass sample container (Figure 15.2), slowly evaporates until 0.5 ml of extract remains. Using an optical sensor allows the evaporation to stop to that point. Multiple extract solvent evaporation can be done sequentially or simultaneously.

15.2.3 Rotary Evaporation

Prior to sample solvent evaporation, turn on the water bath to the desired temperature (e.g. 40 °C) and the coolant water to the condenser. Sample extract volumes can be done in an either a 500- or 1000-ml round-bottomed flask, which allows large extracts to be evaporated. To start the process, a vacuum is applied to the apparatus (this might be done using water or a pump). The sample solvent is removed under reduced pressure by mechanically

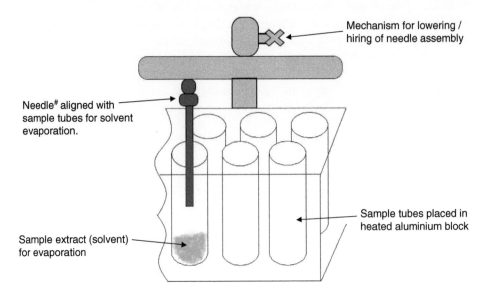

Mechanism for lowering / hiring of needle assembly

Needle# aligned with sample tubes for solvent evaporation.

Sample tubes placed in heated aluminium block

Sample extract (solvent) for evaporation

Figure 15.1 Needle evaporator with an example needle assembly.

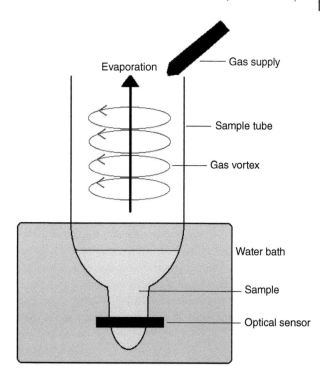

Figure 15.2 Automated evaporator (TurboVap).

rotating the flask containing the sample in the controlled temperature water bath (Figure 15.3). The (waste) solvent is condensed and collected for disposal in the receiver flask. A 'film' of the concentrated extract remains in the original flask. Problems can occur due to loss of volatile analytes, adsorption on to glassware, entrainment of analyte in the solvent vapour, and the uncontrollable evaporation process.

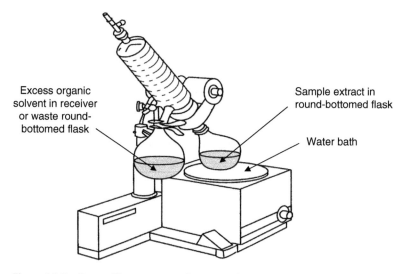

Figure 15.3 Rotary film evaporator (or rotovap).

15.2.4 Kuderna–Danish Evaporative Concentration

The Kuderna–Danish evaporative condenser was developed in the laboratories of Julius Hyman and Co., Denver, Colorado [1, 2]. It consists of an evaporation flask (500 ml) connected at one end to a Snyder column and the other end to a concentrator tube (10 ml) (Figure 15.4). The sample containing organic solvent (200–300 ml) is placed in the apparatus, together with one or two boiling chips, and heated with a water bath. The temperature of the water bath should be maintained at 15–20 °C above the boiling point of the organic solvent. The positioning of the apparatus should allow partial immersion of the concentrator tube in the water bath but also allow the entire lower part of the evaporation flask to be bathed with hot vapour (steam). Solvent vapours then rise and condense within the Snyder column. Each stage of the Snyder column consists of a narrow opening covered by a loose-fitting glass insert. Sufficient pressure needs to be generated by the solvent vapours to force their way through the Snyder column. Initially, a large amount of condensation of these vapours returns to the bottom of the Kuderna–Danish apparatus. In addition to continually washing the organics from the sides of the evaporation flask, the returning condensate also contacts the rising vapours and assists in the process of recondensing volatile organics. This process of solvent distillation concentrates the sample to approximately 1–3 ml in 10–20 minutes. Escaping solvent vapours are recovered using a condenser and collection device. The major disadvantage of this method is that violent solvent eruptions can occur in the apparatus leading to sample losses. Micro-Snyder column systems can be used to reduce the solvent volume still further.

15.2.5 Automated Evaporative Concentration System

Solvent from a pressure-equalized reservoir (500 ml capacity) is introduced, under controlled flow, into a concentration chamber (Figure 15.5) [3]. Glass indentations regulate the boiling of solvent so that bumping does not occur. This reservoir is surrounded by a heater. The solvent reservoir inlet is situated under the level of the heater just above the final concentration chamber. The final concentration chamber is calibrated to 1.0 and 0.5 ml volumes. A distillation column is connected to the concentration chamber. Located near the top of the column are four rows of glass indentations, which serve to increase the surface area. Attached to the top of the column is a solvent recovery condenser with an outlet to collect and hence recover the solvent.

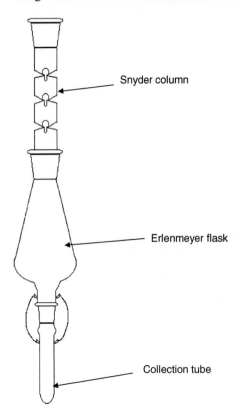

Snyder column

Erlenmeyer flask

Collection tube

Figure 15.4 Kuderna–Danish evaporative concentrator.

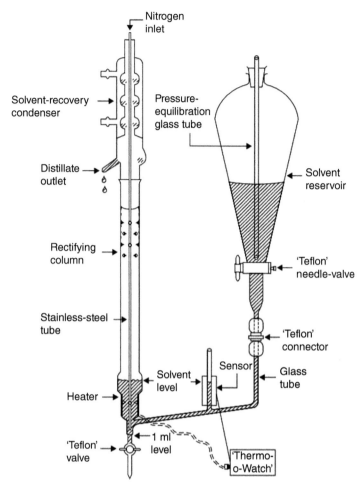

Figure 15.5 Schematic diagram of the automatic evaporative concentration system (EVACS). Based on [3].

To start a sample, the apparatus is operated with 50 ml of high-purity solvent under steady uniform conditions at total reflux for 30 minutes to bring the system to equilibrium. Then the sample is introduced into the large reservoir either as a single volume or over several time intervals. [NOTE: A boiling point difference of approximately 50 °C is required between solvent and compound for the highest recoveries.] The temperature is maintained to allow controlled evaporation. For semi-volatile analytes, this is typically at 5 °C higher than the boiling point of the solvent. The distillate is withdrawn while keeping the reflux ratio as high as possible. During operation, a sensor monitors the level of liquid, allowing heating to be switched off or on automatically (when liquid is present the heat is on and vice versa). After evaporation of the sample below the sensor level, the heating is switched off. After 10 minutes, the nitrogen is started for final concentration from 10 to 1 ml (or less). Mild heat can be applied according to the sensitivity of solvent and compound to thermal decomposition. When the liquid level drops below the tube, stripping nearly stops. The tube is sealed at the bottom, so that the nitrogen is dispersed above the sample and the reduction of volume becomes extremely slow. This prevents the sample from going to dryness even if left for hours. The sample is drained and the column is rinsed with two 0.5 ml aliquots of solvent. Further concentration can take place, if required.

15.3 Post-Extract Evaporation

Subsequently, it is necessary to ensure that all the solvent evaporated extracts are quantitatively transferred to a sample vial for subsequent analysis. Often residues will remain on the extract vial/vessel walls post-evaporation. In this situation, it is necessary to add a small quantity of organic solvent (the choice depends on the selected analytical technique) and manually swirl it around the sides to allow it to settle in the bottom of the sample vial. This process can be aided using a vortex mixer (Figure 15.6) or sonication, in a sonic bath (see Figure 9.6).

15.4 Sample Extract Clean-Up Procedures

A range of clean-up procedures are available and include column chromatography (adsorption, partition, gel permeation, or ion exchange); acid–alkaline partition; acetonitrile–hexane partition; sulphur clean-up; and alkaline decomposition.

15.4.1 Column Chromatography

Adsorption chromatography is often used to separate relatively non-polar organic compounds in environmental sample extracts. Adsorption chromatography separates the extract components according to an equilibrium between the adsorbent (i.e. the stationary phase), the organic compounds (i.e. those under analysis and unwanted matrix

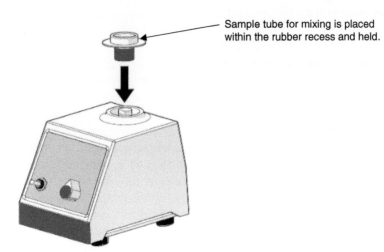

Sample tube for mixing is placed within the rubber recess and held.

Figure 15.6   Vortex mixer.

components), and the eluent (i.e. organic solvent used to elute the organic compounds). Sample extracts are initially concentrated using a procedure outlined in Section 15.4.2. Then, the column is eluted with a range of organic solvents of differing polarity; a non-polar solvent is often used first, e.g. hexane. A range of adsorbents are used for column chromatography including silica gel, florisil, and alumina.

Silica gel: this is weakly acidic, amorphous silica used for clean-up of compounds contain-ing ionic and non-ionic functional groups. To activate the silica gel, heat to 150–160 °C for a few hours prior to use. Silica gel is best if it contains between 3 and 5% w/w water. Beware when using methanol or ethanol as the eluent as issues can arise in the ability of silica gel to function.

Florisil: this is based on magnesium silicate (with an acidic character) and is used for the clean-up of extracts for GC containing pesticides, organochlorine compounds, esters, ketones, phthalic esters, nitrosamines, organophosphate pesticides, as well as ali-phatic and aromatic hydrocarbons. Commercially available Florisil should already be activated, if not it needs to be heated to 667 °C. Also check out batch-to-batch elution solvent consistency. Beware that some pesticides may decompose in ethylether on Florisil.

Alumina: Three types of alumina are available: basic (pH 9–10); neutral; and acidic (pH 4–5). Basic alumina is used for basic and neutral compounds (e.g. alcohols, hydrocarbons, and steroids), which are stable in alkali. Ethyl acetate cannot be used as an eluent, when analysing for esters, as they are unstable in alkali and decom-pose. In addition, acetone cannot be used as an eluent as amidol condensation reac-tions can occur leading to diacetone alcohol formation. Neutral alumina is used for aldehydes, ketones, quinines, and esters while acidic alumina is used for acidic pig-ments (e.g. dyes) or acidic compounds, which are adsorbed by basic or neutral alumina.

15.4.1.1 Partition Chromatography

In reversed-phase column chromatography, the column could be a solid-phase extraction cartridge (see Chapter 4) in which case clean-up can be done using a non-polar stationary phase (e.g. C18) and a polar solvent (e.g. methanol–water). The sample extract is added to the column (in water) and eluted with solvent mixtures (e.g. methanol–water or acetonitrile–water). Reversed-phase chromatography is used for clean-up of polar organic compounds.

15.4.1.2 Gel Permeation Chromatography

In gel permeation chromatography (GPC), sample extracts are separated by molecular size on a column of fixed pore size. Larger molecules elute the quickest. GPC is normally used to remove lipids, proteins, and natural resins from samples.

15.4.1.3 Ion-Exchange Chromatography

Ion-exchange chromatography is used to separate compounds that have fully ionizable functional groups. Sample extracts are added to the column, containing an ion-exchange resin, and eluted using an electrolyte solution.

15.4.2   Acid–Alkaline Partition

Acid–alkaline is used to separate basic, neutral, and acid compounds by adjustment of the pH (of the aqueous sample extract). Phenols can be extracted into organic solvent from an aqueous extract (pH 2), then the phenols are reverse extracted by water (pH 12–13). Finally, the aqueous extract is acidified (<pH 2) and re-extracted by organic solvent. In this situation, only phenols will be extracted. In a similar manner, basic compounds, e.g. amines, can be separated by pH reversal.

15.4.3   Acetonitrile–Hexane Partition

Acetonitrile–hexane partitioning can be used to remove lipids from sample extracts. The compounds of interest partition into acetonitrile, while lipids partition into the hexane phase.

15.4.4   Sulphur Clean-Up

This procedure is used to remove sulphur from the sample extract and is done by addition of copper powder. In Soxhlet extraction, for example, copper powder can be added to the sample in the thimble to affect in situ removal of sulphur.

15.4.5   Alkaline Decomposition

Alkaline decomposition is used to extract organic compounds that are stable in alkaline solution (e.g. PCBs) from biological samples (which contain lipids). The samples are refluxed in an alkaline ethanolic solution, which saponifies the lipids. Then, the extract is extracted using liquid–liquid extraction (see Chapter 3). The addition of salt aids recovery of the target compounds in liquid–liquid extraction.

15.5   Derivatization for Gas Chromatography

Sometimes the organic compounds for GC are not volatile enough or thermally stable at the operating temperatures of the injection port and column oven for analysis. However, it is still possible to analyse these organic compounds by carrying out an additional step of sample pre-treatment, i.e. derivatization. The process of derivatization modifies the functionality of an organic compound to allow separation by GC. The resultant analysis benefits of derivatization are:

- Improved separation of compounds, i.e. resolution (see Section 16.2) and reduced peak tailing (see Section 16.2) of polar compounds; polar compounds contain the following functional groups, e.g. -OH, -COOH, =NH, $-NH_2$, and -SH.
- Improved column efficiency (see Section 16.2).
- Analysis of relatively non-volatile compounds, e.g. higher molecular weight compounds.
- Increased detector sensitivity (see Section 16.3).
- Improved thermal stability of (some) compounds.

The important considerations when selecting a derivatizing agent are:

- The derivatization reaction is 100% complete.
- An appropriate temperature is chosen to allow the reaction to take place; an aluminium heating block is used (Figure 15.7a).
- The selected derivatization reagent will not affect the target organic compound in such a way that any chemical rearrangement or structural alteration takes place.
- The chosen derivatization reagent does not contribute to any loss of the compound during the chemical reaction. Reactions normally take place in a sealed vial, e.g. Reacti-Vial™ (Figure 15.7b).
- The newly derivatized product does not react with the chromatographic column.
- The newly derivatized product does not chemically degrade with storage time.
- Finally, that the newly derivatized product is thermally stable in the GC.

Different derivatizing reagents are available for different organic compounds; the choice of reagent, and its suitability, is dependent upon the organic compounds functional groups. Silylation and acylation are the two main derivatizing reagents (see Section 16.4.6).

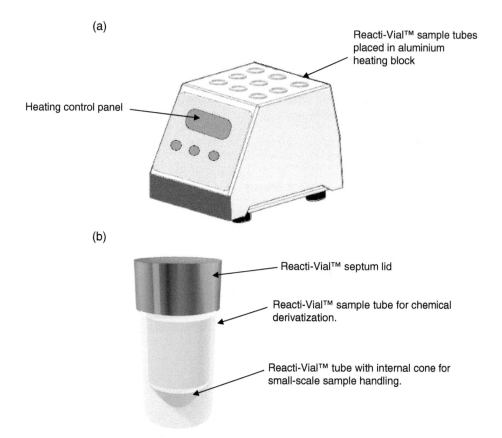

(a)

Reacti-Vial™ sample tubes placed in aluminium heating block

Heating control panel

(b)

Reacti-Vial™ septum lid

Reacti-Vial™ sample tube for chemical derivatization.

Reacti-Vial™ tube with internal cone for small-scale sample handling.

Figure 15.7 Chemical derivatization using (a) aluminium heating block with (b) Reacti-Vial™ glassware.

15.6 Application of Pre-Concentration for Analysis

Case Study A   The Influence of Solvent Evaporation of the Recovery of Polycyclic Aromatic Hydrocarbons

---

[full details are provided in Chapter 10, case study B].

Background: The recovery of PAHs from contaminated soil prior to analysis by gas chromatography-mass spectrometry (GC-MS) was investigated. The influence of alumina (an absorbent) on spiked extract clean-up have been investigated with respect to PAH recovery with and without solvent evaporation.

Experimental

An off-line alumina column (10 g, Sigma-Aldrich, 150 mesh) with a layer of Na_2SO_4 on top was initially washed with 50 ml of hexane, then prior to drying of the column hexane (2 ml) containing the PAH standard (100 µg) was added. The column was again washed with 2 × 15 ml of hexane, prior to elution with approximately 35 ml of dichloromethane. The collected eluant was either analysed directly or evaporated to dryness using needle evaporation, then reconstituted in 2 ml hexane.

Results and Discussion

The results on the influence of 10 PAHs with and without solvent evaporation are shown in Table 15.1. It can be seen that the lightest PAH, i.e. naphthalene, has around 90% lost during the evaporation process. From acenaphthylene to anthracene, less loss of the PAH occurs. Whereas quantitative recovery occurs for the PAHs with four membered rings, i.e. fluoranthene onwards.

Table 15.1   Investigation of solvent evaporation on the recovery of ten polycyclic aromatic hydrocarbons spiked (100 µg) onto an alumina adsorbent.

	% Recovery	
PAH	Without evaporation	With evaporation
Naphthalene	98 ± 1.5	11 ± 8.0
Acenaphthylene	101 ± 2.2	63 ± 2.6
Acenaphthene	105 ± 1.8	77 ± 5.0
Fluorene	101 ± 0.9	83 ± 1.7
Phenanthrene	106 ± 3.8	90 ± 2.3
Anthracene	99 ± 0.8	93 ± 2.1
Fluoranthene	102 ± 3.0	101 ± 4.4
Pyrene	99 ± 3.5	96 ± 6.5
Benzo(a)anthracene	102 ± 1.8	99 ± 2.5
Chrysene	99 ± 1.1	99 ± 2.1

References

1 Karasek, F.W., Clement, R.E., and Sweetman, J.A. (1981). *Anal. Chem.* 53: 1050A.
2 Gunther, F.A., Blinn, R.C., Kolbezen, M.J., and Barkley, J.H. (1951). *Anal. Chem.* 23: 1835.
3 Ibrahim, E.A., Suffet, I.H., and Sakla, A.B. (1987). *Anal. Chem.* 59: 2091.

16

Instrumental Techniques for Environmental Organic Analysis

LEARNING OBJECTIVES
After completing this chapter, students should be able to: • Comprehend the underlying theory of chromatography. • Be aware of the instrumentation and performance of gas chromatography. • Be aware of the instrumentation and performance of high-performance liquid chromatography. • Be able to select the appropriate analytical technique for analysis of extracts.

16.1   Introduction

A wide range of analytical techniques are, and have been used, for the determination of organic compounds in environmental matrices. In this chapter, to be read alongside the knowledge gained in the previous chapters, the operation and function of a wide range of instrumental techniques are described, specifically, those based on chromatography (gas and liquid), as well as other field-based portable analytical techniques.

16.2   Theory of Chromatography

The basic separation process in a column used for chromatography involves interaction between the injected compounds, the carrier gas (or mobile phase) and stationary phase. The generic process of separation is illustrated in Figure 16.1. The compounds in their host organic solvent have been introduced into the column (Figure 16.1a). Under the influence of the carrier gas/mobile phase, the compounds and host organic solvent move through the column. The organic solvent moves much quicker through the column with minimal interactions with the stationary phase (Figure 16.1b). At the same time, and depending upon their physical properties, the compounds interact with the stationary phase for different periods of time and therefore travel in the carrier gas at different rates (Figure 16.1c). The resultant output, the chromatogram, represents the appearance of the organic solvent and compounds, with their appropriate elution times (Figure 16.2). The chromatogram is therefore a plot of the signal (which corresponds to an amount or

Extraction Techniques for Environmental Analysis, First Edition. John R. Dean.
© 2022 John Wiley & Sons Ltd. Published 2022 by John Wiley & Sons Ltd.

(a)

(b)

(c)

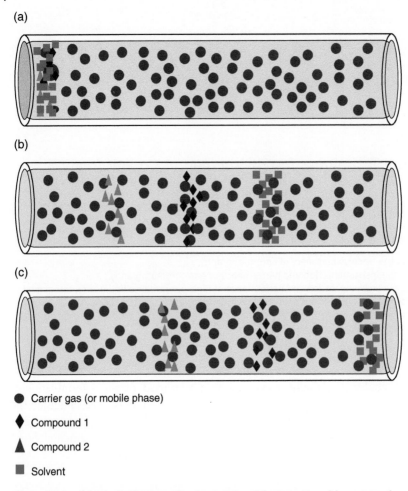

● Carrier gas (or mobile phase)

◆ Compound 1

▲ Compound 2

■ Solvent

Figure 16.1 Schematic diagram of a chromatographic separation: (a) compounds are introduced into column in a solvent, (b) partitioning of compounds between mobile phase (or carrier gas) and stationary phase occurs, (c) solvent reaches end of column.

concentration) of the compounds present as a function of time. Within the chromatogram, it is possible to define some specific terms and measurements (Figure 16.3). The main terms are identified:

- t_0 = the time of elution (minutes) of the unretained compound from the point of sample injection. This sometimes referred to as the column dead time. In practical terms, this is often taken to be the time, from sample injection, when the organic solvent appears in the chromatogram (see Figure 16.3).
- t_r = the time of elution (minutes) of each compound from the point of sample injection to the centre of the peak (its retention time). In the case of the example (see Figure 16.3), two compounds elute from the column; therefore, we can refer to t_{r1} (the time of elution, from the point of injection, of compound 1) and t_{r2} (the time of elution, from the point of injection, of compound 2).

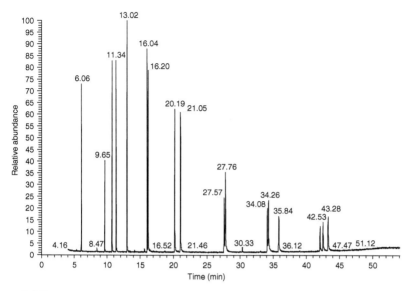

NOTE: This chromatogram is from a gas chromatography-mass spectrometer (GC-MS). You will note that it does not start at 0 minute on the x-axis, but at a (retention) time of 3.5 minutes. This is typical of a GC-MS chromatogram and prevents the detection of the first eluting component, the organic solvent. The organic solvent (in which the compounds have been prepared) will always be in excess compared to the compounds being analysed. It is therefore important to have a 'solvent-window' when operating a GC-MS so that the organic solvent is invisible to the detector.

Figure 16.2 A chromatogram: A plot of signal (relative abundance) (on the y-axis) versus retention time (minutes) (on the x-axis).

- h = the peak height (in units of the y-axis on the chromatogram, e.g. μV). The height of the peak measured from the baseline (i.e. the position with the chromatogram when no compound or solvent is present) to the highest point that the compound attains in the vertical direction.
- A = the peak area (in units that are representative of the y-axis on the chromatogram and the duration of the peak on the x-axis, i.e. time, e.g. μV.seconds or μV.minutes).
- w_b = the width of the peak at its extrapolated base. When a peak is magnified, it is normally observed that some curvature takes place between the baseline and the start of the vertical peak. It is generally accepted that the width of the peak at the base considers this curvature by extrapolating through it. [NOTE: w_{b1} refers to the width of the peak base for compound one and w_{b2} is the width of the peak base for compound 2.]
- $w_{1/2}$ = the width of the peak at half its height. In practical terms, this is done by halving the peak height and measuring the width of the peak at this position.
- $W_{0.6065}$ = the width of the peak at the point of inflection (or curvature as described in w_b) near the peak base. In practical terms, this is done by halving the peak height and measuring the width of the peak at this position.
- k′ = capacity factor (sometimes it is defined by use of a small letter k with a prime, i.e. k′ or simply k. It has no units). To be able to compare the elution time of one compound

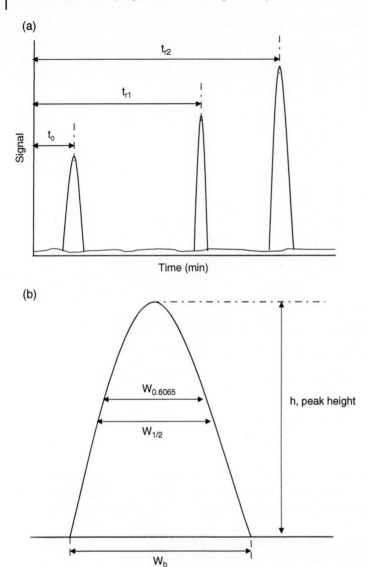

Figure 16.3 Selected chromatographic terms.

between one chromatograph and another (whether in the same laboratory or not) can be done by calculating the capacity factor for that compound. The capacity factor creates a unitless measure of the compounds retention time irrespective of column length or flow rate. It is calculated as follows:

$$k' = \left(t_r - t_o\right)/t_o \tag{16.1}$$

- N = column efficiency. This is calculated using a concept of the number of theoretical plates in a column. Practically, the numerical value provides a measure of the peak narrowness. In principle, therefore, the narrower the peak shape the more peaks (or compounds) can be separated. The number of theoretical plates is therefore a measure of column efficiency (N). It can be determined mathematically in several ways, each of which will provide a number (unitless). The larger the numerical value the more compounds, in theory, can be separated.

[NOTE: The derived number can be compared from column to column provided the same mathematical approach is used. No comparison is possible between alternate mathematical approaches.]

$$N = 16.0 \left(t_r / w_b \right)^2 \tag{16.2}$$

$$N = 5.54 \left(t_r / w_{1/2} \right)^2 \tag{16.3}$$

$$N = 4.0 \left(t_r / w_{0.6065} \right)^2 \tag{16.4}$$

$$N = 2\pi \left(\left(t_r \times h \right) / A \right)^2 \tag{16.5}$$

[NOTE: In practical terms, Eq. (16.5) is the most useful as the relevant information, i.e. retention time (t_r), peak height (h) and peak area (A) are all easily obtained from the chromatographic software data package.]

- L = column length (the dimension needs to be defined in appropriate units, e.g. m, cm or mm).
- HETP = height equivalent to a theoretical plate, expressed as column efficiency (N), in units of mm. In chromatography, different column lengths can be used on different instruments and in different laboratories, therefore, it is possible to normalize the column efficiency, N, by using the term *height equivalent to a theoretical plate*, or HETP (in units of mm).

$$HETP = L / N \tag{16.6}$$

- A_s = asymmetry factor. The plate number assumes that the peak shape is Gaussian (Figure 16.4a), whereas in reality the peak shape can vary due to a range of issues leading to peak fronting (Figure 16.4b) and peak tailing (Figure 16.4c). Peak fronting can be caused by injecting too much sample onto the column, thereby overloading it, whereas

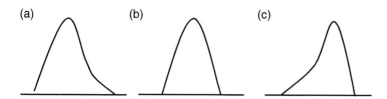

(a) (b) (c)

Figure 16.4   Chromatographic peak shape: (a) peat tailing, (b) Gaussian and (c) peak fronting.

peak tailing is often caused by the compound being separated having too much interaction with the stationary phase. In either case (peak fronting or tailing), it is detrimental to both the ability of the column to separate compounds that elute close to each other and the chromatographic data software package in determining peak area. A measure of the A_s can be done using either Eq. (16.7) at 10% of the peak height and by referring to Figure 16.5a:

$$A_s = b/a \tag{16.7}$$

Or, using either Eq. (16.8) at 5% of peak height and by referring to Figure 16.5b:

$$A_s = (a+b)/2 \times a \tag{16.8}$$

[NOTE: Ideally, the asymmetry factor should have a numerical value (unitless) between 0.9 and 1.2; between these values no issues will arise in the utilization of the peak chromatographic data.]

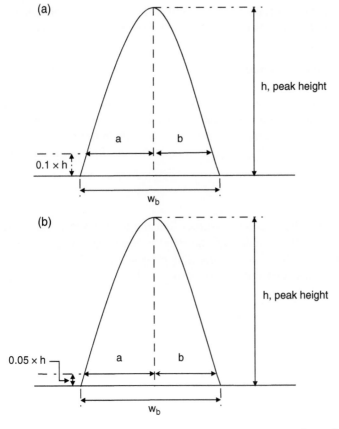

Figure 16.5 Approaches for the estimation of the asymmetry factor. (a) Using the 10% of peak height method and (b) using the 5% of peak height method.

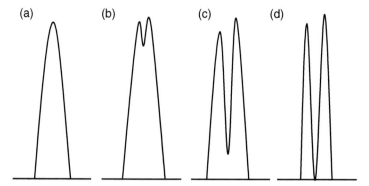

Figure 16.6 An assessment of chromatographic resolution based on the separation of two peaks: (a) no separation, (b) an indication that two peaks are present, (c) integration of two peaks maybe possible using chromatographic software and (d) separation of the two compounds.

- R = a measure of the degree of separation of adjacent compound peaks. Resolution is the ability to separate two adjoining compounds such that their peak bases are distinguishable from each other, i.e. they are separated.

$$R = \left(t_{r2} - t_{r1}\right) / \left(0.5\left(w_{b1} + w_{b2}\right)\right) \tag{16.9}$$

[NOTE: The numerical value for resolution should be >0.9. This allows the chromatographic data software package to be able to distinguish between different compound peaks. An illustration of how values for resolution affect separation is shown in Figure 16.6.]

16.3 Chromatography Detectors: The Essentials

The purpose of the detector is to rapidly respond to a compound passing from the column in either the gas phase (for GC) or liquid phase (for HPLC), then return to its original state and be ready to record the next eluting compound. A range of specific detectors are available for either GC or HPLC. However, all detectors have key performance characteristics:

- Noise: any perturbation of the detector signal not related to an eluting compound is described as detector noise. Ultimately, the presence of this type of signal response will limit the overall sensitivity of the chromatographic system. It can be quantified by determining the average amplitude of the background variation of the baseline in the absence of a known eluting compound.
- Sensitivity: defined as the change in detector signal as a result of the change in concentration (or mass) of an eluting compound. Sensitivity can be calculated by plotting the signal response versus the compound concentration. The slope of the resultant calibration plot is the sensitivity (S).

- Limit of detection: often described as the concentration of compound that produces a signal (e.g. peak area) corresponding to a signal-to-noise (s/n) ratio of 2 (or 3). The LOD can be calculated as follows:

$$LOD = [3 \times N] / [S \times w_{0.5}]$$

(16.10)

where 3 is the proposed basis of the s/n ratio; N = noise; S = sensitivity; and $w_{0.5}$ = peak width at half its height.

- Dynamic range: a measure of the concentration range over which the detector shows an incremental increase in response (signal) for an increase in concentration of the compound. The most useful and significant dynamic range is when the response change occurs in a linear manner, i.e. linear dynamic range. The linear dynamic range for the detector is used to calculate the sensitivity of the detector. An order of magnitude is often applied to dynamic range. One order of magnitude refers to an increasing signal response over, for example, a concentration of between 0.1 and 1.0, i.e. a 10^1 order of magnitude.
- Selectivity: a chromatographic detector can be classified as either selective or universal. In the case of a selective detector, it will produce a heightened response for certain types of atoms in a compound, whereas a universal detector will respond to any eluting compound in the sample.

16.4 Gas Chromatography

A gas chromatograph (Figure 16.7a) consists of the following main instrumental components: a gas supply, sample introduction system, a column located in a temperature-controlled oven, a detector and a PC-controlled data station. Each of these components will now be discussed.

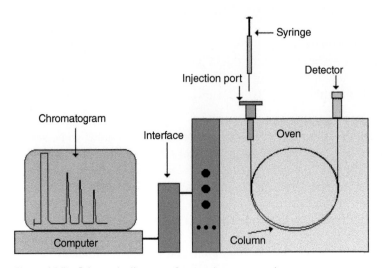

Figure 16.7 Schematic diagram of a gas chromatograph.

16.4.1 Choice of Gas for GC

The choice of carrier gas for a GC is one of the key aspects that ultimately influences the quality of the data obtained via the chromatogram. A comparison of GC performance, i.e. column efficiency, expressed as HETP (see Section 16.3), can be assessed using the van Deemter plot (Figure 16.8). For the best column efficiency, and hence the ability to separate multiple compounds, helium is the carrier gas of choice, whereas nitrogen is often used.

The choice of carrier gas is determined based on two basic principles: availability at a specific cost that is suitable for the analysis or the optimum gas for a specific task that leads to enhanced performance. Normally, the former would result in the use of nitrogen as the carrier gas, particularly when a flame-ionization detector is used (see Section 16.4.5), while the latter would be done using helium when a mass spectrometer is used as the detector (see Section 16.4.5).

[NOTE: an additional gas may be required as fuel for the detector, e.g. hydrogen and air for a flame-ionization detector, see Section 16.4.5.]

As well as the choice of carrier gas, another important quality is its purity. Gas impurities can appear in the resultant chromatogram as either unwanted peaks or result in peak deterioration over time. It is therefore important to use a high-purity carrier gas, e.g. 99.9995% purity. Typical impurities that can occur in the carrier gas are oxygen, water and hydrocarbons. It is possible for impurities (principally oxygen and water) to become entailed with the carrier gas stream downstream of the supply (cylinder or generator) due to miniscule leakages in the connector fittings. One way to reduce their input into the carrier gas stream is to introduce a trap in-line between the carrier gas source and the sample introduction system. Typically, a trap is added in-line and is positioned vertically to prevent channelling with the following: a molecular sieve (to remove moisture), a hydrocarbon trap (to remove hydrocarbons and prevents contamination of the oxygen trap) and an oxygen scrubber (to remove oxygen). The use of electronic pressure devices (EPCs) incorporating mass flow controllers maintains a steady flow of carrier gas. The use of the EPC acts to minimize/ reduce pressure surges as a result of the sample introduction process (see Section 16.4.2) that would lead to chromatogram baseline disturbances and drift. The use of an EPC also compensates for viscosity changes in the carrier gas resulting from the use of temperature programming in the separation process.

Figure 16.8 Van Deemter plot: Influence of carrier gas on column efficiency (expressed as HETP).

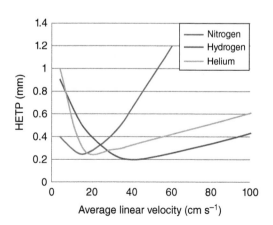

Traditionally, the use of gas cylinders as the source of the carrier gas (and fuel gas) was common. These being purchased or rented from a retail supplier. However, increased safety awareness in laboratories has significantly detracted from locating (multiple) high-pressure (e.g. 2000–3000 psig) cylinders in the laboratory environment. It is now normal practice to use a generator, located either within the laboratory or an adjoining separate room, and connected via pipework, to a GC or a series of GCs. Gas generators are available for nitrogen, hydrogen and air. For helium, however, a cylinder is still the norm, chained to a fitting attached to the interior wall or bench top. Attached to each gas cylinder, or generator, is a regulator that controls the pressure of the gas that is released. In addition, for a cylinder, the regulator also indicates how much gas is left.

16.4.2 Sample Introduction in GC

The most common method of introducing a sample into a GC is using a (precision made) syringe to introduce precisely 1 µl of the sample extract (or calibration standard). The syringe is either operated manually or is incorporated into an autosampler. The syringe introduces the sample extract (or calibration standard) via an independently heated (of the column oven) injection port that allows the volatile components to vapourize, prior to transfer to the column. Various sample injection device options are available.

The split/splitless injector (Figure 16.9) is comprised of a heated chamber containing a glass liner into which the sample is injected through a septum by the syringe. The chamber is heated independently of the column oven. Typically, the injection chamber may be heated to 250 °C, while the column oven may be at 100 °C. The injected sample vapourizes rapidly to form a mixture of carrier gas, solvent vapour and vapourized organic compounds. A portion of this vapour mixture passes onto the column, but the greater volume leaves through a

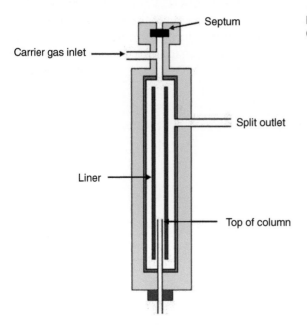

Figure 16.9 Sample injector for GC: a split/splitless injector.

split valve. The ratio of the split flow to the column flow rate is called the split ratio; ratios of 50:1 and 100:1 are common. For example, in a 100:1 split ratio, one part of the injected sample enters the column, while the other 100 parts are vented, via a trap, to waste. A disadvantage of this type of injector is the possibility of discrimination, i.e. production of a chromatogram, which is not truly representative of the actual composition of the mixture.

The on-column injector (Figure 16.10) allows the entire syringe introduced sample extract (or calibration standard) to be introduced directly into the capillary column. Typically, this is done using a special syringe, with a finer needle, that is inserted into the capillary column. On-column injection is a non-vapourizing technique as the sample reaches the column as a liquid. A disadvantage of this type of injector is that the internal surface of the column stationary phase will be damaged by the insertion of the syringe needle unless a retention gap (a short length of silica tubing) is attached to the column.

The programmed temperature vapourization injector (PTV) (or large volume injector) (Figure 16.11) is a combined modified version of the split/splitless and on-column injectors. The sample extract (or calibration standard) is introduced into a cold chamber and is then subjected to rapid heating to affect vapourization of the sample extract. The major advantage of this approach is that the sample volume can be relatively large (up to 250 μl). This large volume injection technique can be used for organic compounds with low concentration in samples. A disadvantage of the PTV is that it requires some method development to allow a reproducible and effective injection.

16.4.3 The GC Oven

The chromatographic column is in an oven (Figure 16.12). The temperature of the oven is controlled accurately and precisely, and its operation is crucial in maintaining

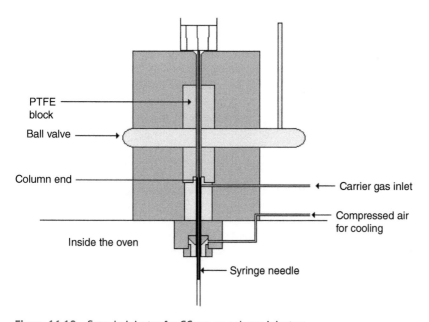

Figure 16.10 Sample injector for GC: an on-column injector.

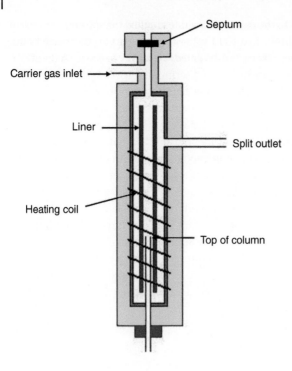

Figure 16.11 Sample injector for GC: a programmed temperature vapourizer or large injection volume injector.

reproducible separation by the column. In addition, the oven must be thermally insulated from both the independently heated injection port (see Section 16.4.2) and the detector and its components (see Section 16.4.5). Typically, the column oven is able to deliver the desired temperature range from ambient (room) temperature up to 400 °C. The column oven can be operated in two modes: isothermal and temperature-programmed GC. In isothermal mode, the column oven maintains a fixed, constant temperature, e.g. 100 °C, for the duration of the chromatographic run. Whereas in temperature-programmed mode, the temperature of the column oven varies throughout the chromatographic run, e.g. 50 °C for 2 minutes followed by a linear temperature gradient at 10 °C

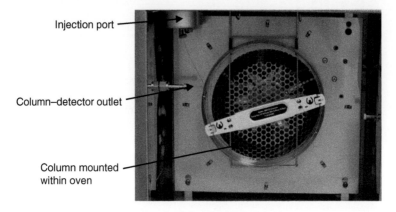

Figure 16.12 The GC oven with *in situ* column.

min^{-1} up to a temperature of 220 °C, with a final hold temperature of 2 minutes. At the end of the temperature-programme run, the column oven can be rapidly cooled allowing the temperature in the oven (and column) to be returned to the starting temperature of 50 °C.

16.4.4 The GC Column

The basic anatomy of a capillary GC column is shown in Figure 16.13. In selecting a column for, either a generic or specific class of compound, separation of four important characteristics need to be considered:

Stationary phase: This is chemically immobilized on the internal surface of a fused silica tube. However, the brittle nature of the fused silica requires that it should be coated in a polymer (i.e. polyimide) that provides rigidity and flexibility to the column, as well as giving the GC column its overall brown colouration. The most used stationary phases are based on polysiloxane (Figure 16.14). It is appropriate to try and match the stationary phase polarity with the polarities of the compounds to be separated, e.g. for non-polar compounds, choose a non-polar stationary phase, i.e. 100% polydimethylsiloxane. The interactions between a non-polar compound and a non-polar stationary phase are mainly governed by van der Waals forces. In contrast, polar compounds and a polar stationary phase are mainly governed by dipole, π-π and/or acid/base interactions. A general 'rule of thumb' is to use the stationary phase, which is least polar to produce the separation required, i.e. satisfactory resolution (see Section 16.2) between neighbouring peaks in the shortest analysis time. A good starting position is to select a DB-1 or DB-5[1] equivalent column. GC column manufacturers produce catalogues, which describe the performance of their different columns with respect to different applications. By comparison of the chromatogram produced by a specific column, under specified operating conditions, it is possible to identify a satisfactory column for a specific application. Manufacturers generally catalogue chromatograms based on the following application

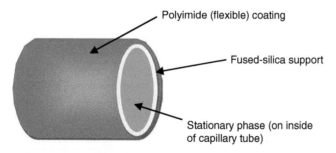

Polyimide (flexible) coating

Fused-silica support

Stationary phase (on inside of capillary tube)

Figure 16.13 A schematic diagram of a capillary GC column.

1 [NOTE: column nomenclature: different manufacturers of GC columns use specific alpha and numeric system designations to identify their brand of column. Fortunately they often retain the same numeric values to allow cross reference from one manufacturer to another. For example, a DB-5 (from J&W) is similar to an HP-5 (from Agilent) as well as an RTX-5 (from Restek), a BP-5 (from SGE), and an SPB-5 (from Supelco)].

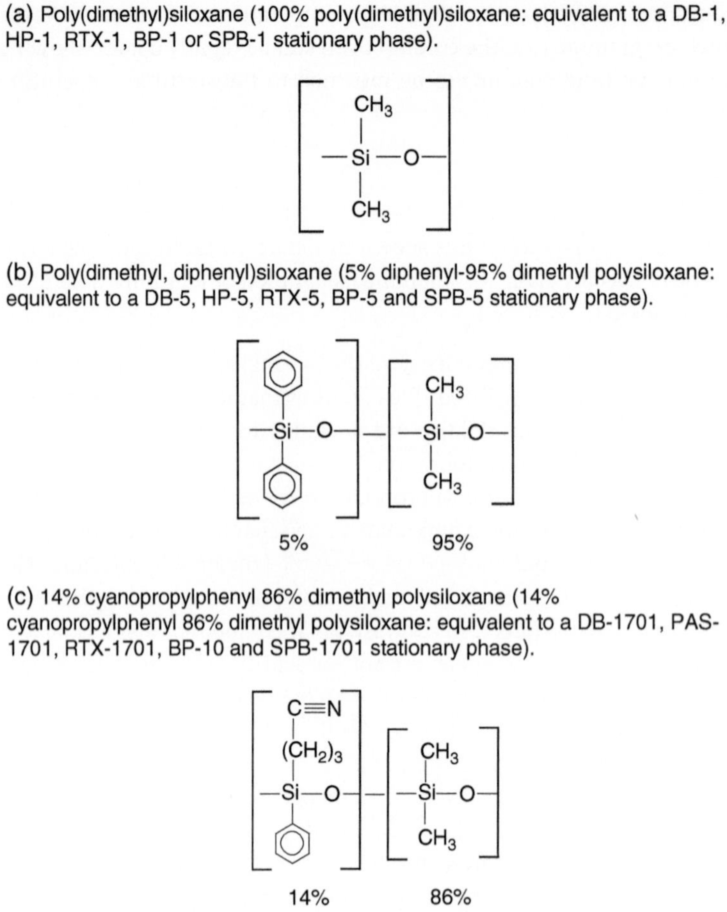

(a) Poly(dimethyl)siloxane (100% poly(dimethyl)siloxane: equivalent to a DB-1, HP-1, RTX-1, BP-1 or SPB-1 stationary phase).

(b) Poly(dimethyl, diphenyl)siloxane (5% diphenyl-95% dimethyl polysiloxane: equivalent to a DB-5, HP-5, RTX-5, BP-5 and SPB-5 stationary phase).

5%   95%

(c) 14% cyanopropylphenyl 86% dimethyl polysiloxane (14% cyanopropylphenyl 86% dimethyl polysiloxane: equivalent to a DB-1701, PAS-1701, RTX-1701, BP-10 and SPB-1701 stationary phase).

14%   86%

Figure 16.14 Examples of three common GC stationary phases: (a) DB-1, (b) DB-5 and (c) DB-1701.

areas: environmental; chemical; food, flavours and fragrances; forensic; and fuels and petrochemicals.

Internal diameter of the column: The internal diameter (i.d.) of a capillary column normally varies between 0.1 and 0.53 mm. Unless a specific application warrants the use of a narrow bore column (e.g. a fast capillary column uses a 0.1 mm i.d. column, or a sample with significantly varying concentrations of its components that requires the use of a >0.25 mm i.d. column to avoid column overload), then a 0.25 mm i.d. column can be used. In general terms, a smaller internal diameter column, e.g. 0.25 mm, will give good resolution of early eluting compounds, but lead to longer analysis times and produce a limited linear dynamic range. In contrast, columns with a larger internal diameter, e.g. 0.53 mm, will result in less resolution for early eluting compounds but allow shorter analysis times with enough resolution for complex mixtures and with a greater linear dynamic range.

Length of the capillary column: This normally varies between 10 and 60 m. Typically, a column length of 30 m will act as a good starting point in developing a separation. For faster analyses, a shorter column may be beneficial provided the compounds are either well separated or few in number. In contrast, a longer column (60 m) may be required when separation of compounds is not possible by using a smaller internal diameter column, using a different stationary phase or altering the column temperature.

Film thickness of the stationary phase: The thickness of the stationary phase (Figure 16.14) normally varies between 0.1 and 5 μm. Typically, by increasing the film thickness, i.e. thickness of the stationary phase, will result in more retention of the compounds, as well as more sample capacity but with an overall lowering in column efficiency (see Section 16.2). In general terms, a thin film thickness is good for separating high boiling point compounds leading to decreased analysis times. In contrast, a thicker film thickness is best for low boiling point compounds resulting in improved resolution of early eluting compounds but with increased overall analysis times. A good starting column for method development would be a film thickness of 0.25 μm.

A typical description of a capillary GC column would be as follows:

$$DB - 530\,m \times 0.25\,mm\,i.d. \times 0.25\,\mu m\,film\,thickness$$

i.e. the manufacturer, as identified by the letters, followed by the number, which identified the stationary phase composition of the polysiloxane, followed by the column length × the internal diameter of the capillary column × the dimensions of the film thickness (i.e. thickness of the stationary phase) as described by the manufacturer and numerical code. In addition, the use of either isothermal or temperature-programmed GC will also influence the separation.

In isothermal analysis, the retention of compounds is more dependent on the column length such that a doubling of column length will double the analysis time, whereas in temperature-programmed mode the retention is more dependent on temperature such that doubling the column length marginally increases analysis time.

16.4.5 GC Detectors

A variety of different detectors can be used in GC. So, as well as the universal flame-ionization detector (FID) and mass spectrometer (MS), others that can be used are the selective detectors: electron capture detector (ECD), nitrogen-phosphorus (or thermionic) detector (NPD) and flame photometric detector (FPD).

Flame-Ionization Detector: The FID is classified as a universal detector and hence will respond to all column eluting compounds; it has an excellent linear dynamic range (up to 10^7 orders of magnitude) and has negligent response to carrier gas impurities such as CO_2 and water. The typical carrier gas for GC-FID is nitrogen, with an additional make-up gas for the detector, e.g. hydrogen. A schematic diagram of an FID is shown in Figure 16.15. It consists of a small hydrogen-air flame located at the end of a jet, which in turn is connected to the end of the chromatographic column. As the eluting compounds exit the column and enter the flame, they will become ionized. The charged species are collected at an electrode producing an increase in electric current proportional to the amount of carbon in the flame (from the eluting organic compound). The resultant electric current is then amplified and recorded as a chromatogram. The FID is the most popular detector for GC.

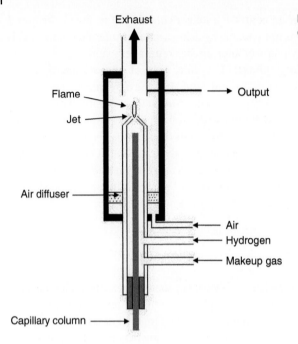

Exhaust

Flame

Jet

Output

Air diffuser

Air
Hydrogen
Makeup gas

Capillary column

Figure 16.15 A flame-ionization detector for gas chromatography.

Mass Spectrometer: The MS provides both quantitative information on the amount of compound present in a sample (as do all other detectors) and identification of an unknown compound by its chemical structure in the form of a mass spectrum. In GC-MS (Figure 16.16), the MS separates ionized compounds based on their mass/charge ratio, while the GC separates unionized compounds. Within the interface between the GC and MS, compounds are ionized; two methods of ionization are possible: electron impact (EI) and chemical ionization (CI). The most popular method is EI ionization due to simpler mass spectra interpretation and requirement for no additional gas to be introduced. In EI ionization, electrons are produced from a heated filament (cathode) (Figure 16.17). As the electrons accelerate towards an anode, they collide with the vapourized sample exiting from the GC column:

$$X_{(g)} + e^- = X^+_{(g)} + 2e^-$$

(16.11)

The generated ions, of specific m/z ratios, are then separated by a mass spectrometer. A range of different mass spectrometers are available for GC. The most common is the quadrupole MS, but others can also be used including the ion trap MS and time-of-flight MS.

Quadrupole Mass Spectrometer: This has four stainless rods located horizontally to each other (Figure 16.16 and insert) such that the same combination of direct current (DC) and radiofrequency (RF) voltages can be applied to opposite rods at the same time. Based on a specific combination of DC/RF voltages an ion with a selected mass/charge (i.e. m/z) ratio will pass through the quadrupole MS and be detected, at that moment all other ions of different m/z ratios are lost. By rapidly altering the combined DC/RF

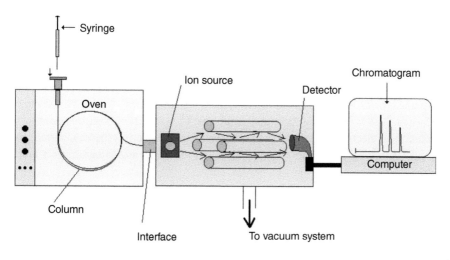

Figure 16.16 A quadrupole mass spectrometer (coupled to a GC) detector for gas chromatography with insert a photograph of a quadrupole mass spectrometer with detector (EMT).

Figure 16.17 A schematic diagram of the electron impact ionization method for GC-MS.

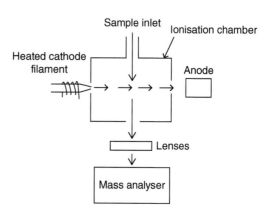

voltages allows ions of different m/z ratios to pass through the mass spectrometer and be detected. For GC-MS, the typical mass range required may extend up to 600 amu.

Detection in Mass Spectrometry: The MS separated ions are detected using an electron multiplier tube (EMT) (Figure 16.18). The ion, of a specified m/z ratio, strikes the surface of a semiconductor where it is converted to an electron. Each electron generated is then cascaded towards an anode. On the way, however, an electron will strike the internal surface of the EMT creating additional electrons. The cascade of electrons generated is collected as an electric current at the anode; the electric current is then converted to a signal and visualized using appropriate software as either a chromatogram or a mass spectrum.

Data Acquisition in Mass Spectrometry: Two modes of operation are possible for the MS (Figure 16.19). In the first all ions, from 0 to600 amu, are monitored in a rapid scanning mode (i.e. full scan or total ion current mode). In total ion current (TIC) mode it is possible to generate a mass spectrum for any eluting compound in the chromatogram. The generated mass spectrum can then be compared to either the mass spectrum generated for the suspected same compound purchased as an authentic standard, from a recognized supplier, or by comparing the generated mass spectrum with a computer-based database of mass spectra. However, once a compound (or range of compounds) has been identified, it is possible for the MS to be operated in single (or sequential)-ion

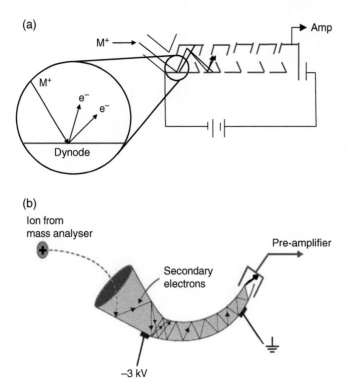

Figure 16.18 Detectors for mass spectrometry: (a) a discrete dynode electron multiplier tube (EMT): mode of operation, and (b) a continuous dynode (or channel) EMT: mode of operation.

(a)

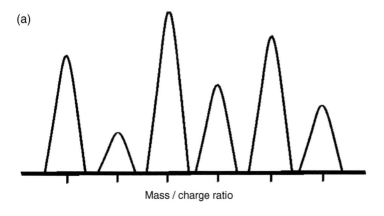

Mass / charge ratio

(b)

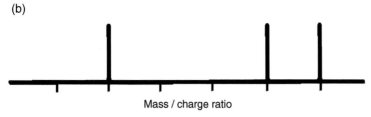

Mass / charge ratio

In full scan mode all ions are monitored by rapid scanning of all mass/charge ratios, whereas in single ion monitoring mode only the identified ions are monitored (e.g. the 3 identified m/z ions).

Figure 16.19 Data acquisition in mass spectrometry: (a) full scan (total ion current mode) and (b) single-(or sequential) ion mode.

mode (SIM). In this mode, it is not possible to obtain a mass spectrum for any eluting compound. However, signal enhancement is evident allowing lower limits of detection to be obtained for the identified compounds. In SIM mode, key ions are selected, characteristic of the compounds separated, in TIC mode first. [NOTE: the same ion (m/z ratio) can be selected for more than one compound. This is because they are eluting from the column at different times.] For example, the ion at m/z ratio, 77 amu is characteristic of C_6H_5 (i.e. a monosubstituted benzene ring), i.e. using the atomic weights of 12C and 1H, which results in $12 \times 6 = 72$ amu plus $1 \times 5 = 5$ amu resulting in a total of 77 amu. Then, instead of the MS rapidly scanning all m/z ratios up to 600 amu in TIC mode, it now spends longer monitoring 77 amu. This results in an enhanced signal in SIM mode.

16.4.6 Compound Derivatization for GC

In some cases, the organic compounds under investigation are not volatile enough to be vapourized in their current chemical form for GC. In this situation, compound derivatization can be applied. Derivatization is used to modify the functionality of an organic compound so that it can be analysed (by GC). Derivatization is normally applied to a compound that has a low volatility and is thermally labile. The most common derivatizing reagents are silylation and

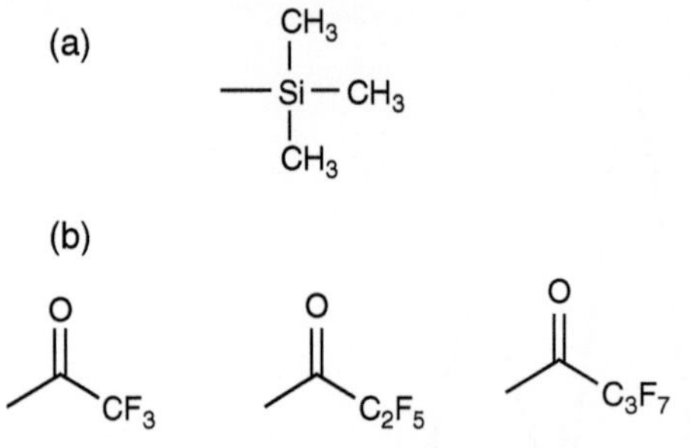

(a)

$$-Si \begin{matrix} CH_3 \\ | \\ -CH_3 \\ | \\ CH_3 \end{matrix}$$

(b)

Figure 16.20 Common derivatization reagents for low volatility or thermally labile compounds in GC (a) silylation using the trimethyl-silyl (TMS) group or (b) acylation using the acyl group.

acylation (Figure 16.20). Silylation involves the addition of either a trimethyl-silyl group or a t-butyldimethylsilyl group to the organic compound using a specific silyating reagent, e.g. N, O-bistrimethylsilyl-acetamide (BSA), N,O-bistrimethylsilyl-trifluoroacetamide (BSTFA), N-methyl-N-trimethylsilyl-trifluoroacetamide (MSTFA) and N-trimethylsilyimidazole (TMSI). In acylation, the addition of acyl derivatives or acid anhydrides is done using a range of acylating reagents including trifluoroacetic acid (TFAA), pentafluoropropionic acid anhydride (PFPA) and heptafluorobutyric acid anhydride (HFBA).

16.5 High-Performance Liquid Chromatography

A high-performance liquid chromatograph (HPLC) is shown in Figure 16.21. It consists of a reciprocating piston pump, injection valve, column located in a temperature-controlled oven and a detector.

16.5.1 The Mobile Phase in HPLC

The mobile phase, typically acetonitrile and water or methanol and water, is used to transport the non-volatile sample through the system. It is stored, on the instrument, in either single solvent or mixed solvent, glass bottles. Occasionally, the water component of the mobile phase may be acidified (e.g. 1% acetic acid or formic acid) or a buffer added. This allows the separated compounds to remain unionized (and charge neutral). The chemical purity of the solvents (acetonitrile, methanol and water) can also be an issue, so that HPLC-grade solvents are required. However, the term HPLC-grade solvents means that any impurities in the solvent are 'invisible' to a UV-Visible detector (i.e. they contain no chromophore compounds). The mobile phase is transported via narrow bore tubing (e.g. stainless steel or PEEK) by a reciprocating piston pump, which is capable of delivering a consistent, and pulseless, flow of mobile phase at a typical flow rate of $1\,ml\,min^{-1}$. The mobile phase is filtered prior to use to remove air bubbles, which can cause cavitation within the pump, and to ensure it is particle-free.

The HPLC can be operated in two modes: isocratic (i.e. a fixed, constant, mobile-phase composition during the HPLC run) (Figure 16.21a) or gradient (i.e. the mobile phase composition varies during the HPLC run) (Figure 16.21b). A typical isocratic mobile phase may consist of a mixture of 45:55%, v/v acetonitrile:water located within one glass bottle on the instrument. Whereas a gradient mobile phase will have individual glass bottles of solvents (e.g. water and acetonitrile), which can be programmed to deliver the mobile phase as follows: 75:25%, v/v water:acetonitrile for 2 minutes hold, then to 25:75%, v/v water:acetonitrile at a gradient of 10 minutes; finally a hold of 2 minutes at 25:75%, v/v water:acetonitrile, with a total chromatographic run time, in this example, of 14 minutes. At the end of the gradient programme, a

(a)

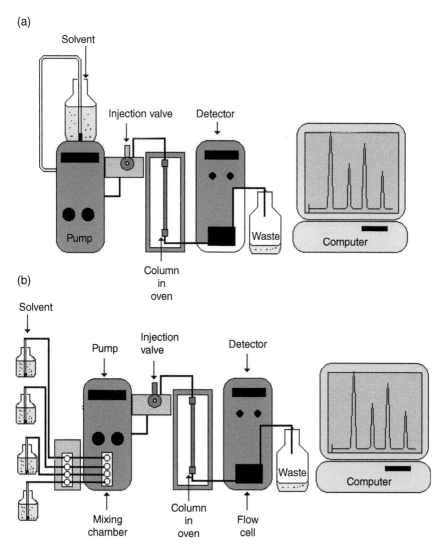

(b)

Figure 16.21 Schematic diagram of a high-performance liquid chromatograph: (a) isocratic system and (b) gradient system.

time is required (a few minutes) to allow the mobile-phase composition to return to its original composition prior to injection of the next sample extract or calibration standard.

16.5.2 Sample Introduction in HPLC

The sample extract (or calibration standard), which consists of organic compounds in a solvent (often like the mobile phase), is introduced using a 'blunt' stainless steel needle syringe, either manually (by hand) or via an autosampler. The syringe allows the sample extract (or calibration standard) to be introduced into the flowing mobile phase via an injection (Rheodyne or 6-port) valve. The injection valve (Figure 16.22) allows a discrete, fixed volume of sample extract (or calibration standard) to be introduced reproducibly into the flowing mobile phase. In the sample load position (Figure 16.22a), the mobile phase is continuously, and uninterruptedly, pumped from the solvent bottle to the column, at the same time the sample extract (or calibration standard)-loaded syringe is inserted into the valve and fills a loop. The loop has a fixed volume, typically in the range 5–50 μl. Then, in the sample inject position (Figure 16.22b), the injection valve is rotated either manually (by hand) or via an automated system, such that the mobile phase is diverted via the loop, which allows the sample extract (or calibration standard) to be introduced into the flow to the column without any interruption.

16.5.3 The HPLC Column

The basic anatomy of an HPLC column is shown in Figure 16.23. In selecting a column for separation of compounds requires consideration of some important characteristics:

Stationary phase: Separation is based on the partitioning of non-volatile compounds with the stationary phase. The stationary phase is chemically bonded to the external surface of a silica particle support. The coated silica particles are located within the physical restraints of a stainless steel column (Figure 16.23). A range of different stationary phases are available. For example, a typical generic stationary phase of a C_{18} hydrocarbon chain (or octadecylsilane, ODS) bonded to silica particles of 3–10 μm in diameter (Figure 16.24). By comparison of the chromatogram produced by a specific column, under specified operating conditions, it is possible to identify a satisfactory column for a specific application.

Dimensions of the column: In physical terms, the column may range in length between 1 and 25 cm (e.g. 10 cm long), with an internal diameter ranging between 1.0 and 4.6 mm (e.g. 4.6 mm internal diameter).

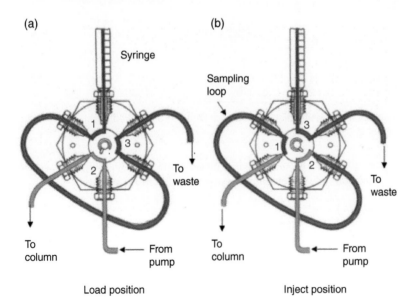

Figure 16.22 A schematic diagram of a 6-port injection valve: (a) sample loading and (b) sample injection.

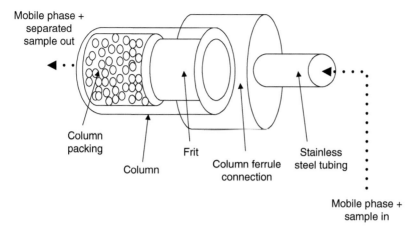

Figure 16.23 A schematic diagram of a typical HPLC column.

Silica particle size: The particle size (of the C_{18} loaded silica particles) can vary between 3 and 10 μm (e.g. 3 μm). The smaller the particle size the better the column efficiency (see Section 16.2).

A typical description of an HPLC column would be as follows:

$$10\,cm \times 0.46\,mm\,i.d. \times 5\,\mu m\,ODS$$

i.e. the column length × the internal diameter of the column × the silica particle size and the stationary phase.

Variation in the physical dimensions and choice of stationary phase will dramatically affect the separation of the compounds. As noted in Section 16.5.1, an HPLC can be operated in either isocratic or gradient HPLC mode. Therefore, it is always a good idea to consult with the column's manufacturer's literature on the capabilities of specific columns for the task required, prior to purchase.

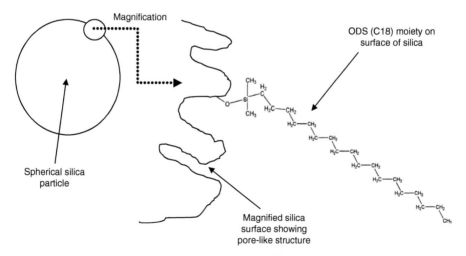

Figure 16.24 An illustration of an HPLC column stationary phase (i.e. octadecylsilane or C18).

In addition, HPLC can be performed in two different formats: reversed-phase HPLC and normal-phase HPLC. As the dominant form is reversed-phase HPLC, the discussion so far has focused on this. However, it is important to appreciate the differences: in reversed-phase HPLC, the stationary phase is non-polar (e.g. C_{18}) and the mobile phase is polar (e.g. methanol–water), whereas in normal-phase HPLC, the stationary phase is polar (e.g. silica) and the mobile phase is non-polar (e.g. isopropanol–heptane).

16.5.4 Detectors for HPLC

A variety of different detectors can be used in HPLC. So, as well as the ultraviolet-visible spectrometer detector (UV/Vis.) and mass spectrometer (MS), others that can be used are the fluorescence detector (FD), and refractive index (RI) detector.

Ultraviolet-Visible Spectrometer: This detector is available in a variety of different formats that include single wavelength, variable (and often programmable) wavelength (Figure 16.25a) or a photodiode array (PDA)/diode array detector (DAD) (that allows multiple wavelength detection) (Figure 16.25b). For visible light measurement, a tungsten lamp is used (380–900 nm), whereas for ultraviolet detection, a deuterium lamp is required (190–380 nm). A key component of the UV/Vis. detector is the flow cell (Figure 16.25c), which has both a low volume (e.g. 8 µl) to allow the separated compounds to retain their discreteness, as well as maintaining a relatively large path length (i.e. 1 cm). The latter maximizes the signal (Absorbance) according to the Beer-Lambert Law, i.e. $A = \varepsilon. c. L$, where A is the absorbance (signal), ε is an extinction coefficient (in appropriate units), c is the concentration (in appropriate units) of the compound and L is the path length of the light through the detector (i.e. 1 cm). The UV/Vis. is the most popular detector for HPLC.

Mass Spectrometer: As in GC (see Section 16.4), the HPLC separates compounds and a mass spectrometer separates the ion of a compound based on its m/z ratio. Therefore, an ion source is required. The function of the ion source is to convert the separated compound into an ion. The most common ion sources for HPLC-MS are electrospray ionization (ESI) and atmospheric pressure chemical ionization (APCI).

In ESI (Figure 16.26a), the mobile phase is pumped through a stainless steel capillary tube, held at a potential of between 3–5 kV. This results in the mobile phase being sprayed from the exit of the capillary tube. As a result of this spraying action, highly charged solvent and solute ions are produced in the form of droplets. By applying a continuous flow of nitrogen carrier gas allows the solvent to evaporate, leading to the formation of solute ions. The resultant ions are transported in to the high-vacuum system of the MS via a sample-skimmer arrangement (often positioned at a right-angle to the sample-skimmer arrangement). By forming a potential gradient between the electrospray and nozzle allows the generated ions to be 'pulled' into the MS while allowing some discrimination between the solute ions and unwanted salts (e.g. from the buffer in the mobile phase).

In APCI (Figure 16.26b), the voltage is applied to a corona pin located in front of (but not in contact with) the stainless steel capillary tubing through which mobile phase from the HPLC passes. To aid the process, the stainless steel capillary tube is heated and surrounded by a coaxial flow of nitrogen gas. The interaction of the nitrogen gas and the mobile phase results in the formation of an aerosol. This aerosol is desolvated due to the presence of the heated nitrogen gas. A voltage (2.5–3 kV) is applied to the corona pin resulting in the

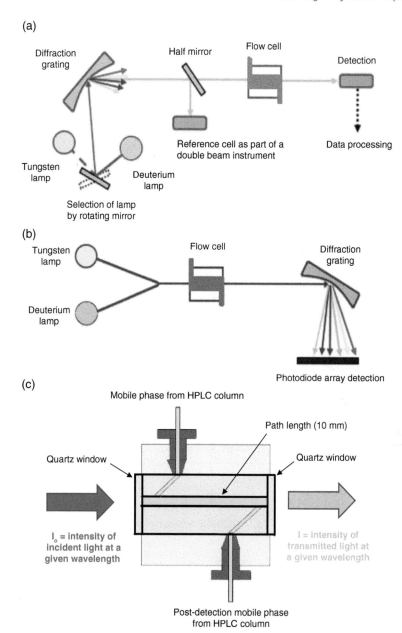

Note: $A = \log_{10} (I_o / I) = \varepsilon \cdot c \cdot L$

Where A is the absorbance, ε is the extinction coefficient, c is the concentration of compound and L is the path length.

Figure 16.25 Schematic diagram of ultraviolet – visible spectrometer for HPLC detection: (a) UV/Visible spectrometer for single or variable wavelength, (b) diode array detector (DAD)/photodiode array (PDA) detector for HPLC and (c) close-up of flow cell of spectrometer for HPLC.

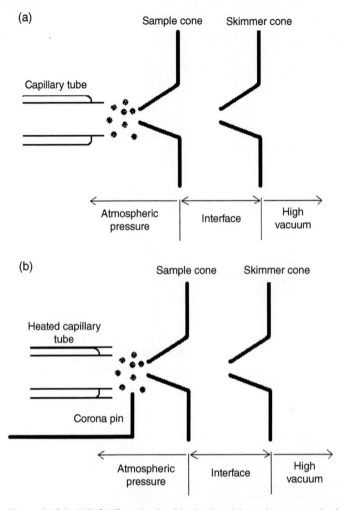

Figure 16.26 HPLC-MS methods of ionization: (a) an electrospray ionization source and (b) an atmospheric pressure chemical ionization source.

formation of a plasma (an ionized gas). The generated ions are transported into the high vacuum system of the MS as described above for ESI.

In ESI or APCI, the molecules form singly charged ions by the loss or gain of a proton (hydrogen atom), e.g. $[M+1]^+$ or $[M-1]^-$, where M = molecular weight of the compound. This allows the MS to operate in either positive ion mode or negative ion mode. In positive ion mode, peaks at mass/charge ratios of $M+1$, i.e. basic compounds, e.g. amines, can be detected, whereas in negative ion mode, peaks at mass/charge ratios of $M-1$, i.e. acidic compounds, e.g. carboxylic acids, can be detected. This process can be complicated in the presence of buffers, e.g. as part of the mobile phase, when adducts can form. For example, in an ammonium buffer solution (a component of the mobile phase), an adduct can form that has a m/z of $M+18$. Similarly, in a sodium buffer solution, an adduct can form that has an m/z of $M+23$.

A range of different mass spectrometers are available including those based on a quadrupole, ion-trap or time-of-flight arrangement (Figure 16.27). Once the ions (mass/

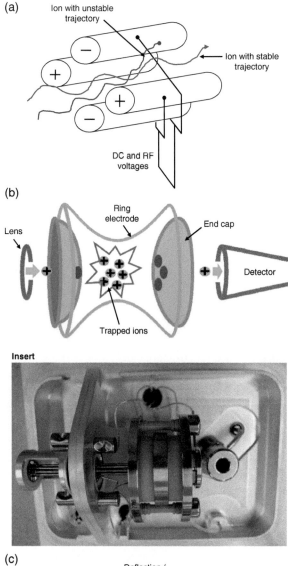

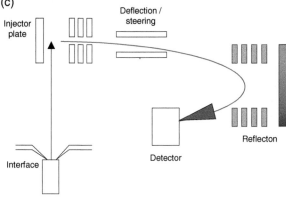

Figure 16.27 Typical mass spectrometers for HPLC: (a) a quadrupole MS, (b) an ion-trap MS with insert a photograph of an ion-trap MS with detector and pre-separation using an octapole MS arrangement and (c) a time-of-flight MS.

charge) are separated by the mass spectrometer, they are detected using an EMT (Figure 16.18). For specific details on the mode of operation of an EMT, see the description in Section 16.4.5. Similarly, the description of the data acquisition for MS is in Section 16.4.5.

16.6 Other Techniques for Environmental Organic Analysis

The chromatography techniques outlined are capable of determining quantitatively the amount of the compound present in a sample. However, other techniques can be used to identify chemical structure information on the organic compounds and include mass spectrometry (MS), nuclear magnetic resonance (NMR) spectroscopy and infrared (IR) spectroscopy. The dominant technique that is used is mass spectrometry, which is often used as the detector for both GC and HPLC (see Sections 16.4 and 16.5) and allows both molecular weight and structural elucidation information to be obtained.

16.6.1 Infrared Spectroscopy

Infrared (IR) spectroscopy is predominantly used for structure elucidation. Infrared spectroscopy is concerned with the energy changes involved in the stretching and bending of covalent bonds in organic compounds. Spectra are represented in terms of a plot of percentage transmittance versus wavenumber (cm^{-1}). In its most common form, infrared spectroscopy is used as a Fourier transformation, a procedure for inter-converting frequency functions and time or distance functions. Fourier transform IR (FTIR) spectra can be obtained of solid, liquid or gaseous samples by the use of an appropriate sample cell. Spectra of liquid samples are normally obtained by placing the pure, dry sample between two sodium chloride disks (plates) and placing it in the path of IR radiation. Solid samples can be prepared as either a Nujol® mull (finely ground solid is mixed with a liquid paraffin) and placed between two sodium chloride disks or a KBr disk (finely ground powder is mixed with potassium bromide and pressed as a pellet). A more recent development, to simplify sample preparation, is the use of an attenuated total reflectance (ATR) diamond anvil sample cell. It consists of a ZnSe focusing element with a small circular diamond crystal mounted on in a stainless steel top plate. The sample (solid, liquid) is placed on the diamond and the IR spectrum obtained.

However, infrared spectroscopy can be used for quantitative analysis of environmental compounds, e.g. petroleum hydrocarbons. The sample is placed in a solution cell, which in turn is located in the path of the IR radiation. For example, the analysis of BTEX in a suitable extract can be determined by observing the IR spectrum at approximately $3000\,cm^{-1}$ (C–H stretching frequencies occur at $>3000\,cm^{-1}$ in unsaturated systems while at $<3000\,cm^{-1}$ C–H stretching frequencies occur for CH_3, CH_2 and CH in saturated systems). By recording the percentage transmittance (signal) for a range of standards and plotting a graph of concentration versus signal, unknown concentrations can be determined. The choice of a suitable solvent is crucial to the use of IR as a quantitative technique. As the FTIR spectra will record signals for C–H bonds in the solvent, it is necessary to use a solvent without any C–H component. One such solvent could be the use of supercritical CO_2 (see Chapter 13) or tetrachloroethene (or perchloroethylene) C_2Cl_4].

16.6.2 Nuclear Magnetic Resonance Spectrometry

Nuclear magnetic resonance (NMR) spectrometry is predominantly used for structure elucidation. NMR spectrometry is concerned with the energy changes involved in atomic nuclei subjected to a magnetic field (generated from radio frequencies in the range 60–600 MHz). The typical magnetic nuclei used to generate spectra from a compound are ^{1}H and ^{13}C. Spectra are represented in terms of a plot of absorption of energy (resonance) versus chemical shift (expressed against an internal standard, based on tetramethylsilane (TMS) (ppm). NMR spectra are obtained of solid or liquid samples by dissolution (or mixing) in a solvent, which does not contain atoms of the nuclei being observed, e.g. $CDCl_3$, deuterated chloroform, the exception to this is when obtaining ^{13}C-NMR spectra. The now deuterated sample solution, filtered, is placed in an NMR tube for analysis. The structure of a compound can be interpretation based on the chemical shift, integration and coupling. The chemical shift is the relative peak positions of the protons (if 1H NMR) to TMS of the compound in their environment (different protons will have different chemical shifts). Integration indicates a relative size of the peak area, depending upon how many protons have the same chemical shift. Coupling is the fine structure on each peak indicating the number of protons on adjacent atoms.

For further information about IR, NMR and MS, please consult the further reading section at the end of the chapter.

16.6.3 Portable Techniques for Field Measurements

Often normal practice is for samples to come to a laboratory for preparation and analysis. But some analytical techniques have a hand-held portable option with batteries for power. While the results may not always be as precise or accurate as those obtained with the more sophisticated laboratory-based instrumentation, they can provide a more 'instant' response that can lead to further sampling and analysis. A range of physical or chemical responses can be obtained using portable meter devices, e.g. a pH meter for water or soil analyses to indicate acidity or alkalinity, as well as dissolved oxygen content for information on water quality and sustaining of aquatic life that may be affected, for example, by blue-green algae. The range of portable analytical instrumentation types is wide and varied and includes FTIR spectroscopy, Raman spectroscopy and NMR for structural identification, as well as quantitative analyses using GC-MS and a photoionization detector. Examples are shown in Figure 16.28 (please note portable/handheld instruments are available for other suppliers: no endorsement of products is inferred. The examples are for illustration purposes only).

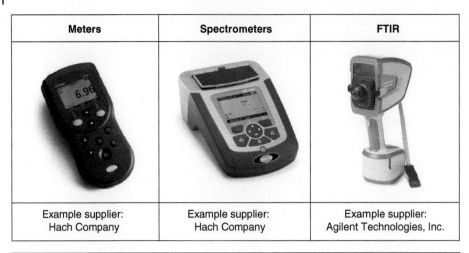

Meters	Spectrometers	FTIR
Example supplier: Hach Company	Example supplier: Hach Company	Example supplier: Agilent Technologies, Inc.

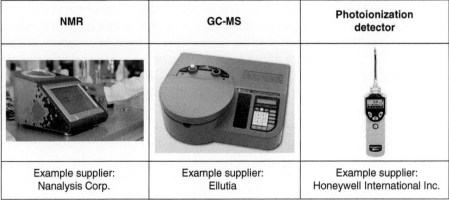

NMR	GC-MS	Photoionization detector
Example supplier: Nanalysis Corp.	Example supplier: Ellutia	Example supplier: Honeywell International Inc.

Figure 16.28 Examples of portable/handheld analytical instruments.

16.7 Applications of Chromatography in Environmental Analysis

Case Study A   Analysis of Polycyclic Aromatic Hydrocarbons by Gas Chromatography Mass Spectrometry – A Comparison of GC Injection Mode

Background: Polycyclic aromatic hydrocarbons are ubiquitous pollutants in the environment. The United States Environmental Protection Agency (US EPA) has identified 16 priority pollutant compounds (Figure 16.29) for analysis.

Experimental
A PAH standard solution was obtained from Thames Restek UK Ltd., Buckinghamshire, UK (2000 µg ml^{-1} in dichloromethane). 4,4′-Difluorobiphenyl (internal standard) was obtained from Sigma-Aldrich Ltd., Dorset, UK. All the solvents were analytical reagent grade and obtained from Fisher Scientific Ltd. (Loughborough, UK).

Instrumentation
The GC-MS instrument was a Trace GC Ultra coupled with a Polaris Q Ion trap MS (Thermo Scientific, UK) and a Triplus autosampler injector. The system was controlled

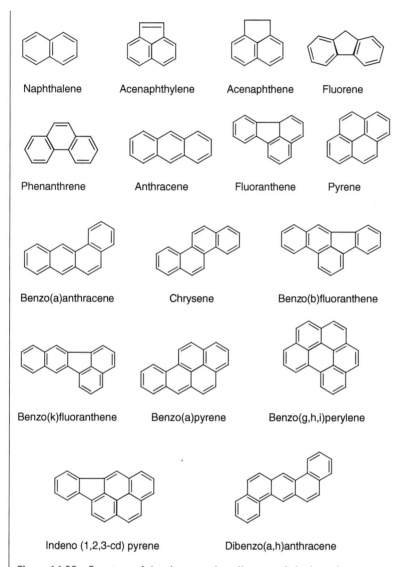

Naphthalene   Acenaphthylene   Acenaphthene   Fluorene

Phenanthrene   Anthracene   Fluoranthene   Pyrene

Benzo(a)anthracene   Chrysene   Benzo(b)fluoranthene

Benzo(k)fluoranthene   Benzo(a)pyrene   Benzo(g,h,i)perylene

Indeno (1,2,3-cd) pyrene   Dibenzo(a,h)anthracene

Figure 16.29 Structure of the sixteen polycyclic aromatic hydrocarbons.

from a PC with Xcalibur software. Separation was performed using a capillary column Rtx®-5MS (5% diphenyl-95% dimethylpolysiloxane, 30 m × 0.25 mm ID × 0.25 μm) supplied from Thames Restek UK Ltd. The temperature programme started at 70 °C for 2 minutes and then 7 °C min^{-1} until 180 °C, then 3 °C min^{-1} until 280 °C, then hold for 3 minutes. Carrier gas flow: 1.5 ml min^{-1}; split flow: 15 ml min^{-1}; split ratio: 10; injector temperature: 280 °C; ion source temperature: 270 °C; start time: 4 minutes; scan mode: selected-ion monitoring; damping gas flow: 0.3 ml min^{-11}. The transfer line temperature was fixed at 300 °C. An injector was used in split mode; alternatively, a PTV (Programme Temperature Vaporizer) with large volume injection was used. The PTV was operated with the following temperature program: injection time: 2 minutes; flow rate 15 ml min^{-1}; no evaporation and cleaning phase; transfer rate 35 °C (for 2 minutes), then

14.5 °C sec^{-1} until 350 °C (then hold for 4 minutes). Finally, the PTV was operated in split mode with a split flow of 15 ml min^{-1} and a carrier gas flow rate of 0.6 ml min^{-1}.

Results and Discussion

Two example chromatograms for the analysis of the 16 PAHs are shown in Figure 16.30. Figure 16.30a shows the chromatogram using the PTV injector for a 0.5 µg ml^{-1} PAH standard, while Figure 16.30b shows the chromatogram using a standard split/splitless

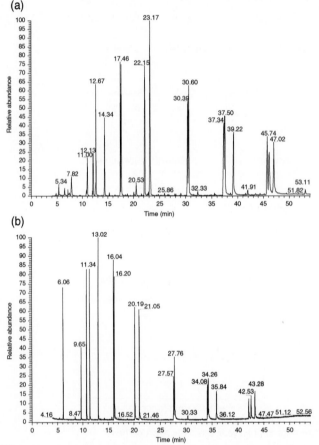

Note:
For PTV-GC-MS analysis of PAHs, retention time minutes: naphthalene, 7.82 minutes; Internal standard (4,4-difluorobiphenyl) 11.00 minutes; acenaphthylene, 12.13 minutes; acenaphthene, 12.67 minutes; fluorene, 14.34 minutes; phenanthrene, 17.40 minutes; anthracene, 17.46 minutes; fluoranthene, 22.15 minutes; pyrene, 23.17 minutes; benzo(a)anthracene, 30.39 minutes; chrysene, 30.60 minutes; benzo(b)fluoranthene, 37.34 minutes; benzo(k)fluoranthene, 37.50 minutes; benzo(a)pyrene, 39.22 minutes; indeno(1,2,3- cd)pyrene, 45.74 minutes; dibenzo(a,h)anthracene, 46.01 minutes; and, benzo(g,h,i)perylene, 47.02 minutes. for split/splitless GC-MS analysis of PAHs, retention time minutes: naphthalene, 6.06 minutes; Internal standard (4,4-difluorobiphenyl) 9.65 minutes; acenaphthylene, 10.76 minutes; acenaphthene, 11.34; fluorene, 13.02 minutes; phenanthrene, 16.04 minutes; anthracene, 16.20 minutes; fluoranthene, 20.19 minutes; pyrene, 21.05 minutes; benzo(a)anthracene, 27.57 minutes; chrysene, 27.76 minutes; benzo(b)fluoranthene, 34.08 minutes; benzo(k)fluoranthene, 34.26 minutes; benzo(a)pyrene, 35.84 minutes; indeno(1,2,3- cd)pyrene, 42.09; dibenzo(a,h)anthracene, 42.53 minutes; and benzo(g,h,i)perylene, 43.28 minutes.

Figure 16.30 GC-MS chromatograms of the polycyclic aromatic hydrocarbons (a) using PTV injector for GC-MS (a 0.5 µg ml^{-1} PAH standard) and (b) using split/splitless mode for GC-MS (a 5 µg ml^{-1} PAH standard).

injector (in split mode) for a 5 μg ml^{-1} PAH standard. The additional peak intensity for the higher molecular weight PAHs (four rings or more) (Figure 16.30) is noted when using the PTV injector. As often the PAH of concern is benzo(a)pyrene due to its carcinogenicity, the use of the PTV can provide enhanced sensitivity (a retention time of 39.22 minutes in Figure 16.30a). In addition, the use of a generic DB-5ms column highlights the lack of baseline separation for the chromatographic peaks for phenanthrene/anthracene, benzo(a)anthracene/chrysene, and benzo(b)fluoranthene/benzo(k)fluoranthene (in both examples).

Case Study B How Should the Instrumental Technique Operating Parameters Be Recorded?

Background: The recording of how the selected instrumentation operates is key to understanding the resultant data obtained. Often with chromatography analyses, some of the operating parameters need to be investigated prior to their final selection. For example, in an experiment using gas chromatography, the temperature programme required to achieve separation of multiple compounds will need to be investigated. Similarly, in an experiment using high-performance liquid chromatography, the wavelength(s) required to achieve optimum performance (absorbance signal) of multiple compounds will need to be investigated.

Activity.
An example template for gas chromatography and high-performance liquid chromatography is shown in Tables 16.1 and 16.2, respectively. In these examples, the key operating parameters are listed as essential. By using these examples as templates, it would be prudent to record your instrumental parameters in your laboratory notebook (whether in paper form or electronically).

Table 16.1   Gas chromatography with flame-ionization or mass spectrometry detection.

Component	Parameter name	Specifics
Injector	Type	split/splitless
	Split flow	[enter value] ml min^{-1}
	Temperature	[enter value] °C
	Split ratio	1 : [enter value]
	Volume	[enter value] μl
Column	Manufacturer	[enter name]
	Type	[enter name]
	Length	[enter value] m
	Internal diameter	[enter value] mm
	Film thickness	[enter value] μm

(Continued)

Table 16.1 (Continued)

Component	Parameter name	Specifics
Carrier gas	Type	helium or nitrogen
	Flow rate	[enter value] ml min^{-1}
Separation	Isothermal (oven temperature)	[enter value] °C
	Temperature programmed:	
	Initial temperature	[enter value] °C
	Hold time:	[enter value] min
	Ramp rate 1:	[enter value] °C min^{-1}
	Temperature:	[enter value] °C
	Ramp rate 2:	[enter value] °C min^{-1}
	Final temperature:	[enter value] °C
	Hold time:	[enter value] min
Detector	**Flame-ionization detector:**	yes/no
	Make-up gas:	[enter name]
	Mass spectrometer:	
	Ionization mode:	Electron impact
	Mass scan range	[enter value] to [enter value] m/z
	Scan mode: full scan	yes/no
	Scan mode: SIM	yes/no
	Ion source temperature	[enter value] °C
	Mass transfer line temperature	[enter value] °C
	Reference library used	[enter name] [enter version]
Quantitation	Peak area	yes/no
	Integration method used	manual/electronic (software)
	Internal standard:	yes/no
	Name of compound	[enter name]
	Purity/supplier	[enter name] [enter name]
	Compound concentration	[enter concentration] [units]
	External standard:	yes/no
	Name of compound	[enter name]
	Purity/supplier	[enter name] [enter name]
	Compound concentration	[enter concentration] [units]
	MS: qualifying ion(s)	[enter ion(s)] m/z
	MS: quantifying ion(s)	[enter ion(s)] m/z

Table 16.1   (Continued)

Component	Parameter name	Specifics
Calibration	Direct or standard additions	[enter method]
	Number of calibration standards	[enter value]
	Calibration standard concentrations	[enter values] [units]
	Linear range of calibration	[enter value] to [enter value][units]
	Calibration equation	[enter values in format y = mx + c]
	Correlation coefficient (r^2)	[enter value to 4 decimal places]
	Limit of detection	[enter value] [units]
	Limit of quantitation	[enter value] [units]

Table 16.2   High-performance liquid chromatography with ultraviolet/visible detection.

Component	Parameter name	Specifics
Injector	Loop volume	[enter value] µl
Column	Manufacturer	[enter name]
	Type	[enter name]
	Length	[enter value] m
	Internal diameter	[enter value] mm
	Particle size	[enter value] µm
	Reversed or normal phase	[enter term]
Mobile phase	Solvent A/Solvent B	[enter name] to [enter name]
	Additional chemical, e.g. buffer	[enter name]
	pH	[enter value]
	Flow rate	[enter value] ml min^{-1}
Separation	Isocratic	[enter % solvent A] [enter % solvent B]
	Gradient:	yes/no
	Solvent composition at start time	[enter name] to [enter name]
	Solvent composition at end time	[enter name] to [enter name]
	Equilibration time	[enter time] [unit]
	Solvent A: purity/supplier	[enter name] [enter name]
	Solvent B: purity/supplier	[enter name] [enter name]

(Continued)

Table 16.2 (Continued)

Component	Parameter name	Specifics
Detector	**UV/vis detector:**	
	Make/model	[enter name] [enter name]
	Fixed or variable wavelength	[enter name]
	Wavelength per compound	[enter value(s)] [units]
Quantitation	Peak area	yes/no
	Integration method used	manual/electronic (software)
	Internal standard:	yes/no
	Name of compound	[enter name]
	Purity/supplier	[enter name] [enter name]
	Compound concentration	[enter concentration] [units]
	External standard:	yes/no
	Name of compound	[enter name]
	Purity/supplier	[enter name] [enter name]
	Compound concentration	[enter concentration] [units]
Calibration	Direct or standard additions	[enter method]
	Number of calibration standards	[enter value]
	Calibration standard concentrations	[enter values] [units]
	Linear range of calibration	[enter value] to [enter value]
	Calibration equation	[enter values in format $y = mx + c$]
	Correlation coefficient (r^2)	[enter value to 4 decimal places]
	Limit of detection	[enter value] [units]
	Limit of quantitation	[enter value] [units]

16.8 Summary

A crossword of the key terms outlined in this chapter can be found in Appendix A3, with the solution in Appendix B3.

Further Readings

1 **NMR:** For information on NMR and its interpretation see, for example: Dean, J.R., Jones, A.M., Holmes, D. et al. (2017). *Practical Skills in Chemistry*, 3e. Harlow, UK: Pearson.
2 **MS:** For information on MS and its interpretation see, for example: Dean, J.R., Jones, A.M., Holmes, D. et al. (2017). *Practical Skills in Chemistry*, 3e. Harlow, UK: Pearson.
3 **IR:** For information on IR and its interpretation see, for example: Dean, J.R., Jones, A.M., Holmes, D. et al. (2017). *Practical Skills in Chemistry*, 3e. Harlow, UK: Pearson.

Section G

Post-Analysis

Decision-Making

17

Environmental Problem Solving

LEARNING OBJECTIVES

After completing this chapter, students should be able to:

- Understand the overall context of environmental analysis.
- Be aware of the initial planning required in any environmental analysis.
- Be aware of how determined analytical data can be interpreted.

17.1   Introduction

This chapter contextualizes the approaches needed to define the overall environmental protocol, i.e. from the initial problem, development of a sampling strategy, the options available for sample pre-treatment, awareness and limitations of analytical techniques, interpretation and data processing, as well as an assessment and utilization of the obtained information (Figure 17.1) using selected case studies including an Escape Room concept. To assist with the investigation of the case studies, it is necessary to consider the overall concept of the analytical protocol for environmental analysis (Figure 17.1).

Extraction Techniques for Environmental Analysis, First Edition. John R. Dean.
© 2022 John Wiley & Sons Ltd. Published 2022 by John Wiley & Sons Ltd.

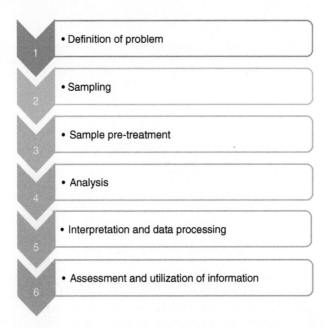

1 • Definition of problem

2 • Sampling

3 • Sample pre-treatment

4 • Analysis

5 • Interpretation and data processing

6 • Assessment and utilization of information

Figure 17.1 Concept of the overall analytical protocol for environmental analysis. In Case Study A, the focus is on defining the problem, whereas Case Study A considers the whole process: stages 1–6 and finally, Case Study C uses an Escape Room that allows decisions to be made along the way.

Case Study A   Defining the Problem: Initial Planning and Considerations

You have been approached to analyse for potential pollutants at a specific site.

At present (Figure 17.2a), the site is an out-of-town shopping, recreational and restaurant site owned by a major retail sales organization. However, concern has been raised by the public about the quality of the water in the adjoining River Urr and Lake Rothersmere from leachate from the site. Both the river and lake are now used for recreational water sport activities involving wild water swimming (lake only), sailing, canoeing, kayaking and stand-up paddleboarding. As the individual responsible for overseeing this work on behalf of the current owner, you need to take the project forward and to a satisfactory conclusion.

As the site is a known former historic industrial site, a good place to start is by undertaking a desk top study. A desk top study, as the name suggests, involves gathering information that is readily available without necessarily having to analyse anything. A desktop study may include the following information:

Outline a physical description of the site. This includes a description of location, map reference, access to site, current land use, along with an overall description of site.

Establish the environmental setting of the site. This might include site geology including a description of surface and below surface geology, e.g. coal seam; site hydrogeology including details of river or stream flows and whether groundwater is abstracted and for what purpose; site hydrology including known rainfall and any river/stream/pond locations; site ecology and archaeology including whether site has any known scheduling,

(a)

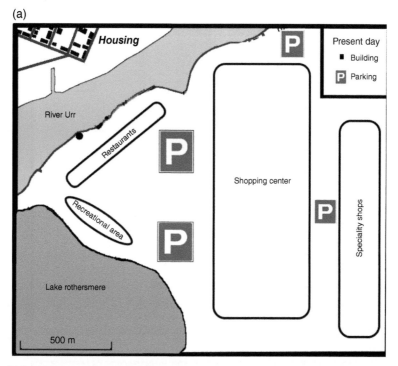

Figure 17.2 (a) Present day map of site to be investigated along with historic maps of the former, contaminated land site in (b) 1898, (c) 1925, (d) 1954 and (e) 1990.

e.g. site of special scientific interest (SSSI), as well as any features of archaeological significance; and mining assessment, e.g. evidence of former quarrying activity;

Identify its former industrial setting and any recent site history. Information available via historic and modern ordnance survey maps including (aerial) photographs of the site, as well as any previous site investigations.

Undertake a qualitative risk assessment. This will include the development of a site-specific conceptual model (see the example in Figure 17.3) that seeks to assess the following: source of contaminants; the pathway by which contaminant could come in to contact with a receptor, e.g. people; and the characteristics and sensitivity of the receptor to the contaminant.

Undertake a site walkover. A visit to the site can assist in understanding of the overall context, as well as enable identification of key issues, major features, position of walkways etc.

Make conclusions and recommendations. This will identify the way forward.

Useful information can be gathered about a former industrial site by obtaining detailed historic ordnance survey maps. By studying these maps, it will be evident what building infrastructure will have been present at set times in history. In this case study, Figure 17.2b shows an historic map from 1898, when the site was largely marsh land and was underdeveloped. However, from the date of the next available maps 1925

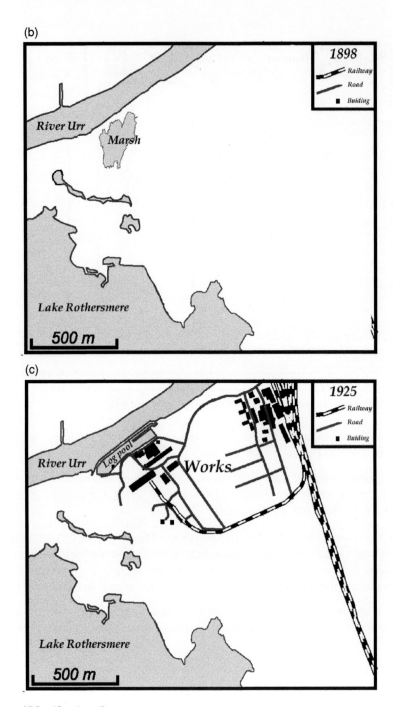

Figure 17.2 (Continued)

(d)

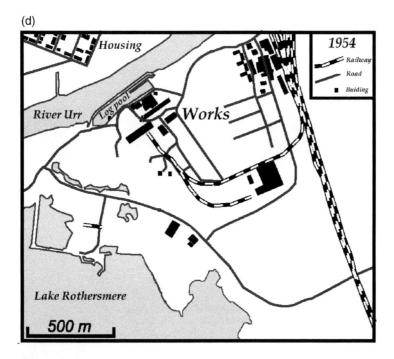

(e)

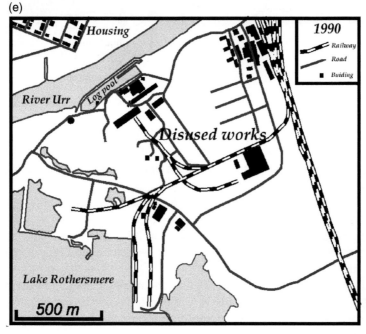

Figure 17.2 (Continued)

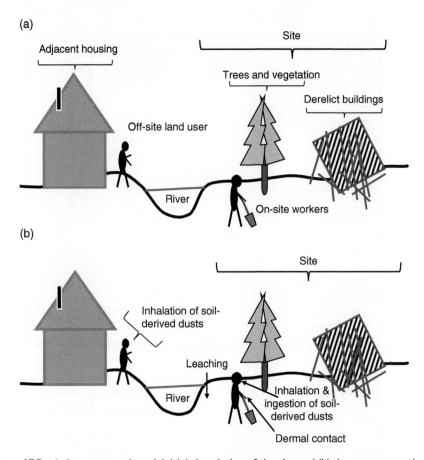

Figure 17.3 A site conceptual model. (a) A description of the site and (b) the exposure pathways.

(Figure 17.2c) and 1954 (Figure 17.2d), they illustrate the growth of the industrial aspects of the site, which ultimately led to its decline as an industrial site by 1990 (Figure 17.2e). Also, it is noted by 1990 the emergent housing development on the north shore of the River Urr.

Other information about the use of the former buildings can be obtained from local archivists, e.g. city/town councils, history societies, who will retain records on historic activities. By gathering this detailed information, it is possible to build up a picture of possible organic contaminants that may have been left on the site and are now leaching into the water courses.

Based on the conclusions and recommendations, it may be necessary to progress further and consider some water sampling (and subsequent analyses). For this type of sampling, it is necessary to obtain answers, in advance, about the following:

1) Is specialized water sampling equipment required? If so, do you have access to it? If not can you obtain the equipment and from whom?

2) How many samples (including replicates) will it be necessary to take?

3) What specific water testing is required?

4) What instrumentation is available to do the analyses?

5) Is the instrumentation limited with respect to sample size (volume)? Does sample size constrain the analytical measurement? Will it be necessary to pre-concentrate/ solvent extract the water sample prior to the analysis.

6) What quality assurance procedures are available? Has a protocol been developed? Does the laboratory have any certified standards for the analysis?

7) What types of container are required to store the samples and do you have enough of them?

8) Do the containers require any pre-treatment/cleaning prior to use?

9) Is any sample preservation or storage conditions required? If so, what is it and how might it impact on the analysis of the contaminants?

Case Study B   A Consideration of the Whole Concept of Environmental Analysis

Background: This case study analyses polycyclic aromatic hydrocarbons from the soil from a current recreational public open space (Figure 17.4). The selected Public Open Space (Figure 17.4) is based in South Tyneside (approximately 4.3 hectares). Historically, the site was used as a landfill site (pre-1920) for commercial and household waste, and subsequently (to 1982) as an allotment garden. However, since 1982 the site has been maintained as a public recreational space. Most of the site is a level-grassed open space, interspersed with overgrown grassed areas. The site is currently used for leisure, although there are no formal picnic facilities available (such as benches or litter bins) and a tarmac path crosses the site allowing access to and from surrounding residential areas and local facilities.

Sampling

Seven shallow soil samples ($0.02-0.20\,cm^2$) were collected from across the site using a stratified sampling grid (Figure 17.4). Soil samples were collected using a stainless-steel trowel. The sampling equipment was cleaned with acetone after each sample was collected to avoid cross contamination. The samples were transferred into suitable containers (i.e. kraft bags) and then transported to the laboratory for subsequent analyses. All soil samples were then subsequently dried (typically <40 °C for a minimum of four days), disaggregated and sieved through a 2 mm nylon mesh followed by a 250 µm nylon mesh and stored in sealed containers, at 4 °C, for subsequent analysis.

Experimental

Chemicals and solvents

Acetone (99.8%), dichloromethane (99.8%) and ethyl acetate (99.5%) were purchased from Fischer Scientific (Loughborough, UK). A PAH Calibration Mix (16 compounds) ($2000\,\mu g\,ml^{-1}$ in methylene chloride was purchased from Restek (Bellefonte, PA, USA). Mixed standard solution of PAHs was prepared in hexane at $1.0\,\mu g\,ml^{-1}$ level by diluting the different commercial standard solutions. All mixed standard solutions were subsequently diluted, as necessary. Hydromatrix was purchased from Thermo Fisher

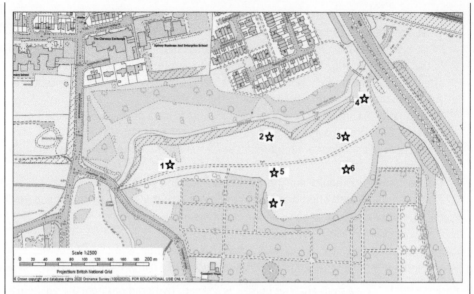

Figure 17.4 The public open space site with identified sampling points.

(Hemel Hempstead, UK). A soil reference material (CRM 172, sandy loam soil) was purchased from AccuStandard Inc. (New Haven, CT, USA).

Procedure for Determination of Soil Properties
Soil pH was determined by taking 10 g (accurately weighed) and suspending in deionized water in the ratio 1:2.5 w/v. After agitation (30 rpm for 10 minutes), the sample was left to stand for 10 minutes and the pH recorded, using a previously calibrated pH probe. Soil organic matter was determined, based on the loss of ignition (LoI) method. In this method, 5 g of the soil sample (W) (accurately weighed) was placed in a pre-weighed crucible. The weight of soil and crucible (W₁) were also recorded. The sample-containing crucible was placed in a pre-heated muffle furnace at 400 °C for 4 hours. After heating the crucible was then allowed to cool in a desiccator. The sample-containing crucible was then re-weighed (W_2); the % LoI was then calculated:

$$\%LoI = \left(\left(W1 - W_2\right)/W\right) \times 100$$

$$(17.1)$$

Procedure for Soil Extraction Using Pressurized Liquid Extraction
Samples were extracted using an ASE 200 Accelerated Solvent Extractor (Thermo Fisher, Sunnyvale, CA, USA) in 11 ml stainless steel cells. A cellulose filter was placed in the bottom of the cell and was filled until a quarter of the cell volume of hydromatrix sorbent. Then, 1.0 g of sample, accurately weighed, was added to the cell. Finally, the cell is fully filled of hydromatrix and is closed. Extraction was performed in a single cycle of a preheating of 5 minutes and 10 minutes static time at 100 °C and 2000 psi using acetone:dichloromethane (50:50, v/v) as extraction solvent. Flush volume of 50%

and purge time of 60 seconds were programmed. Glass ASE collector vial of 60 ml was used to collect the extract. The extract obtained was put into two tubes using a glass pipette. ASE collector vials were cleaned with three portions of extraction solvent (2 ml approximately) and the wash solvent was added to the samples for evaporation. The obtained extract was concentrated to dryness using a sample concentrator (Techne, DB-3, Dri-Block®, Essex, UK) at 35 °C and a N_2 stream of 10 psi. The residue was dissolved in 1.0 ml of ethyl acetate using a vortex and stored in 2 ml vial in a fridge at 4 °C until GC-MS analysis. Each sample, and soil CRM, was extracted in triplicate. All cells were cleaned with acetone and were pre-extracted with cellulose filters inside using the same extraction program in the ASE. Glassware were cleaned with soap, rinsed several times using water and finally rinsed with acetone.

Analysis of Extracts Using GC-MS
Chromatographic separation was achieved using a TG-5MS column (30 m × 0.25 mm ID, 0.25 μm film thickness) from Thermo Scientific (Hemel Hempstead, UK). The chromatographic system consisted of a Trace 1300 gas chromatograph, TriPlus RSH with liquid sampling tool and an ISQ 7000 Single Quadrupole Mass Spectrometer (Thermo Scientific). Sample volume injection was 1.0 μl in splitless mode with a split flow of 30 ml min^{-1}, 1.0 minutes of splitless time and 10 ml min^{-1} of purge flow. The injector temperature was maintained at 300 °C. Helium was used as a carrier gas with a constant flow of 1.0 ml min^{-1}. The oven temperature program was 70 °C held for 1.0 minutes, followed by an increase by 40 °C min^{-1} to 110 °C, and held 2.0 minutes. The temperature was then increased to 170 °C by 5 °C min^{-1} and increased to 200 °C by 2.50 °C min^{-1} and held for 3.0 minutes. Finally, an increase to 310 °C by 5 °C min^{-1} was performed, and the temperature is held for 5.0 minutes with a total analysis time of 58 minutes. The temperatures of the source and MS transfer line were 300 and 280 °C, respectively. The MS was operated in selective-ion monitoring (SIM) mode using electron impact ionization (EI).

Results and Discussion
Calibration graphs for the 16 PAHs were constructed (Table 17.1). All calibration graphs were linear with correlation coefficients (R^2) greater than 0.99. The limit of detection (LOD) and limit of quantitation (LOQ) were determined, using the slope of the calibration graph and the standard deviation of the intercept, based on the following equations: LOD = $3.3\sigma/s$ and LOQ = $10\sigma/s$, where σ is standard deviation of intercept and s is the slope (Table 17.1). Typical LOD data varied between 6 ng g^{-1} for anthracene and dibenzo(a,h)anthracene to 322 ng g^{-1} for benzo(a)anthracene, with corresponding LOQ data of 17–19 ng g^{-1} and 975 ng g^{-1}, respectively. The accuracy of the data was assessed by analysis of a CRM (CRM172, a sandy loam soil). The concentrations found for the 8 PAHs (Table 17.2) are in good agreement with the certified values after statistical evaluation by applying a t-test at 95% confidence level for two degrees of freedom. The t_{cal} values for all PAHs (Table 17.2) are lower than the t_{tab} value of 4.30. These validation results indicated that the PLE - GC-EI-MS method developed in this work was acceptable. An example chromatogram for the separation of the PAHs is shown in Figure 17.5.

Table 17.1 Example, analytical figures of merit for the PAHs in soil.

Compound	Retention time (min)	Qualitative m/z	Quantitative m/z	Calibration range (ng ml^{-1})	N° of data points	Calibration graph	R^2	LOD (ng g^{-1})	LOQ (ng g^{-1})
Naphthalene	7.21	127, 102	128	0–500	5	Y = 211.12x − 6652.7	0.9908	31	94
Acenaphthylene	10.98	153, 76	152	0–250	5	Y = 1188.3x + 171618	0.9943	132	399
Acenaphthene	12.37	76, 80	153	0–500	6	Y = 1946x − 12786	0.9993	17	52
Fluorene	14.57	165, 139	166	0–500	6	Y = 1342.1x − 14772	0.9979	50	153
Phenanthrene	19.06	160, 176	178	0–500	5	Y = 1278.7x − 15257	0.9981	43	130
Anthracene	19.21	160, 176	178	0–500	6	Y = 960.4x − 20679	0.9917	6	17
Fluoranthene	26.56	101, 106.5	202	0–500	5	Y = 984.8x − 23588	0.9935	220	667
Pyrene	28.07	101, 106.5	202	0–500	5	Y = 1124.9x − 26961	0.9946	29	89
Benzo(a) anthracene	38.06	113.5, 236	228	0–500	5	Y = 326.73x − 14585	0.9924	322	975
Chrysene	38.29	113.5, 236	228	0–500	5	Y = 840.54x − 21553	0.9919	142	432
Benzo(b) fluoranthene + Benzo(k) fluoranthene	44.01	101, 141	252	0–500	5	Y = 1366.5x − 61960	0.9958	103	312
Benzo(a)pyrene	45.34	125, 132	252	0–500	5	Y = 804.64x − 37335	0.9974	134	405
Benzo(g,h,i) perylene	50.67	228, 138	276	0–500	5	Y = 474.79x − 21269	0.9918	18	54
Indeno(1,2,3-c,d) pyrene	49.82	276, 138	292	0–500	4	Y = 239.63x − 12987	0.9936	16	49
Dibenzo(a,h) anthracene	50.00	138, 126	278	0–500	5	Y = 394.3x − 16402	0.9982	6	19

Table 17.2 Analysis of a soil certified reference material (CRM 172).

| | CRM172 (Sandy loam soil) | | |
Compound	Certified value (ng g^{-1})	Found value (ng g^{-1}) (n = 3)	$|t_{exp}|^{a}$
Acenaphthylene	55.6 ± 18.1	<LOQ	
Acenaphthene	94.9 ± 24.7	<LOQ	
Fluorene	66.4 ± 11.2	<LOQ	
Phenanthrene	168 ± 7.6	170.4 ± 4.6	0.74
Anthracene	17.7 ± 2.7	13.6 ± 0.2	4.14
Fluoranthene	634 ± 82.4	670.2 ± 30.2	1.70
Pyrene	86.5 ± 13	73.9 ± 4.8	3.71
Benzo(a)anthracene	303 ± 47.7	<LOQ	
Chrysene	154 ± 20.8	146.5 ± 3.1	3.42
Benzo(b)fluoranthene + Benzo(k)fluoranthene	NC	<LOQ	
Benzo(a)pyrene	33.9 ± 10.9	<LOQ	
Benzo(g,h,i)perylene	452 ± 81.2	392.4 ± 21.2	4.00
Indeno(1,2,3-c,d)pyrene	150.7 ± 30.5	151.3 ± 4.9	0.17
Dibenzo(a,h)anthracene	284 ± 30.5	277.8 ± 16.1	0.55

NC = not certified.

<LOQ = less than limit of quantitation.

^{a} t_{exp} calculated as follows: $t_{exp} = \left| [\ ]_{certified} - [\ ]_{found} \right| \times \dfrac{\sqrt{n}}{SD}$, $[\ ]_{found}$ and SD are the mean and standard deviation values (n = 2) after PLE-GC-MS and $[\ ]_{certified}$ is the certified concentration.

Analysis of Pollutants in Soil Samples

The public open space site was selected for consideration due to its use as a recreational site with a historic context. It can be observed (Table 17.3) that the highest concentrations (>10 µg g^{-1}) of individual PAH were identified as fluoranthene (24.1, 11.5 and 14.7 µg g^{-1} in sites 5, 6 and 7) and benzo(a)anthracene (13.6 and 18.5 µg g^{-1} in sites 3 and 6), while the ΣPAHs across all seven sampling sites varied between 6.2 and 63 µg g^{-1} (Figure 17.6)

For the Public Open Space Residential, POS$_{resi}$, identified as an area with green open space close to housing, and that includes the possibility of tracking back soil, and of relevance to the two areas considered in this study, the Environment Agency (EA) has identified a provisional Category 4 Screening Level (C4SL) of 10 mg kg^{-1} for BaP [1]. Fortunately, in this POS the highest concentration determined was 2.0 µg g^{-1} BaP in sample 5. On that basis, the risk to the public from exposure to the soil via the three main pathways of dermal, inhalation and ingestion is minimal.

It is also possible to consider the cancer potency of each PAH based on its BaP equivalent (BaP$_{eq}$) concentration. This BaP$_{eq}$ concentration (ng g^{-1}) was calculated in

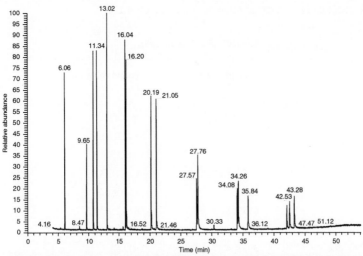

PAH (and retention time, minutes): naphthalene (6.06 minutes); acenaphthylene (10.76 minutes); acenaphthene (11.34 minutes); fluorene (13.02 minutes); phenanthrene (16.04 minutes); anthracene (16.20 minutes); fluoranthene (20.19 minutes); pyrene (21.05 minutes); benzo(a)anthracene (27.57 minutes); chrysene (27.76 minutes); benzo(b)fluoranthene (34.08 minutes); benzo(k)fluoranthene (34.26 minutes); benzo(a)pyrene (35.84 minutes); indeno(1,2,3-cd)pyrene (42.09 minutes); dibenzo(a,h)anthracene (42.53 minutes); benzo(g,h,i)perylene (43.28 minutes); and internal standard (4,4-difluorobiphenyl) (9.65 minutes).

Figure 17.5 Example chromatograph of polycyclic aromatic hydrocarbon (PAH) separation using GC-MS.

soil, based on the measured concentrations of the individual PAH compounds, by using the following equation [2]:

$$\left[BaP_{eq}\right] = \Sigma\left(\left[PAH_i\right] \times TEF_i\right)$$

(17.2)

where $[PAH_i]$ represents the concentration (ng g^{-1}) of an individual PAH in samples, and TEF_i (ng g^{-1}) is the toxic equivalence factor of a given PAH (PAH_i) relative to BaP [3]. The TEF approach adopts BaP as the reference compound because of its highly potent carcinogenic effect. Calculated TEFs for acenaphthene, phenanthrene, anthracene, fluoranthene, pyrene, benzo(a)anthracene, chrysene, benzo(b)fluoranthene, benzo(k) fluoranthene, benzo(a)pyrene, benzo(g,h,i)perylene, indeno(1,2,3)pyrene and dibenz(a,h) anthracene are 0.001, 0.01, 0.01, 0.01, 0.001, 0.1, 0.01, 0.1, 1, 0.01, 0.1 and 0.1, respectively, according to the USEPA [3, 4]. The BaP$_{eq}$ concentration values were calculated for the soil samples (Table 17.4) and their values varied from 396 to 5964 ng g^{-1}, with an average of 2706 ng g^{-1}. The maximum BaP$_{eq}$ concentration value was found at site 5.

The incremental lifetime cancer risk (ILCR) model has been explored to assess the carcinogenic potential of the measured PAHs in soil to humans, through the three main

Table 17.3 Analysis and characterization of soil at public open space.

Compound	#1	#2	#3	#4	#5	#6	#7
Sample identifier							
Soil pH	7.2	6.7	5.7	7.1	6.7	7.0	6.7
%LoI	17.0	27.5	41.4	18.9	15.9	14.9	15.9
Compound	**Concentration (ng g^{-1})$\pm$SD (n = 3)**						
Naphthalene	<LOQ	<LOQ	<LOQ	<LOQ	<LOQ	<LOQ	<LOQ
Acenaphthylene	<LOQ	<LOQ	<LOQ	<LOQ	<LOQ	<LOQ	<LOQ
Acenaphthene	20.8 ±4.6	82.3 ±12	18.2 ±1.2	<LOQ	21.2 ±1.8	<LOQ	29.4
Fluorene	<LOQ	<LOQ	<LOQ	<LOQ	<LOQ	<LOQ	<LOQ
Phenanthrene	2250 ±63	2730 ±195	3090 ±175	867 ±23	5980 ±1150	2720 ±133	4040 ±317
Anthracene	88.7 ±3.1	72.7 ±5.5	150 ±15	27.3 ±0.5	184 ±13	91.7 ±2.4	101 ±15
Fluoranthene	9590 ±440	4390 ±270	<LOQ	2280 ±66	24100 ±1550	11500 ±268	14700 ±1080
Pyrene	1360 ±47	672 ±22	2510 ±140	291 (298, 285)^a	<LOQ	1360 ±24	1760 ±144
Benzo(a) anthracene	6990 ±311	4400 ±79	13600 ±1250	1700 ±97	18500 ±1320	6980 ±99	8680 ±661
Chrysene	1000 ±182	771 ±20	1960 ±149	<LOQ	2840 ±188	1040 ±30	1360 ±76
Benzo(b) fluoranthene + Benzo(k) fluoranthene	633 ±64	336 ±27	1080 ±34	<LOQ	1330 ±58	545 ±22	747 ±25
Benzo(a)pyrene	773 ±16	493 ±20	1080 ±34	<LOQ	2010 ±156	708 ±16	973 ±62
Benzo(g,h,i) perylene	801 ±13	647 ±14	1380 ±88	278 ±19.1	1680 ±109	666 ±19	934 ±50
Indeno(1,2,3-c,d) pyrene	2100 ±92	1660 ±33	4080 ±33	681 ±52.5	5030 ±314	1800 ±85	2630 ±67
Dibenzo(a,h) anthracene	366 ±20	431 ±6.8	750 ±151	123 ±4.3	1120 ±66	339 ±6.8	372 (381, 363)$^{\#}$
ΣPAHs	25973	16685	29698	6247	62795	27750	36326
BaP$_{eq}$	2250	1650	3774	396	5964	2141	2764

^{a} n = 2; mean (individual values).

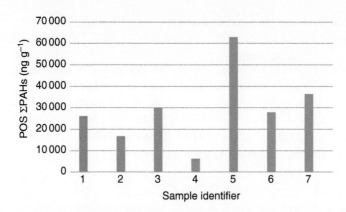

Figure 17.6 A summary of total PAH data across the public open space site.

exposure pathways, namely ingestion, inhalation and dermal absorption. The ILCR model estimates the probability of an individual who is exposed to PAHs, during his or her lifetime, from developing cancer [5]. This ILCR model was applied along with TEF values using the equations [6, 7]:

$$ILCR_{ing} = CSF_{ing} \times \sqrt[3]{\frac{BW}{70}} \times \frac{\left[BaP_{eq}\right] \times IR_{ing} \times \left(EF \times ED \times 10^{-6}\right)}{BW \times AT}$$
(17.3)

$$ILCR_{inh} = CSF_{inh} \times \sqrt[3]{\frac{BW}{70}} \times \frac{\left[BaP_{eq}\right] \times \left(\frac{IR_{inh}}{PEF}\right) \times \left(EF \times ED\right)}{BW \times AT}$$
(17.4)

and

$$ILCR_{derm} = CSF_{derm} \times \sqrt[3]{\frac{BW}{70}} \times \frac{\left[BaP_{eq}\right] \times \left(SA \times AF\right) \times \left(EF \times ED \times 10^{-6}\right) \times ABS}{BW \times AT}$$
(17.5)

where $ILCR_{ing}$, $ILCR_{inh}$ and $ILCR_{derm}$ are the incremental lifetime cancer risks resulting from the three exposure pathways (ingestion, inhalation and dermal absorption pathways, respectively); CSF_{ing}, CSF_{inh} and CSF_{derm} are the carcinogenic slope factor (7.3, 3.85 and 25 mg^{-1} kg day) for ingestion, inhalation and dermal absorption pathways, respectively) [8]. While IR_{ing} is the soil/dust ingestion rate (10 and 200 mg day^{-1} for adult and children, respectively) [9]; PEF is the particle emission factor (1.36 × 10^9 m^3 kg^{-1}); BW is the body weight (70 and 15 kg for adult and children, respectively); IR_{inh} is the inhalation rate (20 and 10 mg day^{-1} for adult and children, respectively); EF is the exposure frequency (180 day year^{-1}); ED is the exposure duration (24 and 6 year for adult and children, respectively); AT is the average life span (25 550 days); SA is the surface area of dermal exposure (5700 and 2800 cm^2 day^{-1} for adult and children, respectively); AF is the skin adherence factor (0.07 and 0.2 mg cm^{-2} for adult and children, respectively); and ABS is the dermal adsorption fraction (0.13, unitless) [7, 10, 11].

The sum of the individual ILCR values (i.e. ingestion, inhalation and dermal absorption) are then defined as incremental lifetime cancer risk sum (ILCR$_s$):

$$ILCR_s = \sum \left(ILCR_{ing} + ILCR_{inh} + ILCR_{derm} \right) \tag{17.6}$$

where an ILCR$_s \leqslant 10^{-6}$ suggests a negligible risk under most regulatory programmes; an ILCR$_s$ between 10^{-6} and 10^{-4} suggests a potential risk; and, an ILCR$_s > 10^{-4}$ implies a potentially high risk [6].

The ILCR values through dust ingestion, inhalation and dermal absorption pathways for adults and children are shown in Table 17.4. The results (Table 17.4) indicate that the exposure pathways of PAHs, for both adults and children, follow the general order: dermal > ingestion >>> inhalation (i.e. ingestion and dermal absorption are the most dominant exposure pathways when compared with the inhalation pathway).

The ILCR through inhalation varies between 5.4×10^{-8} to 2.3×10^{-7} (for an adult) and 7.9×10^{-8} to 1.0×10^{-7} (for a child). This indicates that the inhalation-induced cancer risk was almost negligible. Whereas high ILCR values were obtained through the ingestion pathway ($7.0 \times 10^{-5} - 2.9 \times 10^{-3}$ (for an adult) and $9.8 \times 10^{-4} - 1.5 \times 10^{-2}$ (for a child). Similarly, high ILCR$_{derm}$ values were obtained, $8.6 \times 10^{-3} - 1.2 \times 10^{-2}$ (for an adult) and $8.5 \times 10^{-3} - 1.2 \times 10^{-2}$ (for a child). The total cancer risk (ILCR$_s$) obtained was

Table 17.4 PAH incremental lifetime cancer risk for adult and child in soil samples.

	#1	#2	#3	#4	#5	#6	#7
ILCR$_{ing}$^{adult}	4.0×10^{-4}	2.9×10^{-4}	6.7×10^{-4}	7.0×10^{-5}	7.0×10^{-5}	3.8×10^{-4}	4.9×10^{-4}
ILCR$_{inh}$^{adult}	3.1×10^{-7}	2.3×10^{-7}	5.2×10^{-7}	5.4×10^{-8}	5.4×10^{-8}	2.9×10^{-7}	3.8×10^{-7}
ILCR$_{derm}$^{adult}	7.0×10^{-3}	5.2×10^{-3}	1.2×10^{-2}	1.2×10^{-3}	1.2×10^{-3}	6.7×10^{-3}	8.6×10^{-3}
ILCR$_s$^{adult}	**7.4×10^{-3}**	**5.5×10^{-3}**	**1.3×10^{-2}**	**2.5×10^{-3}**	**2.0×10^{-2}**	**7.1×10^{-3}**	**9.2×10^{-3}**
ILCR$_{ing}$^{child}	5.5×10^{-3}	4.1×10^{-3}	9.3×10^{-3}	9.8×10^{-4}	1.5×10^{-2}	5.3×10^{-3}	6.8×10^{-3}
ILCR$_{inh}$^{child}	1.1×10^{-7}	7.9×10^{-8}	1.8×10^{-7}	1.9×10^{-8}	2.9×10^{-7}	1.0×10^{-7}	1.3×10^{-7}
ILCR$_{derm}$^{child}	6.9×10^{-3}	5.1×10^{-3}	1.2×10^{-2}	1.2×10^{-3}	1.8×10^{-2}	6.6×10^{-3}	8.5×10^{-3}
ILCR$_s$^{child}	**1.2×10^{-2}**	**9.1×10^{-3}**	**2.1×10^{-2}**	**2.2×10^{-3}**	**3.3×10^{-2}**	**1.2×10^{-2}**	**1.5×10^{-2}**

PAH incremental lifetime cancer risk via ingestion (ILCR$_{ing}$); PAH incremental lifetime cancer risk via inhalation (ILCR$_{inh}$); PAH incremental lifetime cancer risk via dermal absorption (ILCR$_{derm}$); total incremental lifetime cancer risk (ILCR$_s$): negligible risk $\leq 10^{-6}$; potential risk 10^{-6} to 10^{-4}; **potentially high risk $> 10^{-4}$**.

$9.2 \times 10^{-3} - 1.3 \times 10^{-2}$ (for an adult) and $9.1 \times 10^{-3} \times 1.2 \times 10^{-2}$ (for a child), exceeding the safety level and posing a high potential cancer risk to exposed adults and children. However, this analysis needs some further consideration. The main exposure pathways have been identified as dermal absorption and ingestion for this POS site. The calculated incremental lifetime cancer risk values need to be considered against the habits of individuals visiting the POS site including their physical contact with the soil in what is a recreational site used for leisure (e.g. dog walking). So, while the potential risk using the ILCR model is high, the reality is that an individual's exposure via dermal absorption and ingestion are minimal. For dermal absorption to occur an individual would need, for example, to be handling the soil on a very regular basis with their hands (or other exposed skin areas), while for the ingestion exposure pathway an individual would need to be consuming between 10 and 200 mg day^{-1} (adult and child, respectively) principally by hand-to-mouth contact (e.g. soil under finger nails). It is also important to recall that the ILCR model assumes an exposure frequency of 180 days per annum, over several years (24 years for an adult and six years for a child) and an individual who will live to the age of 70.

Case Study C   Environmental Chemistry Escape Room

This case study was developed (autumn 2019) as a student team activity in the form of an Environmental Chemistry Escape Room in which participants would solve problems to obtain 2–4 digit codes to unlock boxes in which subsequent clues were provided. It was delivered in spring 2020, but due to COVID-19, it was converted into an on-line activity in a virtual learning environment. This is a considerably modified version for text format.

Scenario: You are a chemist who is an independent environmental consultant employed to advise on contamination issues relating to the land-use of a specific site. You are required to consider all the evidence and advise your client on whether there is a risk to people using this site.

Activity 1: Desk Top Study of Site Over Time
Initially a desk-top study is required to check out the history of the site to obtain clues as to the likely contaminants. By considering the available maps (Figure 17.7), obtain information to help you develop an action plan. Note: the area to be considered is highlighted within the oval shape.

Activity 1: Response
By looking at the maps in chronological order, you can see that in 1860 the highlighted site labelled St Anthony's is a hive of industrial activities being located on the northern shore of the River Tyne, Newcastle upon Tyne. To the left of centre (west) is the named 'St Anthony's Lead Works', whereas to the right of centre (or East) is the 'St Anthony's Chemical Works'. And at this stage (1860) in the site's development, it has the 'Old Copperage Works' with multiple ferry links across the River Tyne to provide additional access to the site for the workforce from the south shore-side.

   In 1890, the centre of the site has been cleared of any major vegetation trees and the 'St Anthony's Lead Works' looks to be fully developed. Whereas the 'St Anthony's Chemical Works' site looks to have developed as an on and off loading and storage site with multiple cranes evident.

   By the 1940s, major changes have occurred on the site. The 'St Anthony's Lead Works' has disappeared, and the land now appears to be landscaped with a redundant slipway. In the centre of the site has appeared some allotments for local people to grow fruit and vegetables for personal consumption. The use of allotments would have been very popular (as they are now) as harsh rationing of food due to WW2 precluded a ready supply with 'Dig for Victory' being a popular slogan promoted by the British Ministry of Agriculture. To the east, the original 'St Anthony's Chemical Works' site has expanded considerably with evidence of multiple storage tanks for fuel/oil and a jetty for transport links, via the river. The area now has only one ferry link across the river.

(a)

(b)

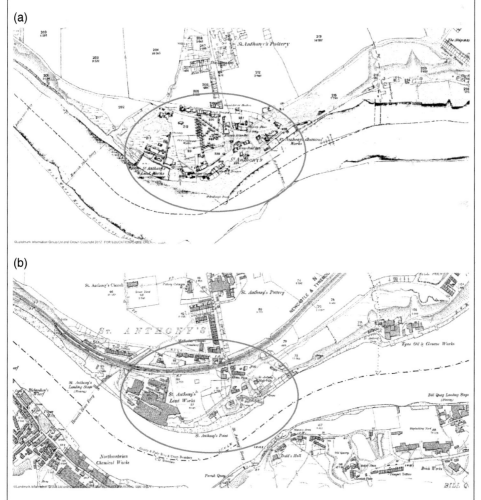

Figure 17.7   Historical maps of a former, historic contaminated land site known as St Anthony's (a) 1860, (b) 1890, (c) 1940s and (d) 2017.

(c)

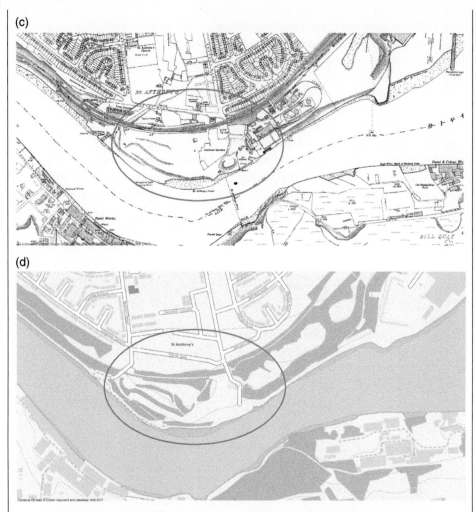

(d)

Digimap Historic, Ordnance Survey, St. Anthony's, Walker, Newcastle upon Tyne, using Digimap
Ordnance Survey Web Map Service (http://digimap.edina.ac.uk) (accessed 30 October 2019)

Figure 17.7 (Continued)

And by 2017, any evidence, on the maps, of their former industrial heritage has gone. The former 'St Anthony's Lead Works' site is now a public open space consisting of pathways and grassed areas. The former 'St Anthony's Chemical Works' site is less formal and more an area of public open space dominated by scrubland.

Action Plan: some suggestions

- The former industrial site is likely to be contaminated with lead (from the 'St Anthony's Lead Work'), while the 'St Anthony's Chemical Works' site is likely to be contaminated with potentially volatile organic compounds, petroleum hydrocarbons, polycyclic aromatic hydrocarbons and other compounds and materials from leakages and spillages.
- Exposure to environmental contaminants is a major concern of regulatory agencies in countries worldwide. The regulatory agencies are concerned with the risk to humans from environmental contaminants such as organic compounds, metals,

metalloids, inorganic chemicals and others (including asbestos). The exposure (e.g. Figure 17.3) results from a variety of pathways:

- Direct soil and dust ingestion
- Consumption of homegrown produce
- Dermal contact outdoors
- Inhalation of dust indoors
- Inhalation of vapours indoors

Regulatory agencies often use a risk-based approach based on the source – pathway – receptor concept to understand the implications to the exposure pathways to humans.

- In this scenario, the sources most likely are Pb and a range of organic compounds, e.g. polycyclic aromatic hydrocarbons (PAHs). In the case of the PAHs, the most important due to its potential carcinogenicity (by the International Agency for Research on Cancer (IARC)) and as an indicator compound for other PAHs is benzo(a)pyrene (Figure 17.8).
- In this scenario, the pathways most likely are ingestion of soil from hand-to-mouth by individuals sitting on the grassed areas; and airborne dust (and soil particles) in dry weather leading to inhalation risk or deposition on hands, linked to hand-to-mouth ingestion.
- In this scenario, the receptors most likely are people (young and old) undertaking recreational activities on the site, e.g. dog walking, picnics in fine weather, sporting activities on the level grassed areas.

Activity 2: Identification of Type of Site

By considering the map provided (Figure 17.9) what description best fits the type of site it is today (as defined in the UK Contaminated Land Exposure Assessment, CLEA)?

- Public Open Space: two scenarios have been identified
  - The scenario of green space close to housing that includes tracking back of soil (POS_{resi}); and
  - A park-type scenario where the park is at a sufficient distance that there is negligible tracking back of soil (POS_{park}).
- Allotment Site: a place where fruit and vegetables are grown for own consumption (normally)
- Commercial Site: a place where commercial activity takes place, e.g. a car park, an office block

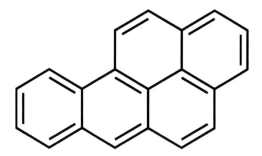

Figure 17.8 Chemical structure of benzo(a)pyrene.

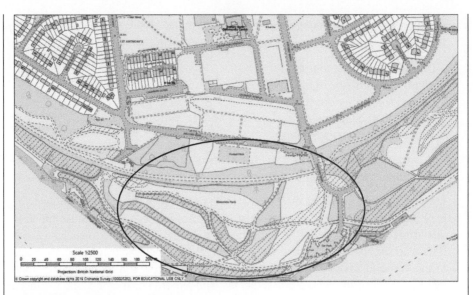

Figure 17.9   A modern map of the St Anthony's site.

- Residential Area: two scenarios have been identified
  - With home-grown produce
  - Without home-grown produce

Activity 2: Response

The site is clearly a public open space and neither of the other options apply. However, within the public open space are two scenarios. The question is which is best applied to this site. The POS_{resi} is defined as the predominantly grassed areas (up to 500 m²) adjacent to high-density housing and the central green area around which houses are located, as on many housing estates from the 1930s to 1970s. It is also anticipated that this land-use would include the smaller areas commonly incorporated in newer developments as informal grassed areas or more formal landscaped areas with a mixture of open space and covered soil with planting. The site is likely to be extensively used by children for playing and may be used for informal sports activities such as a football 'kickabout'. Whereas a public park (POS_{park}) is an area of open space provided for recreational use and usually owned and maintained by the local authority. It is anticipated that POS_{park} could be used for a wide range of activities, including family visits and picnics; children's play area; sporting activities on an informal basis; and dog walking.

Hopefully you will concur that this site (Figure 17.9) is a POS_{park}.

Activity 3: Pollutant Concentrations

Using the x and y axes, respectively, in Figure 17.10, identify the concentration ($\mu g\ g^{-1}$) of lead and benzo(a)pyrene as provided from an independent consultancy company.

- 6 and 4
- 5 and 3
- 6 and 2
- 7 and 3

Now from within the now outlined sampling grid ◇ identify the concentration of Pb and BaP, respectively.

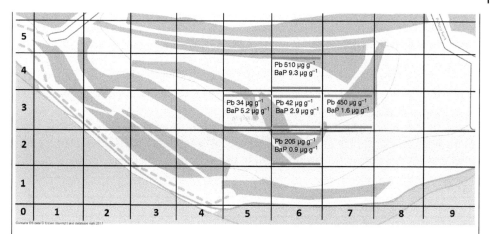

Figure 17.10 Chemical analyses of soil from selected sampling grid areas.

Activity 3: Response
Hopefully, you will agree that the concentration at the site (Figure 17.10) for Pb is 42 µg g^{-1} and for BaP is 2.9 µg g^{-1}.

Activity 4: Analytical Techniques
[Note: As this book is concerned with organic compounds, the determination of Pb is out of scope. Nevertheless, the activity is included for completeness.]

Identify which is the most appropriate analytical technique to determine Pb, in solution, at low concentration by answering the following multiple-choice questions (MCQs).

- Flame atomic absorption spectroscopy (FAAS)
- Energy-dispersive X-ray fluorescence spectroscopy (EDXRF)
- Inductively coupled plasma-mass spectrometry (ICP-MS)

Question	True	False
I require a hollow cathode lamp to operate		
My operating temperature is around 6000 K		
I require an X-ray tube to operate		
I can use a range of nebulizers to introduce samples into the plasma		
I can operate with a quadrupole mass spectrometer		
I have a radio frequency generator that operates at 40 MHZ		
I can analyse pressed sample pellets		
I use an electron multiplier tube to detect ions		
My plasma is sustained with argon		
I can suffer from isobaric interferences		
I use a flame as my atomization cell		
An Echelle spectrometer allows me to identify multiple elements		

Now, simply add the total number of TRUE responses and the total number of FALSE responses and record the two-digit number generated.

Activity 4: Response

From the *Activity 3: Response*, you know that Pb was determined in soil at a concentration of 42 μg g^{-1}. However, in this activity we know that Pb was determined in solution, so some dissolution/digestion, with heat and acid must have taken place to 'extract' the Pb from the soil. No details are provided, but we could assume a dilution ratio of a weight of soil (g):solution (ml) of 50–100, w/v, so the Pb concentration in solution would be either 0.84 or 0.42 μg ml^{-1}. By prior knowledge (or further reading), you might well conclude that inductively coupled plasma-mass spectrometry would be ideal and be able to (easily) determine Pb in this concentration range, in solution.

Question	True	False
I require a hollow cathode lamp to operate		x
My operating temperature is around 6000 K	x	
I require an X-ray tube to operate		x
I can use a range of nebulizers to introduce samples into the plasma	x	
I can operate with a quadrupole mass spectrometer	x	
I have a radio frequency generator that operates at 40 MHZ	x	
I can analyse pressed sample pellets		x
I use an electron multiplier tube to detect ions	x	
My plasma is sustained with argon	x	
I can suffer from isobaric interferences	x	
I use a flame as my atomization cell		x

The answer is therefore 7 True and 4 False or 74.

Activity 5

Identify which is the most appropriate analytical technique to determine BaP, in a dichloromethane (DCM) extract, by answering the following multiple-choice questions (MCQs).

- Gas chromatography-mass spectrometry (GC-MS)
- High-performance liquid chromatography-UV-visible spectroscopy (HPLC-UV)

Question	True	False
I require an organic solvent as the mobile phase		
I operate with a carrier gas of helium		
Samples are introduced via a 1 μl syringe		
Samples are introduced using a fixed volume loop		
My column can be 60 m long		
My column can be 25 cm long		
My column stationary phase (as an abbreviation) is ODS		
My column stationary phase (as an abbreviation) is DB-5		
I can operate with a quadrupole mass spectrometer		
My cell path length is 1 cm long		

Question	True	False
My separation is controlled (mainly) by a combination of solvent		
My separation is controlled (mainly) via oven temperature		

Now, simply add the total number of TRUE responses and the total number of FALSE responses and record the two-digit number generated.

Activity 5: Response

From the **Activity 3: Response**, you know that BaP was determined in soil at a concentration of 2.9 µg g^{-1}. However, in this activity, we know that BaP was determined in a DCM extract, so an extraction has taken place to recover the BaP from the soil. No experimental details are provided, but we could assume a dilution ratio of a weight of soil (g):solution (ml) of 10–25, w/v, so the BaP concentration in the DCM extract would be either 0.29 or 0.12 µg ml^{-1}. By consideration of the information in earlier chapters or prior knowledge, you might well conclude that gas chromatography-mass spectrometry would be ideal and be able to determine BaP in this concentration range, in a DCM extract.

Question	True	False
I require an organic solvent as the mobile phase		x
I operate with a carrier gas of helium	x	
Samples are introduced via a 1 µl syringe	x	
Samples are introduced using a fixed volume loop		x
My column can be 60 m long	x	
My column can be 25 cm long		x
My column stationary phase (as an abbreviation) is ODS		x
My column stationary phase (as an abbreviation) is DB-5	x	
I can operate with a quadrupole mass spectrometer	x	
My cell path length is 1 cm long		x
My separation is controlled (mainly) by a combination of solvent		x
My separation is controlled (mainly) via oven temperature	x	

The answer is therefore 6 True and 6 False or 36.

Activity 6

You now need to compare all five data sets for Pb and BaP obtained from the Independent consultancy company with the *relevant* Environment Agency Category 4 Screening Level (C4SL) [12]. The C4SLs (in England) consist of *cautious estimates of contaminant concentrations in soil that are still considered to present an acceptable level of risk*, within the context of the Environmental Protection Act (1990) (Part 2A), by combining information on human health toxicology, exposure assessment and normal ambient levels of contaminants in the environment. The identified generic land-uses for consideration of the contaminant concentrations are:

- Public Open Space (POS_{resi} or POS_{park}).
- Allotment
- Commercial
- Residential (with or without home-grown produce)

So, is the site safe to the public based on your evaluation of all the data available and the relevant C4SL?

Activity 6: Response

Consider the C4SLs, for the different land-uses in Table 17.5.

Earlier, we identified that the type of site we are considering was POSpark (**Activity 2: Response**); an area used for a wide range of activities, including family visits and picnics; children's play area; sporting activities on an informal basis; and dog walking.

Now consider all the determined Pb and BaP concentration data (Figure 17.10):

Pb = 510, 34, 42, 450 and 205 $\mu g\ g^{-1}$ and BaP = 9.3, 5.2, 2.9, 1.6 and 0.9 $\mu g\ g^{-1}$ against the selected C4SL data (Pb = 1300 mg kg^{-1}; BaP = 21 mg kg^{-1}).

On that basis for Pb, all the determined soil concentrations are below the C4SL, and the same applies for BaP. It is concluded therefore that the site, for these two pollutants, is considered well within the acceptable risk, i.e. safe, against the C4SL (which was determined by combining information on human health toxicology, exposure assessment and normal ambient levels of contaminants in the environment).

Table 17.5 Category 4 screening levels (mg kg^{-1}) based on the risk management of the site.

| Substance | Residential | | Allotments | Commercial | Public open space (POS_{resi})[a] | Public open space (POS_{park})[b] |
	With home-grown produce	Without home-grown produce				
Arsenic	37	40	49	640	79	170
Benzene	0.87	3.3	0.18	98	140	230
Benzo(a) pyrene	5.0	5.3	5.7	77	10	21
Cadmium	22	150	3.9	410	220	880
Chromium (VI)	21	21	170	49	21	250
Lead	200	310	80	2300	630	1300

[a] The scenario of green space close to housing that includes tracking back of soil (POS_{resi}).
[b] A park-type scenario where the park is considered to be at a sufficient distance that there is negligible tracking back of soil (POS_{park}).

References

1 CL:AIRE (2014). Development of Category 4 Screening Levels for Assessment of Land Affected by Contamination: Appendix E Provisional C4SLs for Benzo(a)pyrene as a surrogate marker for PAHs.

2 Gao, P., Liu, D., Guo, L. et al. (2019). *Sci. Total Environ.* 659: 1546–1554.

3 USEPA (1994). *Benzo[a]pyrene (BaP)*. Washington, DC: US: Environmental Protection Agency.

4 Samburova, V., Zielinska, B., and Khlystov, A. (2017). *Toxics* 5: 17.

5 USEPA (2005). *Human Health Risk Assessment Protocol for Hazardous Waste Combustion Facilities*. Washington, DC: US Environmental Protection Agency.

6 Ali, N., Ismail, I.M.I., Khoder, M. et al. (2017). *Sci. Total Environ.* 601–602: 478–484.

7 Cao, Z., Zhao, L., Shi, Y. et al. (2017). *Hum. Ecol. Risk Assess. An Int. J.* 23: 1072–1085.

8 Ma, Y., Liu, A., Egodawatta, P. et al. (2017). *Sci. Total Environ.* 575: 895–904.

9 USEPA (2017). *Chapter 5 of the Exposure Factors Handbook: Soil and Dust Ingestion*, 1–100. USEPA.

10 Cao, Z., Wang, M., Chen, Q. et al. (2019). *Sci. Total Environ.* 653: 423–430.

11 USEPA (2002). Supplemental Guidance for developing soil screening levels for Superfund sites. Office of Solid Waste and Emergency Response (OSWER).

12 Department for Environment, Food and Rural Affairs (2014). SP1010: Development of Category 4 Screening Levels for Assessment of Land Affected by Contamination – Policy Companion Document.

Section H

Historical Context

18

A History of Extraction Techniques and Chromatographic Analysis

LEARNING OBJECTIVES
After completing this chapter, students should be able to:
• Understand the longstanding tradition that encompasses the extraction techniques described.   • Be aware of the historical developments related to chromatography.

18.1   Introduction

The history of the development of extraction techniques is long and varied, linked to other scientific developments and advances. The following listings are a history of both the extraction techniques used and their accompanying chromatographic analytical techniques. Perhaps the birth of solid–liquid extraction can be attributed to Franz von Soxhlet (Figure 18.1) whose invention of the Soxhlet extractor (Figure 18.2) revolutionized the extraction of compounds from solid matrices. Whereas the birth of chromatography is often being attributed to M. Tswett (Figure 18.3). His early experimental set-up (Figure 18.4) used solid particles, packed in a column, for the separation of leaf pigments. Tswett is also attributed with coining the phrase chromatography, from the Greek chroma (colour) and graphein (to write). The term chromatography was first used by Tswett in his paper of 1906 [3] where he wrote 'Ich solches Präparat nenne ich eine Chromatogramm und die entsprechende Methode, die chromatographische Methode' or 'I call such a preparation a chromatogram and the corresponding method the chromatographic method'. What follows is an historical viewpoint on the developments in chromatography and extraction techniques (Table 18.1).

Extraction Techniques for Environmental Analysis, First Edition. John R. Dean.
© 2022 John Wiley & Sons Ltd. Published 2022 by John Wiley & Sons Ltd.

Brief biography: Franz von Soxhlet was born (1848) in Brno, Czech Republic. He was the son of a Belgian immigrant. He obtained a PhD, in 1872, from Leipzig University, Germany. He is credited with the invention of the Soxhlet extractor in 1979, the same year he became Professor of Agricultural Chemistry at the Technical University in Munich.

Figure 18.1   The German agricultural chemist Franz Ritter von Soxhlet (12 January 1848 – 5 May 1926). Franz Ritter von Soxhlet / Wkimedia Commons / CC BY-SA 3.0.

Figure 18.2   Diagram of the Soxhlet extractor [1].

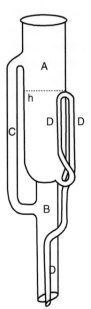

Also spelled as Tsvett, Tswett, Tswet, Zwet and Cvet.

Brief biography: Mikhail Tsvet was born (1872) in Asti, Italy, to an Italian mother and a Russian father. He was raised by his father, due the death of his mother, in Geneva, Switzerland. Hegraduated from the University of Geneva (in 1893) after being awarded a BSc degree from the Department of Physics and Mathematics. This was followed by a PhD in botany that focused on cell physiology (1896). That same year he accompanied his father to St. Peterburg, Russia. He started work in the Biological Laboratory for the Russian Academy of Science. He subsequently moved to Warsaw University, Poland (in 1902) and became a laboratory assistant, and later assistant professor,at the Institute of Plant Physiology. Due to the start of World War I, he was evacuated, with his institution and colleagues, to Moscow, Russia. Later, in 1917, he became Professor of Botany and Director of theBotanical Gardens, at the University of Tartu, Estonia. Again, this university was re-located (in 1918) to Voronezh, Russia. He died due to a chronic inflammation of the throat on 26 June 1919 (aged 47).

Figure 18.3 The Russian–Italian botanist Mikhail Semyonovich Tsvet (14 May 1872–26 June 1919). Mikhail Semyonovich Tsvet / Wkimedia Commons / CC BY-SA 3.0.

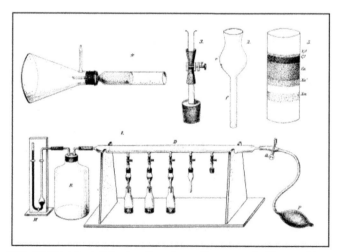

Note: Apparatus for multiple adsorption columns. The manometer M, uses a 3 l bottle R which serves as a pressure reservoir, in which a certain air pressure can be produced through the tube D by means of a rubber bulb P. By means of the pinchcock Q, P is closed off from the rest of the apparatus. Tube D serves as a pressure distributor; it is provided with several tube-shaped attachments to which the actual filtration vessels are to be attached (3 shown attached). The small filtration funnel is connected with the pressure distributor D by means of a tight-fitting stopper, which is provided with a glass tube and rubber tube in a suitably movable connection.

Figure 18.4 The original chromatographic apparatus developed by M. Tsvett [2].

Table 18.1 Historical developments in chromatography and extraction techniques.

Year	Discovery	Comments
AD 23–79	Dye separation on papyrus	Pliny (*Naturalis Historia*) outlined a test for dyes by spotting solutions on papyrus [4]
1794	Liebig condenser	The first apparatus which used a counter-current cooling stream of water as part of the distillation process [5]
1822	Supercritical phase	The discovery of the supercritical phase is attributed to Baron Cagniard de la Tour [6]. He observed that the boundary between a gas and a liquid disappeared for certain substances when the temperature was increased in a sealed glass container
1850	Paper chromatography	Friedrich Runge (1795–1868) discovered that coloured substances, when spotted on filter paper, would spread out in concentric rings
1861	Capillary analysis	Christian F. Schönbein (1799–1868) observed during qualitative analysis of ozone that when an aqueous solution is spotted on a filter, the water precedes the dissolved substances, and also that different substances are drawn up the paper to varying degrees
1879	Soxhlet	Soxhlet extraction introduced by Franz Ritter von Soxhlet [1]
1889	Thin layer chromatography	M.W. Beyerinck developed an approach to separate acids through a thin layer of gelatin [5]
1893	Compressed paper column	Fischer separated inorganic ions using a compressed paper column, through which the solution was allowed to flow
1893	Adsorption chromatography	L. Reed used columns packed with powdered absorbents [4]
1895	Cavitation	The principle of (acoustic) cavitation, the basis of ultrasound (sonication) was discovered; cavitation leading to bubble formation, severe vibration and surface disruption (on an ocean going destroyer, HMS Daring, resulting in damage to its propellers) reported [7]
1897	Adsorption chromatography	D.T. Day reported separation of different fractions of crude petroleum on passage down a column of powdered limestone [4]
1903	Adsorption chromatography	The discovery of column chromatography is attributed to Mikhail Tsvett; he separated plant pigments using mixtures of ether and alcohol on a column of calcium carbonate [2]
1903	Liquid–liquid extractor	A continuous liquid–liquid extractor for solvents lighter than water was designed (Kutscher and Stendel) [5]
1912	Mass spectrometry	First example of mass spectrometry; the separation of neon isotopes by J.J. Thomson. He used magnetic and electric fields to deflect neon ions; detection was done suing a photographic plate
1917	Ion exchange resin	First use of an ion exchange resin for clean-up/separation demonstrated [8]
1918	Mass spectrometry	A.J. Dempster constructed the first focusing magnetic mass spectrometer with an electron ionization source
1919	Mass spectrometry	F.W. Aston, working with J.J. Thomson, credited with development of the first workable mass spectrometer

Table 18.1 (Continued)

Year	Discovery	Comments
1921	Magnetron	The description of the microwave source is first described [9]
1922	Theory of HETP	K. Peters introduced the concept of HETP (height equivalent to a theoretical plate); this allowed a comparison of the efficiency of separations [5]
1935	Mass spectrometry	Dempster built the first double-focusing MS
1935	Synthetic ion exchange resin	First practical demonstration of a synthetic ion exchange resin [10]
1937	Liquid–liquid extractor	A continuous liquid–liquid extractor for solvents heavier than water was designed (Wehrli) [11]
1938	Thin layer chromatography	N.A. Ismailof and M.S. Schraiber used 'chromatostrips' for separation of terpene derivatives [4]
1940	Mass spectrometry	Nier's first precision (60°) mass spectrometer for investigation of isotopic abundances and isotopic ratios
1941	Gas chromatography/ high-performance liquid chromatography	Martin and Synge describe the principle of liquid-phase partitioning; they described the chromatography of volatile substances in a column using a gas (as carrier) (GC), as well as the concept of using a liquid mobile phase (HPLC) and the concept of theoretical plates [12]
1943	Paper chromatography	The first paper on paper chromatography published [13]
1944	Ion exchange chromatography	First use of ion exchange chromatography; published data in 1947 on separation of fission products [14]
1946	Microwaves	The heating effect of microwaves was attributed to Dr Percy Spencer of Raytheon Corp., USA
1946	Mass spectrometry	First time-of-flight mass spectrometer (Stephens)
1947	Microwave oven	The first commercial microwave ovens for cooking in the home are launched into the marketplace (Radarange)
1948	Mass spectrometry	First commercial mass spectrometers built by Consolidated Engineering Co. (USA) and Metropolitan Vickers Electric Co. (UK).
1950–1951	Kuderna-Danish solvent evaporator	J. Kuderna Jr. and A. Danish develop a method to concentrate down pesticide extracts (to measure their residues)
1952	Gas chromatography	James and Martin separated volatile fatty acids by partition chromatography using N_2 as the carrier gas and a stationary phase of silicone oil/stearic acid supported on diatomaceous earth; first reported description and application of GC [15]
1952	Temperature-programming for GC	Aiding separation in GC by increasing the temperature of the column during a GC run [16]; by 1966 had become critical to application of GC [17]
1953	Mass spectrometry	First quadrupole mass filter and quadrupole ion trap (Paul)
1954	Thermal conductivity detector	N.H. Ray applied thermal conductivity detector for gas–liquid partition chromatography [5].

(Continued)

Table 18.1 (Continued)

Year	Discovery	Comments
1955	Gas chromatography	First commercial gas–liquid partition chromatography apparatus [5]
1955–1956	Gas chromatograhy-Mass spectrometry	First coupling of gas chromatography to a mass spectrometer by F. McLafferty and R. Gohlke
1956	Thin layer chromatography	E. Stahl first reported the use of a very thin layer of absorbent on a glass plate as the stationary phase [4]
1956	Van Deemter plot	J.J. van Deemter, F.J. Zuiderweg and A. Klinenberg developed the relationship that links HETP to the flow velocity, particle diameter and solute diffusivity [5]
1957–1958	Capillary column for GC	Invention of the capillary column for GC by M.J.E. Golay [18, 19]
1958	Flame-ionization detector (for GC)	The FID as a universal detector was discovered; the FID uses the thermal energy of a hydrogen/air flame to bring about emission of electrons from organic molecules (in the flame) to produce ions; these ions are measured as an ion current, which is proportional to the amount of organic molecules, per unit time, entering the flame [20, 21]
1958	Chromatographic retention index system	E. Kovats proposed a retention index system to express column retention [5]
1958	UV/visible detector	First use of visible detector for chromatography [22]
1958	High-precision microsyringe	C.H. Hamilton (USA) developed a high-precision microsyringe for sample introduction for GC [5]
1957–1959	Directly coupled GC-MS	First direct coupling of the GC eluent into a mass spectrometer [23]
1959	Capillary column for GC	Rapid expansion in high-resolution gas chromatography using capillary columns reported by Desty, Scott, Condon, Zlatkis, Lipsky and Lovelock [5]
1960	Electron capture detector for GC	Invention of the electron capture detector, by J.E. Lovelock, for measurement of halogen-containing compounds; it is based on the ability of an organic molecule to capture free electrons from a β-radiation source, e.g. ^{63}Ni [24]
1961–1967	Thermionic detector for GC	A detector based on a flame-ionization detector whose flame contained alkali metal ions to selectively interact with heteroatoms developed; selectivity identified for nitrogen, phosphorus and halogens [5]
1964	Commercial SFE	A patent filed to use supercritical carbon dioxide to decaffeinate coffee [25].
1965	Chemically bonded support for HPLC columns	G. Nickless et al. developed a support (n-$C_{16}H_{33}$), which was chemically bonded to the support phase [5]
1966	Flame photometric detector for GC	S.S. Brody and J.E. Chaney developed this detector, based on optical filters, to monitor the emission wavelengths for phosphorus and sulphur [5].

Table 18.1   (Continued)

Year	Discovery	Comments
1966	High-pressure pump (for HPLC)	Development (by R. Jentoft and T.H. Gouw) of a reciprocating piston pump for HPLC developed that could deliver pulseless flow at high pressure [5]
1968	Selected ion monitoring for MS	The ability to selectively identify m/z ions was developed; enhanced sensitivity possible when operating in SIM mode [26].
1968	Electrospray ionization	First used electrospray ionization with mass spectrometry described by M. Dole
1964–1969	High-performance liquid chromatography	An awakening of the possibilities offered by HPLC; led by pioneers in the practice and application of separation using a liquid mobile phase [27, 28]
1968	LC-MS interface	The first reported attempt at coupling a flowing liquid stream (from an LC) with a mass spectrometer [29]
1969	High-pressure pump (for HPLC)	First commercial reciprocating pump for HPLC; delivered a pulse flow at high pressure (Bombaugh et al.) [5]
1969–1971	HPLC column development	J.J. Kirkland used uniform microspheres as the packing in a narrow column of steel (or glass) through which liquids are forced at high pressure [4]
1972	Chemically bonded support for HPLC columns	D.C. Locke et al. developed methods in which, via a carbon-silicon bond, the silicon atoms on the surface of the core material could be bonded [5]
1974	Soxtec	An alternative for Soxhlet developed by Randall [30]; it is based on a two-stage process involving boiling and rinsing
Mid-1970s	Thermal desorption (for GC)	Development of thermal desorption as an adaption for injection in gas chromatography developed. Originally used for occupational monitoring in the workplace, in the form of personal badge-type monitors
1979	Fused-silica GC column	Invention of the modern capillary GC column [31]
1979–1980	Programmed temperature vaporizer (PTV) injector for GC	Developed of large volume injection for capillary gas chromatography [32]
1980s	SFE instrumentation	Instrumentation and applications of supercritical fluid chromatography are first reported in 1986 [25, 33]
1982	APCI source for LC-MS	The first reported paper on the use of atmospheric pressure chemical ionization, as the interface, for LC-MS [34]
1984	Electrospray ionization (ESI) source for LC-MS	The first reported paper on the use of atmospheric pressure electrospray as a source for LC-MS [35]; based on earlier pioneering developments [36].
1985	MALDI source	First reported publication on matrix-assisted laser desorption ionization (MALDI) [37]

(Continued)

Table 18.1 (Continued)

Year	Discovery	Comments
1985	Purge and Trap (for GC)	Invention of the procedure of passing a stream of gas through a water sample to release volatile organic compounds reported [38]; purge and trap
1986	Microwave-assisted extraction	A microwave oven used for the first time to extract organic compounds from matrices [39]
1990	Solid-phase microextraction	The first paper published on SPME [40]
1990	Semipermeable membrane devices (for passive sampling)	The semipermeable membrane device (SPMD) introduced for passive sampling of non-polar organic pollutants in water [41]
1991	96-well plate for SPE	The development of the 96-well plate for automated solid-phase extraction invented [42]
1992	Automation of solid-phase microextraction	The first paper published on the automation of SPME [43]
1993	Headspace solid-phase microextraction	The first paper published on the theory and practice of headspace SPME [44]
1993	Solid-phase microextraction	The first commercial SPME device (Supelco)
1993	Passive in situ concentration–extraction sampler (for passive sampling)	The passive in situ concentration–extraction sampler (PISCES) for passive sampling of organic compounds in water [45]
1994	Diffusive gradients in thin films (for passive sampling)	Diffusive gradients in thin films (DGT) introduced for passive sampling of metals ions in water; later modified for organic compounds in water [46]
1995	Accelerated solvent extraction	Accelerated solvent extraction, first launched commercially, by Dionex Corp [47]. It has been generically re-branded as pressurized liquid extraction (PLE)
1996	Single-drop microextraction	First reported paper on single-drop microextraction [48]
1997	Inside needle capillary adsorption trap	First reported paper on the inside needle capillary adsorption trap (INCAT) for microextraction of volatile organic compounds [49]
1999	Hollow-fibre liquid-phase microextraction	First reported paper on hollow-fibre liquid-phase microextraction, HF-LPME [50]
1999	Stir bar sorptive extraction (SBSE)	First reported publication on the use of SBSE by Baltussen et al. [51]
2000	Solid-phase dynamic extraction	SPDE commercially launched by Chromtech (Idstein, Germany); commercial version of INCAT device
2000	Orbitrap mass spectrometer	First reported publication on the orbitrap mass analyser by Makarov [52]
2000	Chemcatcher® (for passive sampling)	Chemcatcher® introduced to measure time-averaged concentrations of organic micropollutants in aquatic environments [53]

Table 18.1 (Continued)

Year	Discovery	Comments
2001	Membrane-enclosed sorptive coating (for passive sampling)	Membrane-enclosed sorptive coating (MESCO) introduced for passive sampling, in water, of organic compounds [54]
2001	Thin-film SPME (for passive sampling)	Thin-film solid-phase microextraction (TF-SPME) for time-weighted average water sampling introduced [55]
2002	QuEChERS	The phrase 'Quick, Easy, Cheap, effective, Rugged and Safe' was coined by Anastassiades et al. [56]
2003	Needle-type concentrator	First reported publication on the needle-type concentrator [57]
2004	Microextraction by packed sorbent	First reported paper on MEPS [58]
2005	Needle trap extraction	First reported publication on needle trap extraction device [59]
2005	Orbitrap mass spectrometer	Commercialization of orbitrap mass analyser by Thermo Fisher Scientific
2006	Dispersive liquid-liquid microextraction	First reported papers on dispersive liquid–liquid microextraction [60, 61]
2006	In-needle extraction device	First reported publication on in-needle extraction device [62]
2006	In-tube extraction	ITEX system commercially available via CTC Analytics AG (Zwingen, Switzerland)
2008	Polar organic chemical integrative sampler (for passive sampling)	The polar organic chemical integrative sampler (POCIS) introduced for passive sampling of organic compounds in water [63]
2019	HECAM (for passive sampling)	Passive sampling of hydrophilic/hydrophobic organic compounds using a hydrophilic–lipophilic balance sorbent-embedded cellulose acetate membrane (HECAM) [64]

18.2 Application

Case Study A   A Comparison of Extraction Methods for Solid and Liquid Matrices

Background. In comparing different extraction methods, it is necessary to identify some key parameters on which to make the judgement, in addition to their actual purpose and function.

Activity. The following may be the appropriate criteria:

- *Sample mass/volume:* the amount of sample that an extraction method requires is an important aspect and can directly influence the sensitivity of the measurement component, i.e. the more organic compound that can be extracted from a larger sample will allow its determination to be made at a lower concentration.
- *Extraction time*: the length of time that the extraction methodology takes is one important consideration. However, while it may be obvious to link the extraction

time (faster is better) with the analysis step, the argument does not always hold. Just as multiple samples can be extracted simultaneously, using some approaches, so the use of autosamplers on chromatographic systems means that multiple sample extracts can be pre-loaded ready for analysis overnight, if necessary. Perhaps the faster extraction is better assessed in terms of the customer requirements/needs.

- *Solvent type and consumption:* Not all extraction techniques require organic solvent usage as part of the process. If solvent is required, it would be beneficial if the type of solvent used could be environmentally friendly, cheap to purchase with minimal disposal cost and use small quantities.

- *Sequential or simultaneous extraction:* This criterion could be taken alongside the 'extraction time' criterion above. However, the question is more fundamental. Is it better to extract a sample using a 'one-at-a-time' approach or to extract samples 'several-at-a-time'? The latter is undoubtedly important once any experimental variation in the influence of the extraction technique is known and can be simply repeated multi-fold. The sequential approach does provide some investigation of the important operating variables of the extraction technique/methodology. An understanding of these variables could have long-term benefits, if properly understood.

- *Equipment cost:* No one wants to pay a large amount of money for the extraction approach adopted, provided the chosen one is effective and in-line with other customer/client criteria. Nevertheless, all approaches have an inherent capital cost that needs to be assessed as part of their selection criteria. In addition to the initial capital outlay cost, it is also important to consider the routine and regular cost for maintaining the extraction technique in consumables and maintenance costs.

- *Level of automation:* The greater the level of automation, the undoubted higher the initial capital cost and possibly the higher routine running costs. However, these costs may be overcome by (a) the lower costs in terms of staffing that may be required or (b) the deployment of staff on more productive aspects rather than routine activities.

- *Extraction method approval:* Several organizations worldwide produce 'methods' that have been tested and 'approved' for use in extraction analytes from matrices. The most comprehensive list of 'official' environmental methods has been produced by the US Environmental Protection Agency (US-EPA). Other organizations that produce 'approved' methods include: Association of Official Analytical Chemists (AOAC); Deutsches Institut fur Normung (DIN); National Metrology Institute of Japan (NMIJ); American Society of Testing and Materials (ASTM).

Tables 18.2 and 18.3 summarize the key criteria for extraction of organic compounds from solid and liquid matrices.

Table 18.2 A comparison of the extraction methods for solid matrices.

	Soxhlet	Soxtec	Shake flask	Ultrasonic	PFE	MAE	MSPD	SFE
Description of method	Utilizes cooled condensed solvents to pass over the sample contained in a thimble to extract analytes. Uses specialist glassware and heating apparatus	Also known as automated Soxhlet. Soxtec places the sample into the boiling solvent and then flushes clean solvent over the sample. Faster than Soxhlet	Sample is covered with organic solvent in a container, then shaken	Sample is covered with organic solvent, then a sonic horn is placed inside the beaker with the solvent and sample	Utilizes high temperature (100 °C) and pressure (2000 psi) to extract analytes. Solvent and analytes are flushed from the extraction vessel using a small volume of fresh solvent and a N_2 purge. Fully automated	Utilizes microwave radiation to heat solvent. Either done under pressure or at atmospheric pressure	Sample is mixed with a dispersant e.g. C18 media, and then placed in an empty solid phase extraction cartridge. Analytes eluted with appropriate solvent	Utilizes supercritical CO_2 with or without organic modifier to extract analytes. Pressures up to 680 atm and temperature up to 250 °C can be used. Analytes collected in solvent
Sample mass (g)	10	10	1–5	1–5	Up to 30g	2–5	1–5	1–10
Extraction time	6, 12 or 24 h	Reduced time compared to Soxhlet i.e. 2–4 h	Typically 30 min, but repeated for up to three times.	Typically 5–15 min, but repeated for up to three times.	12 min	20 min (plus 30 min cooling and pressure reduction)	Should be possible in under 30 min	30 min–1 h
Solvent type	Acetone:hexane (1:1, v/v); acetone: DCM (1:1, v/v); DCM only; or, toluene:methanol (10:1, v/v). Or a solvent system of your choice	As Soxhlet	As Soxhlet	As Soxhlet	Acetone:hexane (1:1, v/v) or acetone: DCM (1:1, v/v) for OCPs, semivolatile organics, PCBs or OPPs; acetone: DCM:phosphoric acid (250:125:15, v/v) for chlorinated herbicides	Typically, acetone:hexane (1:1, v/v). The solvent(s) is/are required to be able to absorb microwave energy	Requires optimization	CO_2 (plus organic modifier). Tetrachloroethene used as the collection solvent for TPHs for determination by FTIR, otherwise DCM

(Continued)

Table 18.2 (Continued)

	Soxhlet	Soxtec	Shake flask	Ultrasonic	PFE	MAE	MSPD	SFE
Solvent consumption (ml) per extraction	150–300	40–50	30–100	50–100	25	25–45	20–50	10–20
Sequential or simultaneous	Sequential (but multiple assemblies can operate simultaneously)	Systems available for 2 or 6 samples simultaneously	Simultaneous if an automated shaken is used	sequential	Sequential	Simultaneous (up to 14 vessels can be extracted simultaneously)	sequential	sequential
Equipment cost	Low	Low-moderate	Low-moderate	Low-moderate	High	Moderate	Low-moderate	High
Level of automation	Minimal	Minimal	Minimal	Minimal	Fully automated up to 24 samples can be extracted	Minimal	Minimal	Minimal to high
EPA method	3540	3541	None	3550	3545	3546	None	3560 for TPHs, 3561 for PAHs and 3562 for PCBs and OCPs

TPHs – total petroleum hydrocarbons; PAHs – polycyclic aromatic hydrocarbons; OCPs – organochlorine pesticides.

Table 18.3 A comparison of extraction methods for liquid matrices.

	Liquid–liquid extraction (LLE)	Purge and trap	Headspace	Liquid-phase microextraction	Solid-phase extraction (SPE)	Solid-phase microextraction (SPME)	In-tube extraction	Stir-bar sorptive extraction	Membrane extraction
Description of method	Sample is partitioned between two immiscible solvents; continuous and discontinuous operation possible	Volatile analytes are recovered from aqueous sample by purging with N_2. Recovered analytes are then trapped (concentrated) prior to thermal desorption directly into GC	Volatile analytes are recovered by heating the sample (solid or liquid); after equilibration a sample is recovered and introduced directly into GC	A miniature version of LLE; a single drop used to recover analytes from solution. Recovered analytes are injected directly into GC	Analyte retained on a solid absorbent; extraneous sample material washed from sorbent. Desorption of analyte using organic solvent	Analyte retained on a sorbent-containing fibre attached to a silica support. Fibre protected by a syringe barrel when not in use. Most commonly found for GC applications	Analytes are recovered from aqueous solution or headspace by retention on a sorbent located within the syringe barrel; analytes recovered using either organic solvent or heat; directly into GC or HPLC	Analytes are retained on a coated stir bar in solution. Analytes desorbed by either thermal desorption or organic solvent	Analytes are retained on a sorbent as part of a passive sampling system. Analytes desorbed by organic solvent
Sample size	1–2 l	5–25 ml	2 g or 10 ml	5–50 ml	1–1000 ml	1–1000 ml	1–20 ml	10–100 ml	In situ sampling in natural waters
Extraction time	Discontinuous: 20 min; continuous: up to 24 h	20–24 min per sample	5–20 min per sample	30 min	10–20 min	10–60 min (requires optimization)	5–20 min	30–240 min	Hours, days to weeks

(Continued)

Table 18.3 (Continued)

	Liquid–liquid extraction (LLE)	Purge and trap	Headspace	Liquid-phase microextraction	Solid-phase extraction (SPE)	Solid-phase microextraction (SPME)	In-tube extraction	Stir-bar sorptive extraction	Membrane extraction
Solvent consumption (mL) per extraction	30–60 ml for discontinuous; up to 500 ml for continuous	No organic solvent required; N_2 required	No organic solvent required; N_2 required	Minimal organic solvent	Organic solvent required for wetting sorbent and elution of analyte (10–20 ml)	No solvent required	Minimal organic solvent required	No organic solvent for thermal desorption; 20 ml of organic solvent	Minimal organic solvent required
Sequential or simultaneous extraction	Can be simultaneous using multiple set-ups	Sequential	Sequential	Sequential	Can be simultaneous	Sequential	Sequential	Sequential	Can be simultaneous using multiple set-ups
Equipment cost	Low	Moderate cost (capital cost of Purge and Trap)	Moderate cost (capital cost of static or dynamic headspace)	Low cost for syringe	Low–high (depends on degree of automation)	Low cost (but also available as an automated system)	Low cost for MEPS/moderate cost for ITEX	Moderate cost (capital cost of thermal desorption system)	Low to moderate cost
Level of automation	Can be automated on a moving bed roller	Automated	Automation using autosampler	Manual	Automation using dedicated systems; also with on-line possibilities as well	Automation using autosampler	Automation using autosampler	Automation using autosampler	Manual
EPA method	3510 and 3520	5030C	5021	None	3535	None	None	None	None

References

1 Soxhlet, F. (1879). *Dingler's Polytech. J.* 232: 461–465.

2 Tswett, M. (1903). *Trudy warsavsk. Obst. Jestesvoisoitat, Otd. Biol.* 14: 20–39.

3 Tswett, M. (1906). *Ber. Dtsch. Bot. Ges.* 24: 316–323.

4 Irving, H.M.N.H. (1974). The techniques of analytical chemistry. A short historical survey. HMSO. ISBN: 0 11 290203 0.

5 Laitinen, H.A. and Ewing, G.W. (1977). *A History of Analytical Chemistry*. York, PA, USA: American Chemical Society, Maple Press Co.

6 de la Tour, C. (1822). *Annales de Chimie* 21: 127–132.

7 Thorncroft, J. and Barnaby, S.W. (1895). *Proc. Inst. Civ. Eng.* 122: 51–69.

8 Folin, O. and Bell, D.R. (1917). *J. Biol. Chem.* 29: 329–335.

9 Papoutsis, D. (1984). *Photonics Spectra* 53.

10 Adams, B.A. and Holmes, E.L. (1935). *J. Soc. Chem. Ind.* 54: 1–6.

11 Wehrli, S. (1937). *Helv. Chim. Acta* 20: 927–931.

12 Martin, A.J.P. and Synge, R.L.M. (1941). *Biochem. J.* 35: 1358–1368.

13 Gordon, A.H., Martin, A.J.P., and Synge, R.L.M. (1943). *Biochem. J.* 37: 79–86.

14 Tompkins, E.R., Khym, J.X., and Cohn, W.E. (1947). *J. Am. Chem. Soc.* 69: 2769–2777.

15 James, A.T. and Martin, A.J.P. (1952). *Biochem. J.* 50: 679–690.

16 Griffiths, J.H., James, D.H., and Phillips, C.S.G. (1952). *Analyst* 77: 897–904.

17 Harris, W.E. and Habgood, H.W. (1966). *Programmed Temperature Gas Chromatography*. New York, USA: Wiley.

18 Golay, M.J.E. (1958). *Gas Chromatography* (ed. V.J. Coates), 1–13. New York, USA: Academic Press.

19 Golay, M.J.E. (1958). *Gas Chromatography* (ed. D.H. Desty), 36–55. London: Butterworths.

20 Harley, J., Nel, W., and Pretorius, V. (1958). *Nature* 181: 177–178.

21 McWilliam, I.G. and Dewar, R.A. (1958). *Nature* 181: 760.

22 Spackman, D.H., Stein, W.H., and Moore, S. (1958). *Anal. Chem.* 30: 1190–1206.

23 Gohlke, C. (1959). *Anal. Chem.* 31: 535–541.

24 Lovelock, J.E. and Lipsky, S.R. (1960). *J. Amer. Chem. Soc.* 82: 431–433.

25 Dean, J.R. (ed.) (1993). *Applications of Supercritical Fluids in Industrial Analysis*. Glasgow: Blackie Academic and Professional.

26 Hammer, G., Holmstedt, B., and Ryhage, R. (1968). *Anal. Biochem.* 25: 532–548.

27 Horvath, C. and Lipsky, S. (1966). *Nature* 211: 748–749.

28 Csaba, G., Horvath, C., Preiss, B.A., and Lipsky, S.R. (1967). *Anal. Chem.* 39: 1422–1428.

29 Tal'roze, V.L. and Karpov, G.V. (1968). *Russian J. Phys. Chem.* 42: 1658–1664.

30 Randall, E.L. (1974). *J. Assoc. Offic. Anal. Chem.* 57: 1165–1168.

31 Dandeneau, R.D. and Zerenner, E.H. (1979). *J. High Res. Chromatogr.* 2: 351–356.

32 Vogt, W., Jacob, K., Ohnesorge, A.B., and Obwexer, H.W. (1980). *J. Chromatogr.* 186: 197–205.

33 Hawthorne, S.B. (1990). *Anal. Chem.* 62: 633A–642A.

34 Henion, J.D., Thomson, B.A., and Dawson, P.H. (1982). *Anal. Chem.* 54: 451–456.

35 Yamashita, M. and Fenn, J.B. (1984). *J. Phys. Chem.* 88: 4671–4675.

36 Dole, M., Ferguson, L., and Alice, M. (1968). *J. Chem. Phys.* 49: 2240–2249.

37 Karas, M., Bachmann, D., and Hillenkamp, F. (1985). *Anal. Chem.* 57: 2935–2339.

38 Badings, H.T. and Cooper, R.P.M. (1985). *J. High Res. Chromatogr. Chromatogr. Comm.* 8: 755–763.

39 Ganzler, K., Salgo, A., and Valko, K. (1986). *J. Chromatogr.* 371: 299–306.

40 Arthur, C.L. and Pawliszyn, J. (1990). *Anal. Chem.* 62: 2145–2148.

41 Huckins, J., Tubergen, M., and Manuweera, G. (1990). *Chemosphere* 20: 533–552.

42 Merson, J.R. and Bojanic, D. (1991). An assay tray assembly. European Patent EP0454315, Pfizer Limited, 8 April 1991.

43 Arthur, C.L., Killam, L.M., Buchholz, K.D. et al. (1992). *Anal. Chem.* 64: 1960–1966.

44 Zhang, Z. and Pawliszyn, J. (1993). *Anal. Chem.* 65: 1843–1852.

45 Litten, S., Mead, B., and Hassett, J. (1993). *Environ. Toxicol. Chem.* 12: 639–647.

46 Davison, W. and Zhang, H. (1994). *Nature* 367: 546–548.

47 Richter, B.E., Jones, B.A., Ezell, J.L. et al. (1996). *Anal. Chem.* 68: 1033–1039.

48 Jeannot, M.A. and Cantwell, F.F. (1996). *Anal. Chem.* 68: 2236–2240.

49 McComb, M.E., Oleschuk, R.D., Giller, E., and Gesser, H.D. (1997). *Talanta* 44: 2137–2143.

50 Pedersen-Bjergaard, S. and Rasmussen, K.E. (1999). *Anal. Chem.* 71: 2650–2656.

51 Baltussen, E., Sandra, P., David, F., and Cramers, C. (1999). *J. Microcolumn Sep.* 11: 737–747.

52 Makarov, A. (2000). *Anal. Chem.* 72: 1156–1162.

53 Kingston, J.K., Greenwood, R., Mills, G.A. et al. (2000). *J. Environ. Monit.* 2: 487–495.

54 Vrana, B., Popp, P., Paschke, A., and Schuurmann, G. (2001). *Anal. Chem.* 73: 5191–5200.

55 Wilcockson, J.B. and Gobas, F.A.P. (2001). *Environ. Sci. Technol.* 35: 1425–1431.

56 Anastassiades, M., Lehotay, S.J., Stajnbaher, D., and Schenck, F.J. (2003). *J. AOAC Intern.* 86: 412–431.

57 Berezkin, V.G., Makarov, E.D., and Stolyarov, B.V. (2003). *J. Chromatogr. A* 985: 63–65.

58 Abdel-Rehim, M. (2004). *J. Chromatogr. B: Anal. Technol. Biomed. Life Sci.* 801: 317–321.

59 Wang, A., Fang, F., and Pawliszyn, J. (2005). *J. Chromatogr. A* 1072: 127–135.

60 Rezaee, M., Assadi, Y., Hosseini, M.R.M. et al. (2006). J. Chromatogr. A. 1116: 1–9.

61 Berijani, S., Assadi, Y., Anbia, M. et al. (2006). *J. Chromatogr. A* 1123: 1–9.

62 Saito, Y., Ueta, I., Kotera, K. et al. (2006). *J. Chromatogr. A* 1106: 190–195.

63 Seethapathy, S., Gorecki, T., and Li, X.J. (2008). *J. Chromatogr. A* 1184: 234–253.

64 Gao, X.Z., Xu, Y.P., Ma, M. et al. (2019). *Ecotoxicol. Environ. Saf.* 178: 25–32.

Appendices

Crossword Puzzles to Aid Learning and Understanding

Appendix A1: A Crossword of Key Terms in Chapters 4–8: Extraction Techniques for Aqueous Samples

Extraction Techniques for Environmental Analysis, First Edition. John R. Dean.
© 2022 John Wiley & Sons Ltd. Published 2022 by John Wiley & Sons Ltd.

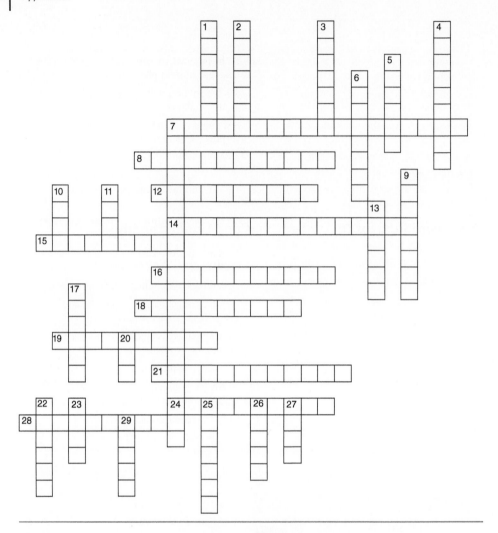

Across

7 A type of medium-to-high polarity coating used in stir-bar sorptive extraction.

8 A type of polar phase used in solid-phase microextraction.

12 A mechanism for physical or chemical trapping of compounds in solid-phase microextraction.

14 By packed sorbent.

15 A sampling location to retain compounds from a coated stir-bar in stir-bar sorptive extraction.

Down

1 A desorption process to recover compounds from a coated stir-bar in stir-bar sorptive extraction.

2 What solid-phase extraction does to the compound.

3 The opposite of active sampling in membrane extraction.

4 Where the solvent and sample are located in the solid-phase extraction cartridge.

5 Trap extraction.

Across	Down
16 Sounds like it retains chemicals that escape in membrane extraction.	6 The place where compounds are thermally desorbed in solid-phase microextraction.
18 A mixed sorbent phase in solid-phase extraction.	7 A type of non-polar phase used in solid-phase microextraction.
19 The process that allows multiple samples to be extracted routinely in solid-phase extraction.	9 A property of the stir bar used in stir-bar sorptive extraction.
21 The step that is immediately before the loading of the sample in solid-phase extraction.	10 A type of in-tube extraction.
24 A mechanism for partitioning of compounds in solid-phase microextraction.	11 A timeframe for sampling in membrane extraction.
28 A type of aggressive agitation used in solid-phase microextraction.	13 The twelfth astrological sign in the zodiac used in membrane extraction.
	17 A pump is used to create this in solid-phase extraction.
	20 A model used in membrane extraction.
	22 A type of aggressive agitation movement used in solid-phase microextraction.
	23 A weighted average concentration in membrane extraction.
	25 The active phase in solid-phase extraction.
	26 An acronym for a passive sampling device, with solid sorbent, used in membrane extraction.
	27 Automated in-tube extraction.
	29 A type of sorbent trap used in in-tube extraction.

Appendix A2: A Crossword of Key Terms in Chapters 10–13: Extraction Techniques for Solid Samples

Across

1 A common support material in matrix solid-phase dispersion.

6 Only solvents with a permanent one gets heated in microwave-assisted extraction.

7 Where the sample is placed ready for extraction in pressurized liquid extraction.

Down

2 Conventional heating source for organic solvents.

3 An alternate name for in situ clean-up in pressurized liquid extraction.

4 The generic name of a substance added to supercritical carbon dioxide to increase its overall polarity.

Across	Down
8 The name for the unwanted waste material in matrix solid-phase dispersion.	5 The microwave generator.
10 A type of localized heating in microwave-assisted extraction.	9 A polar organic solvent.
11 The dielectric and dipolar processes by which the heating effect occurs in microwave-assisted extraction.	13 A type of diagram used to look at the boundaries between a solid, liquid and gas.
12 Having a high one of these allows penetration of the sample by the solvent in supercritical fluid extraction.	16 An instrumental component in supercritical fluid extraction used to create pressure within the system.
14 A type of inert packing material used in instrumental extraction techniques.	
15 An increase in this parameter can enhance compound solubility in pressurized liquid extraction.	
17 Having a low one of these allows penetration of the sample by the solvent in supercritical fluid extraction.	
18 The type of expansion that allows supercritical carbon dioxide to create cardice.	
19 A tool used to grind the sample in matrix solid-phase dispersion.	
20 An elevation in this parameter can enable the organic solvent to remain in liquid form in pressurized liquid extraction.	

Appendix A3: A Crossword for Key Terms on Instrumental Techniques for Environmental Organic Analysis

Across

2 Alternate name to chromatography.

4 Name of valve used to introduce samples in HPLC.

6 The seal on inlet of GC injection port.

7 A type of high-performance liquid chromatography that uses a polar mobile phase and a non-polar stationary phase.

Down

1 The same mobile phase during the entire high performance liquid chromatography run.

3 A stationary phase for GC.

5 Silica support of GC column.

10 The 'guide' within the GC injection port.

Across	Down
8 I can separate volatile compounds.	12 Apparatus to move solvent in HPLC.
9 I can separate non-volatile compounds.	13 I am used as a carrier gas for gas chromatography.
11 A detector for gas chromatography.	14 Name of time when a compound elutes.
17 Name of detector for GC.	15 Alternate name of a C18 HPLC sorbent.
19 Location of column in GC and HPLC.	16 The abbreviated name of a detector for high-performance liquid chromatography.
20 HPLC column packing material.	18 A type of high-performance liquid chromatography that uses a non-polar mobile phase and a polar stationary phase.
22 I am used as a carrier gas for gas chromatography mass spectrometry.	21 A fixed temperature gas chromatography.
23 The changing mobile phase in high-performance liquid chromatography.	
24 Name of device to inject samples into GC and HPLC.	

Crossword Solutions

Appendix B1: Solution to Crossword of Key Terms in Chapters 4–8: Extraction Techniques for Aqueous Samples

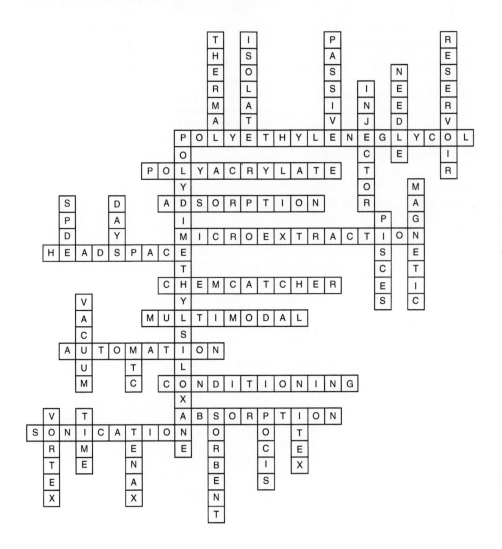

Appendix B2: Solution to Crossword of Key Terms in Chapters 10–13: Extraction Techniques for Solid Samples

Appendix B3: Solution for Crossword for Key Terms on Instrumental Techniques for Environmental Organic Analysis

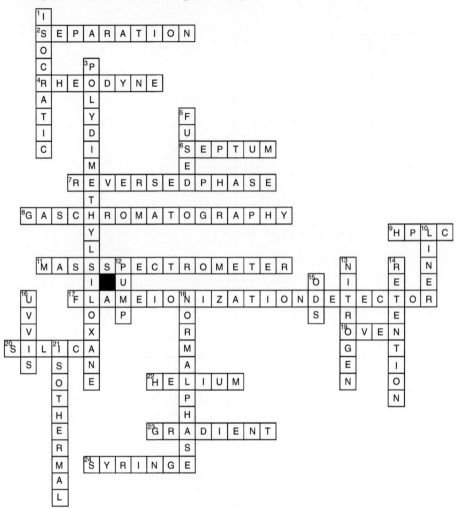

SI Units and Physical Constants

SI Units

The SI system of units is generally used throughout this book. It should be noted, however, that according to present practice, there are some exceptions to this, for example, wavenumber (cm^{-1}) and ionization energy (eV).

Base SI units and physical quantities

Quantity	Symbol	SI unit	Symbol
Length	L	metre	m
Mass	M	kilogram	kg
Time	T	second	s
Electric current	I	ampere	A
Thermodynamic temperature	T	kelvin	K
Amount of substance	N	mole	mol
Luminous intensity	I_v	candela	cd

Prefixes used for SI units

Factor	Prefix	Symbol
10^{12}	tera	T
10^{9}	giga	G
10^{6}	mega	M
10^{3}	kilo	k
10^{2}	hecto	h
10	deca	da
10^{-1}	deci	d
10^{-2}	centi	c
10^{-3}	milli	m

Extraction Techniques for Environmental Analysis, First Edition. John R. Dean.
© 2022 John Wiley & Sons Ltd. Published 2022 by John Wiley & Sons Ltd.

10^{-6}	micro	μ
10^{-9}	nano	n
10^{-12}	pico	p

Some derived SI units with special names and symbols

Physical quantity	SI unit		Expression in terms of base or derived SI units
	Name	Symbol	
Frequency	Hertz	Hz	$1\,Hz = 1\,s^{-1}$
Force	Newton	N	$1\,N = 1\,kg\,ms^{-2}$
Pressure	Pascal	Pa	$1\,Pa = 1\,Nm^{-2}$
Energy	Joule	J	$1\,J = 1\,Nm$
Power	Watt	W	$1\,W = 1\,J\,s^{-1}$
Electric charge	Coulomb	C	$1\,C = 1\,As$
Electric potential	Volt	V	$1\,V = 1\,JC^{-1}$
Electric capacitance	Farad	F	$1\,F = 1\,CV^{-1}$
Electric resistance	Ohm	Ω	$1\,\Omega = 1\,VA^{-1}$
Electric conductance	Siemens	S	$1\,S = 1\,\Omega^{-1}$
Celsius temperature	Degree Celsius	°C	$1\,°C = 1\,K$

Physical Constants

Recommended values of selected physical constants[a]

Constant	Symbol	Value
Acceleration of free fall (acceleration due to gravity)[b]	g_n	$9.806\,65\,m\,s^{-2}$
Atomic mass constant (unified atomic mass unit)	m_u	$1.660\,540\,2(10) \times 10^{-27}\,kg$
Avogadro constant	L, N_A	$6.022\,136\,7(36) \times 10^{23}\,mol^{-1}$
Boltzmann constant	k_B	$1.380\,658(12) \times 10^{-23}\,J\,K^{-1}$
Electron specific charge (charge-to-mass ratio)	$-e/m_e$	$-1.758\,819 \times 10^{11}\,C\,kg^{-1}$
Electron charge (elementary charge)	E	$1.602\,177\,33(49) \times 10^{-19}\,C$
Faraday constant	F	$9.648\,530\,9(29) \times 10^4\,C\,mol^{-1}$

Ice-point temperature[b]	T_{ice}	273.15 K
Molar gas constant	R	8.314 510(70) J K^{-1} mol^{-1}
Molar volume of ideal gas (at 273.15 K and 101 325 Pa)	V_m	22.414 10(19)×10^{-3} m^3 mol^{-1}
Planck constant	H	6.626 075 5(40)×10^{-34} J s
Standard atmosphere[b]	Atm	101 325 Pa
Speed of light in vacuum[b]	C	2.997 924 58×10^8 ms^{-1}

[a] Data are presented in their full precision, although often no more than the first four or five significant digits are used; figures in parentheses represent the standard deviation uncertainty in the least significant digits.
[b] Exactly defined values.

The Periodic Table

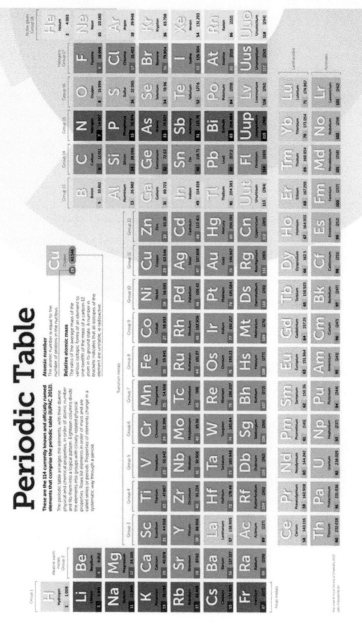

Index

Extraction Techniques for Environmental Analysis, First Edition. John R. Dean.
© 2022 John Wiley & Sons Ltd. Published 2022 by John Wiley & Sons Ltd.